J. Casillas, O. Cordón, F. Herrera, L. Magdalena (Eds.)

Accuracy Improvements in Linguistic Fuzzy Modeling

Springer-Verlag Berlin Heidelberg GmbH

Studies in Fuzziness and Soft Computing, Volume 129

http://www.springer.de/cgi-bin/search_book.pl?series=2941

Editor-in-chief
Prof. Janusz Kacprzyk
Systems Research Institute
Polish Academy of Sciences
ul. Newelska 6
01-447 Warsaw
Poland
E-mail: kacprzyk@ibspan.waw.pl

Further volumes of this series can be found on our homepage

Vol. 110. E. Fink l. 112. Y. Jin
Advanced Fuzzy Systems Design and Applications, 2003
ISBN 3-7908-1523-3

Vol. 111. P.S. Szcepaniak, J. Segovia, J. Kacprzyk and L.A. Zadeh (Eds.)
Intelligent Exploration of the Web, 2003
ISBN 3-7908-1529-2

Vol. 112. Y. Jin
Advanced Fuzzy Systems Design and Applications, 2003
ISBN 3-7908-1537-3

Vol. 113. A. Abraham, L.C. Jain and J. Kacprzyk (Eds.)
Recent Advances in Intelligent Paradigms and Applications", 2003
ISBN 3-7908-1538-1

Vol. 114. M. Fitting and E. Orowska (Eds.)
Beyond Two: Theory and Applications of Multiple Valued Logic, 2003
ISBN 3-7908-1541-1

Vol. 115. J.J. Buckley
Fuzzy Probabilities, 2003
ISBN 3-7908-1542-X

Vol. 116. C. Zhou, D. Maravall and D. Ruan (Eds.)
Autonomous Robotic Systems, 2003
ISBN 3-7908-1546-2

Vol 117. O. Castillo, P. Melin
Soft Computing and Fractal Theory for Intelligent Manufacturing, 2003
ISBN 3-7908-1547-0

Vol. 118. M. Wygralak
Cardinalities of Fuzzy Sets, 2003
ISBN 3-540-00337-1

Vol. 119. Karmeshu (Ed.)
Entropy Measures, Maximum Entropy Principle and Emerging Applications, 2003
ISBN 3-540-00242-1

Vol. 120. H.M. Cartwright, L.M. Sztandera (Eds.)
Soft Computing Approaches in Chemistry, 2003
ISBN 3-540-00245-6

Vol. 121. J. Lee (Ed.)
Software Engineering with Computational Intelligence, 2003
ISBN 3-540-00472-6

Vol. 122. M. Nachtegael, D. Van der Weken, D. Van de Ville and E.E. Kerre (Eds.)
Fuzzy Filters for Image Processing, 2003
ISBN 3-540-00465-3

Vol. 123. V. Torra (Ed.)
Information Fusion in Data Mining, 2003
ISBN 3-540-00676-1

Vol. 124. X. Yu, J. Kacprzyk (Eds.)
Applied Decision Support with Soft Computing, 2003
ISBN 3-540-02491-3

Vol. 125. M. Inuiguchi, S. Hirano and S. Tsumoto (Eds.)
Rough Set Theory and Granular Computing, 2003
ISBN 3-540-00574-9

Vol. 126. J.-L. Verdegay (Ed.)
Fuzzy Sets Based Heuristics for Optimization, 2003
ISBN 3-540-00551-X

Vol 127. L. Reznik, V. Kreinovich (Eds.)
Soft Computing in Measurement and Information Acquisition, 2003
ISBN 3-540-00246-4

Vol 128. J. Casillas, O. Cordón, F. Herrera, L. Magdalena (Eds.)
Interpretability Issues in Fuzzy Modeling, 2003
ISBN 3-540-02932-X

J. Casillas
O. Cordón
F. Herrera
L. Magdalena (Eds.)

Accuracy Improvements in Linguistic Fuzzy Modeling

Springer

Dr. Jorge Casillas
casillas@decsai.ugr.es
Dr. Oscar Cordón
E-mail: ocordon@decsai.ugr.es
Dr. Francisco Herrera
E-mail: herrera@decsai.ugr.es
Dpto. Ciencias de la Computación
e Inteligencia Artificial
Escuela Técnica Superior
de Ingeniería Informática
Universidad de Granada
E - 18071 Granada
Spain

Dr. Luis Magdalena
E-mail: llayos@mat.upm.es
Dpto. Matemáticas Aplicadas
a las Tecnologías de la Información
Escuela Técnica Superior de Ingenieros
de Telecomunicación
Universidad Politécnica de Madrid
E - 28040 Madrid
Spain

DOI 10.1007/978-3-540-37058-1

Library of Congress Cataloging-in-Publication-Data applied for

A catalog record for this book is available from the Library of Congress.

Bibliographic information published by Die Deutsche Bibliothek
Die Deutsche Bibliothek lists this publication in the Deutsche Nationalbibliographie; detailed bibliographic data is available in the internet at <http://dnb.ddb.de>.

http://www.springer.de

Originally published by Springer-Verlag Berlin Heidelberg New York in 2003.
MyCopy version of the original edition 2003

Typesetting: camera-ready by editors
Cover design: E. Kirchner, Springer-Verlag, Heidelberg
Printed on acid free paper 62/3020/M - 5 4 3 2 1 0
www.springer.com/mycopy

Foreword

When I accepted the editors' invitation to write this foreword, I assumed that it would have been an easy task. At that time I did not realize the monumental effort that went into the organization and compilation of these chapters, the depth of each contribution, and the thoroughness with which the book's theme had been covered.

A foreword usually tries to impress upon the reader the importance of the book's main topic, placing the work within a comparative framework, and identifying the new trends or ideas that are pushing the state-of-the-art. While doing this, one also tries to relate the book's main theme to some personal experience that will help the reader understand the usefulness and applicability of the various contributions. I will do my best to achieve at least some of these lofty goals.

The need for trading off interpretability and accuracy is intrinsic to the use of fuzzy systems. Before the advent of soft computing, and in particular of fuzzy logic, accuracy was the main concern of model builders, since interpretability was practically a lost cause. In a recent article in which I reviewed hybrid Soft Computing (SC) systems and compared them with more traditional approaches [1], I remarked that the main reason for the popularity of soft computing was the synergy derived from its components. In fact, SC's main characteristic is its intrinsic capability to create hybrid systems that are based on the integration of constituent technologies. This integration provides complementary reasoning and searching methods that allow us to combine domain knowledge and empirical data to develop flexible computing tools and solve complex problems.

Soft Computing provides a different paradigm in terms of representation and methodologies, which facilitates these integration attempts. For instance, in classical control theory the problem of developing models is usually decomposed into system identification (or system structure) and parameter estimation. The former determines the order of the differential equations, while the latter determines its coefficients. In these traditional approaches, the main goal is the construction of accurate models, within the assumptions used for the model construction. However, the models' interpretability is very limited, given the rigidity of the underlying representation language.

The equation "**model = structure + parameters**"[1], followed by the traditional approaches to model building, does not change with the advent of soft computing. However, with soft computing we have a much richer repertoire to represent the structure, to tune the parameters, and to iterate this process. This repertoire enables us to choose among different trade-

[1] It is understood that the search method used to postulate the structures and find the parameter values is an important and implicit part of the above equation, and needs to be chosen carefully for efficient model construction.

offs between the model's interpretability and accuracy. For instance, one approach aimed at maintaining the model's transparency might start with knowledge-derived linguistic models, where the domain knowledge is translated into an initial structure and parameters. Then the model's accuracy could be improved by using global or local data-driven search methods to tune the structure and/or the parameters. An alternative approach aimed at building more accurate models might start with data-driven search methods. Then, we could embed domain knowledge into the search operators to control or limit the search space, or to maintain the model's interpretability. Post-processing approaches could also be used to extract more explicit structural information from the models.

This book provides a comprehensive yet detailed review of all these approaches. In the introduction the reader will find a general framework, within which these approaches can be compared, and a description of alternative methods for achieving different balances between models' interpretability and accuracy. The book is mainly focused on the achievement of the mentioned tradeoff by improving the accuracy while preserving interpretability in linguistic fuzzy modeling. Thus, it presents constrained optimization methods, as well as extensions to the modeling process and model structures to do so.

These topics are germane to many applications and resonate with recent issues that I have addressed. Therefore, I would like to illustrate the pervasiveness of this book's main theme by relating it to a personal experience. By virtue of working in an industrial research center, I am constantly faced with the constraints derived from real-world problems. There are situations in which the use of black-box models is not acceptable, due to legal or compliance reasons. On the other hand, the same situations require a degree of accuracy that is usually prohibitive for purely transparent models.

An example of such a situation is the automation of the insurance underwriting process, which consists in evaluating an applicant's medical and personal information to assess his/her potential risk and determine the appropriate rate class corresponding to such risk. To address this problem, we need to maintain full accountability of the model decisions, i.e. full transparency. This legal requirement, imposed by the states insurance commissioners, is necessary since the insurance companies need to notify their customers and explain to them the reasons for issuing policies that are not at the most competitive rates. Yet, the model must also be extremely accurate to avoid underestimating the applicants' risk, which would decrease the company's profitability, or overestimating it, which would reduce the company's competitive position in the market.

We solved this problem by creating several hybrid SC models, some of them transparent, for use in production, and some of them opaque, for use in quality assurance. The commonalities among these models are the tight integration of knowledge and data, leveraged in their construction, and the loose integration of their outputs, exploited in their off-line use. In different parts of this project we strived to achieve different balances between interpretability and accuracy. This project exemplifies the pervasiveness of the theme and highlights the timeliness of this book, which fills a void in the technical literature and describes a topic of extreme relevance and applicability.

Piero P. Bonissone
General Electric Global Research Center
Schenectady, New York, 12308, USA

[1] "Hybrid Soft Computing Systems: Industrial and Commercial Applications", P. P. Bonissone, Y-T Chen, K. Goebel and P. S. Khedkar, Proceedings of the IEEE, pp 1641-1667, vol. 87, no. 9, September 1999.

Preface

System modeling with fuzzy rule-based systems, i.e. fuzzy modeling, usually comes with two contradictory requirements in the obtained model: the *interpretability*, capability to express the behavior of the real system in an understandable way, and the *accuracy*, capability to faithfully represent the real system.

To obtain high degrees of interpretability and accuracy is a contradictory purpose and, in practice, one of the two properties prevails over the other. While linguistic fuzzy modeling (mainly developed by linguistic fuzzy systems) is focused on the interpretability, precise fuzzy modeling (mainly developed by Takagi-Sugeno-Kang fuzzy systems) is focused on the accuracy.

Analyzing the research made from the former approach (linguistic fuzzy modeling), a large number of publications are found being oriented towards the use of new techniques and structures to extend the classical, rigid linguistic fuzzy modeling, whith the main aim of improving its accuracy. Thus, more flexible model structures with a larger number of freedom degrees (based on tools such as weights, hierarchical knowledge, or linguistic hedges) and advanced modeling processes (such as multicriteria optimizations or membership function learning) are performed. Of course, the flexibilizations made to enhance the precision should be performed under the assumption of preserving a good interpretability, otherwise, one of the most interesting features of linguistic fuzzy models would be ignored: its good capability to describe its intrinsic knowledge. From this perspective, this book focuses on showing a state-of-the-art on the recent proposals that attempt to obtain linguistic fuzzy models with a good *interpretability-accuracy trade-off* by improving their accuracy.

The book is organized as follows. Section 1 introduces an overview of the different accuracy improvement mechanisms existing in the recent literature. Section 2 collects a set of contributions focused on using different accuracy improvements performed under some constrains that avoid an excessive interpretability loss; restrictions such as rigid structures, comprehensibility criteria of the membership functions, or compactness of the rule set are considered. Section 3 contains a set of contributions that propose more sophisticated modeling processes to attain a good accuracy while preserving interpretability. Finally, Section 4 introduces a different approach that performs the accuracy improvement extending the traditional model structure by using different methodologies such as importance factors for each rule, knowledge bases with different granularities, etc.

We believe that this volume presents an up-to-date state of the current research that will be useful for non expert readers, whatever their background, to easily get some knowledge about this area of research. Besides, it will also support those specialists who wish to discover the latest results as well as the latest trends in research work in fuzzy modeling.

Finally, we would like to express our most sincere gratitude to Springer-Verlag (Heidelberg, Germany) and in particular to Prof. J. Kacprzyk, for having given us the opportunity to prepare the text and for having supported and encouraged us throughout its preparation. We would also like to acknowledge our gratitude to all those who have contributed to the books by producing the papers that we consider to be of the highest quality. We also like to mention the somehow obscure and altruistic, though absolutely essential, task carried out by a group of referees (all the contributions have been reviewed by two of them), who, through their comments, suggestions, and criticisms, have contributed to raising the quality of this edited book.

Granada and Madrid (Spain)
January 2003

Jorge Casillas, Oscar Cordón,
Francisco Herrera, and Luis Magdalena

Table of Contents

3. EXTENDING THE MODELING PROCESS TO IMPROVE THE ACCURACY

4. EXTENDING THE MODEL STRUCTURE TO IMPROVE THE ACCURACY

SECTION 1

OVERVIEW

Accuracy Improvements to Find the Balance Interpretability-Accuracy in Linguistic Fuzzy Modeling: An Overview

Jorge Casillas[1], Oscar Cordón[1], Francisco Herrera[1], and Luis Magdalena[2]

[1] Department of Computer Science and Artificial Intelligence,
University of Granada, E-18071 Granada, Spain
e-mail: {casillas,ocordon,herrera}@decsai.ugr.es

[2] Department of Mathematics Applied to Information Technologies,
Technical University of Madrid, E-28040 Madrid, Spain
e-mail: llayos@mat.upm.es

Abstract. System modeling with fuzzy rule-based systems (FRBSs), i.e. fuzzy modeling (FM), usually comes with two contradictory requirements in the obtained model: the *interpretability*, capability to express the behavior of the real system in an understandable way, and the *accuracy*, capability to faithfully represent the real system. While linguistic FM (mainly developed by linguistic FRBSs) is focused on the interpretability, precise FM (mainly developed by Takagi-Sugeno-Kang FRBSs) is focused on the accuracy. Since both criteria are of vital importance in system modeling, the balance between them has started to pay attention in the fuzzy community in the last few years.

The chapter analyzes mechanisms to find this balance by improving the accuracy in linguistic FM: deriving the membership functions, improving the fuzzy rule set derivation, or extending the model structure.

1 Introduction

System modeling is the action and effect of approaching to a model, i.e., to a theoretical scheme that simplifies a real system or complex reality with the aim of easing its understanding. Thanks to these models, the real system can be explained, controlled, simulated, predicted, and even improved. The development of *reliable* and *comprehensible* models is the main objective in system modeling. If not so, the model loses its usefulness.

There are at least three different paradigms in system modeling. The most traditional approach is the *white box modeling*, which assumes that a thorough knowledge of the system's nature and a suitable mathematical scheme to represent it are available. As opposed to it, the *black box modeling* [74] is performed entirely from data using no additional a priori knowledge and considering a sufficiently general structure. Whereas the white box modeling has serious difficulties when complex and poorly understood systems are considered, the black box modeling deals with structures and associated parameters that usually do not have any physical significance [2]. Therefore, generally the former approach does not adequately obtain reliable models while the latter one does not adequately obtain comprehensible models.

A third, intermediate approach arises as a combination of the said paradigms, the *grey box modeling* [37], where certain known parts of the system are modeled considering the prior understood and the unknown or less certain parts are identified with black box procedures. With this approach, the mentioned disadvantages are palliated and a better balance between reliability and comprehensibility is attained.

Nowadays, one of the most successful tools to develop grey box models is *fuzzy modeling* (FM) [50], which is an approach used to model a system making use of a descriptive language based on fuzzy logic with fuzzy predicates [75]. FM usually considers model structures (fuzzy systems) in the form of fuzzy rule-based systems (FRBSs) and constructs them by means of different parametric system identification techniques. Fuzzy systems have demonstrated their ability for control [28], modeling [63], or classification [18] in a huge number of applications. The keys for their success and interest are the ability to incorporate human expert knowledge – which is the information mostly provided for many real-world systems and is described by vague and imprecise statements – and the facility to express the behavior of the system with a language easily interpretable by human beings. These interesting advantages allow them to be even used as mechanisms to interpret black box models such as neural networks [16].

As a system modeling discipline, FM is mainly characterized by two features that assess the quality of the obtained fuzzy models:

- *Interpretability* — It refers to the capability of the fuzzy model to express the behavior of the system in a understandable way. This is a subjective property that depends on several factors, mainly the model structure, the number of input variables, the number of fuzzy rules, the number of linguistic terms, and the shape of the fuzzy sets. With the term interpretability we englobe different criteria appeared in the literature such as *compactness*, *completeness*, *consistency*, or *transparency*.
- *Accuracy* — It refers to the capability of the fuzzy model to faithfully represent the modeled system. The closer the model to the system, the higher its accuracy. As closeness we understand the similarity between the responses of the real system and the fuzzy model. This is why the term approximation is also used to express the accuracy, being a fuzzy model a fuzzy function approximation model.

As Zadeh stated in its *Principle of Incompatibility* [87], "*as the complexity of a system increases, our ability to make precise and yet significant statements about its behavior diminishes until a threshold is reached beyond which precision and significance (or relevance) become almost mutually exclusive characteristics.*"

Therefore, to obtain high degrees of interpretability and accuracy is a contradictory purpose and, in practice, one of the two properties prevails over the other one. Depending on what requirement is mainly pursued, the FM field may be divided into two different areas:

- *Linguistic fuzzy modeling (LFM)* — The main objective is to obtain fuzzy models with a good interpretability.
- *Precise fuzzy modeling (PFM)* — The main objective is to obtain fuzzy models with a good accuracy.

The relatively easy design of fuzzy systems, their attractive advantages, and their emergent proliferation have made FM to suffer a deviation from the seminal purpose directed towards exploiting the descriptive power of the concept of a linguistic variable [87,88]. Instead, in the last few years, the prevailing research in FM has focused on increasing the accuracy as much as possible paying little attention to the interpretability of the final model.

Nevertheless, a new tendency in the FM scientific community that looks for a good balance between interpretability and accuracy is increasing in importance [3,13,72,79]. The aim of this chapter is to review some of the recent proposals that attempt to address this issue using mechanisms to improve the accuracy of fuzzy models with a good interpretability.

The chapter is organized as follows. Section 2 analyzes the different existing lines of research related to the improvement of interpretability and accuracy to find a good balance in FM, Sect. 3 introduces the most useful kinds of FRBSs to improve their accuracy, Sect. 4 presents mechanisms to increase the accuracy of linguistic fuzzy models, and, finally, Sect. 5 points out some conclusions.

2 Major Lines of Work

The two main objectives to be addressed in the FM field are *interpretability* and *accuracy*. Of course, the ideal thing would be to satisfy both criteria to a high degree but, since they are contradictory issues, it is generally not possible. In this case, more priority is given to one of them (defined by the problem nature), leaving the other one in the background. Hence, two FM approaches arise depending on the main objective to be considered: LFM (interpretability) and PFM (accuracy).

Regardless of the approach, a common scheme is found in the existing literature to perform the FM:

1. Firstly, the main objective (interpretability or accuracy) is tackled defining a specific model structure to be used, thus setting the FM approach.
2. Then, the modeling components (model structure and/or modeling process) are improved by means of different mechanisms to define the desired ratio interpretability-accuracy.

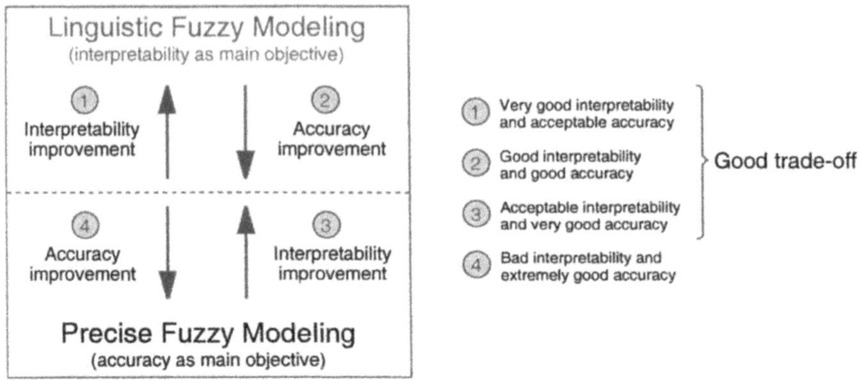

Fig. 1. Improvements of interpretability and accuracy in fuzzy modeling

This procedure results in four different possibilities (see Fig. 1): LFM with improved interpretability, LFM with improved accuracy, PFM with improved interpretability, and PFM with improved accuracy.

Although historically more priority has been given to the accuracy, currently the search of a good balance between both criteria is increasing in importance. Indeed, a significative effort is being performed by several researchers proposing *improvement mechanisms* to compensate for the initial difference. Among the four said lines of work, clearly this philosophy is pursued by two of them: *LFM with improved accuracy* and *PFM with improved interpretability* (approaches 2 and 3 in Fig. 1, respectively).

Moreover, another interesting proposal is *LFM with improved interpretability* (approach 1 in Fig. 1). Although LFM uses a model structure with a high description power by itself, there are some problems (curse of dimensionality, excessive number of input variables or fuzzy rules, garbled fuzzy sets, etc.) that make it not to be as interpretable as desired and the need of interpretability improvements to restore the searched balance is justified.

Finally, the modus operandi of obtaining more accuracy in PFM (approach 4 in Fig. 1) does not pay attention to the comprehensibility of the model and acts close to black box techniques. This approach does not follow the original objective of FM and does not profit from the advantages that distinguish it from other modeling techniques. Although the approach is useful when only accuracy is required, it goes away from the aim of the present book.

This chapter is devoted to review different accuracy improvements that have been proposed to attain the desired balance. Thus, Sect. 4 shows some mechanisms found in the recent literature to do so.

3 Types of Fuzzy Rule-Based Systems

Before presenting the search of a balance interpretability-accuracy in FM by improving the accuracy, it seems that there is need to introduce the different kinds of FRBSs usually employed. It is a significant aspect to consider since depending on the rule structure used, an FRBS has itself a specific capability of description and approximation. The section is only focused on the FRBS types usually considered to improve their accuracy for the sake of a good trade-off, thus having a good description capability themselves.

3.1 Linguistic Fuzzy Rule-Based System

Also known as Mamdani-type FRBS [57,58], the linguistic FRBS constitutes the main tool to develop LFM. A crucial reason why this approach is worth considering is that it may remain verbally interpretable, playing the concept of linguistic variable [88] a central role. Linguistic FRBSs are formed by linguistic rules with the following structure:

IF X_1 is A_1 and ... and X_n is A_n
THEN Y_1 is B_1 and ... and Y_m is B_m ,

with X_i and Y_j being input and output linguistic variables respectively, and with A_i and B_j being linguistic labels with fuzzy sets associated defining their meaning. These linguistic labels will be taken from a global *semantic* defining the set of possible fuzzy sets used for each variable (Fig. 2 shows an example with triangular membership functions). This structure provides a natural framework to include expert knowledge in the form of fuzzy rules.

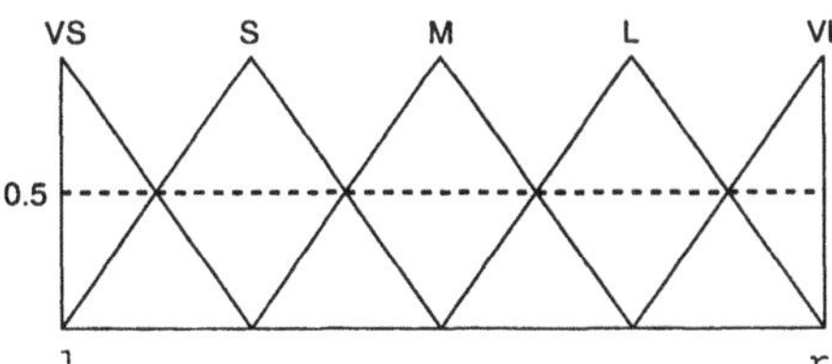

Fig. 2. Graphical representation of an example of the semantic considered for a variable, standing VS for *very small*, S for *small*, M for *medium*, L for *large*, and VL for *very large*, with $[l, r]$ being the corresponding variable domain

In these systems, the knowledge base (KB) – the component of the FRBS that stores the knowledge about the problem being solved – is composed of:

- the *rule base* (RB), constituted by the collection of linguistic rules themselves joined by means of the connective *also*, and

- the *data base* (DB), containing the term sets and the membership functions defining their semantics.

3.2 Singleton Fuzzy Rule-Based System

The singleton FRBS, where the rule consequent takes a single real-valued number, may be considered as a particular case of the linguistic FRBS (the consequent is a fuzzy set where the membership function is one for a specific value and zero for the remaining ones). Its rule structure is the following:

IF X_1 is A_1 and ... and X_n is A_n
THEN Y_1 is y_1 and ... and Y_m is y_m ,

with y_j being real-valued values.

Compared with the linguistic FRBS, the fact of having a different consequent value for each rule (no global semantic is used for the output variable) slightly worsens the interpretability. Nevertheless, the singleton FRBS may be used to develop LFM.

3.3 Fuzzy Rule-Based Classification Systems

A fuzzy rule-based classification system is an automatic classification system that uses fuzzy rules as knowledge representation tool. Therefore, the fuzzy classification rule structure is as follows:

IF X_1 is A_1 and ... and X_n is A_n
THEN Y is C ,

with C being the class label.

Other alternative representations that consider a certainty degree for each rule or that include all the possible class labels with their corresponding certainty degrees in the consequent part are usually also considered.

4 Improving the Accuracy in Linguistic Fuzzy Modeling

LFM has certain inflexibility due to the use of a global semantic that gives a general meaning to the used fuzzy sets. In fact, the use of linguistic variables imposes the following constrains [4,10]:

1. There is a lack of flexibility in the FRBS because of the rigid partitioning of the input and output spaces.
2. When the system input variables are dependent themselves, it is very hard to fuzzy partition the input spaces.

3. The homogeneous partitioning of the input and output spaces when the input-output mapping varies in complexity within the space is inefficient and does not scale to high-dimensional spaces.
4. The size of the KB directly depends on the number of variables and linguistic terms in the system. The derivation of an accurate linguistic FRBS requires a significant granularity amount, i.e., it needs of the creation of new linguistic terms. This granularity increase causes the number of rules to rise significantly, which may make the system lose the capability of being interpretable by human beings. In the most of the cases, it would be possible to obtain an equivalent FRBS having a very lesser number of rules if there would not exist that input space rigid partitioning.

However, it is possible to make some considerations to face this drawback [13]. Basically, two ways of improving the accuracy in LFM can be considered by performing the improvement in

- the *modeling process*, extending the model design to other components different from the RB such as the DB or considering more sophisticated derivations of it, or in
- the *model structure*, slightly changing the rule structure to make it more flexible.

The following three subsections introduce the improvements existing in the literature for designing the DB, learning the RB with sophisticated methods, and extending the model structure.

4.1 Data Base Design

Basic LFM methods are exclusively focused on determining the set of fuzzy rules composing the RB of the model [78,82,85]. In these cases, the DB is usually obtained from expert information (if available) or by a normalization process and it remains fixed during the RB derivation process.

However, the automatic design of the DB has shown to be a very suitable mechanism to increase the approximation capability of the linguistic models. Generally speaking, the procedure involves either defining the most appropriate shapes for the membership functions that give meaning to the fuzzy sets associated to the considered linguistic terms or determining the optimum number of linguistic terms used in the variable fuzzy partitions, i.e., the granularity (e.g., [25,29,68,70]).

In the following, we show different approaches of the DB design regarding the way of integrating it in the derivation process and the effects caused to the membership functions. Some considerations on the interpretability preservation are also discussed.

Integration of the DB design in the whole FRBS derivation process

The design of the DB may be integrated within the whole derivation process of a FRBS with different schemata:

- *Preliminary design* (learning the DB) — It involves extracting the DB a priori by induction from the available data set. This process is usually performed by non supervised clustering techniques [53,64].
- *Embedded design* (learning the KB) — This approach derives the DB using an embedded basic learning method [23,25,26,30,40,68,70]. The technique involves having a simple learning method that designs, from a specific DB, other components of the fuzzy linguistic model (e.g., the RB). Following a meta-learning process, the method generates different DBs and samples its efficacy running the basic learning method.
- *Simultaneous design* (learning the DB together with other components) — The process of designing the DB is jointly developed with the derivation of other components such as the RB in a simultaneous procedure [31] [39,45,48,56,73,81,83].
- *A posteriori design* (tuning the DB) — This approach, usually called DB *tuning*, involves refining the DB from a previous definition once the remaining components have been obtained. It is one of the most common procedures. Usually, the tuning process changes the membership function shapes [7,21,38,46,52,60,76] and the main requirement is to improve the accuracy of the linguistic model. Nevertheless, as shown in the previous section, sometimes another kind of a posteriori DB design is made to improve the interpretability (e.g., merging membership functions [29]).

Of course, several of these approaches can also jointly be considered. For example, in [51] a two-stage DB design is made by first deriving simultaneously the DB and the RB and then performing an a posteriori tuning. In [44], an initial generation of the RB with a subsequent three-stage DB design (input variable selection, simultaneous DB tuning and RB reduction, and DB fine tuning) is developed.

Preliminary, embedded, and a posteriori design approaches are usually combined with other methods to perform the whole derivation process in several sequential stages. Instead, the simultaneous design of the DB together with other components constitutes a whole derivation process itself.

The sequential derivation has the main advantage of reducing the search space since confined spaces are tackled at each stage. On the other hand, in the simultaneous derivation, the strong dependency of the components is properly addressed. However, the process becomes significantly more complex in the latter case because the search space significantly grows making the selection of an appropriate search technique crucial.

Recently, the cooperative coevolutionary paradigm [69] has shown an increasing interest thanks to its high ability to manage with huge search spaces and decomposable problems, and new simultaneous derivation methods are currently emerging using this technique [15,66,67].

Finally, we should say that the DB design gives more flexibility to the modeling process but it runs the risk of losing interpretability and overfitting the problem, wherefore this task must be carefully made. Some mechanisms to keep a good interpretability are discussed later on.

Learning/tuning the membership function shapes

Another interesting aspect to consider when designing the DB is the way of defining the membership function shapes. The most usual approaches are the following:

- *Learning/tuning the membership function parameters* — The most common way of deriving the membership functions is to change their definition parameters [7,21,31,38,39,44–46,48,51,60,67,70,76,81]. For example, if the following triangular-shape membership function is considered:

$$\mu(x) = \begin{cases} \frac{x-a}{b-a}, & \text{if } a \leq x < b \\ \frac{c-x}{c-b}, & \text{if } b \leq x \leq c \\ 0, & \text{otherwise} \end{cases},$$

  changing the basic parameters – a, b, and c – will vary the shape of the fuzzy set associated to the membership function (see Fig. 3), thus influencing the FRBS performance. The same yields for other shapes of membership functions (trapezoidal, gaussian, sigmoid, etc.).

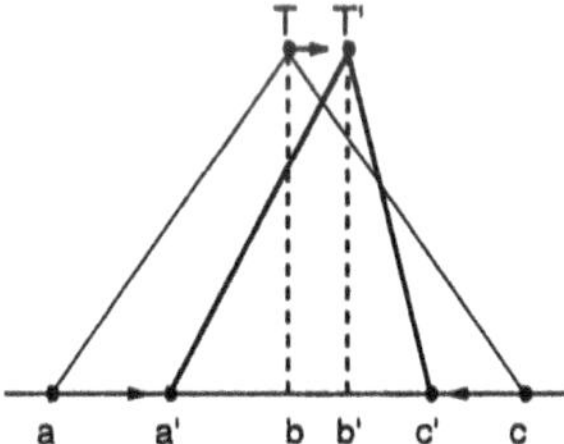

Fig. 3. Definition of the membership function shapes by their parameters

- *Using linguistic modifiers* — Another way to define the membership function shapes of the DB is to use linguistic modifiers [20,52]. Section 4.3 describes them in depth. They involve considering more flexible alternative expressions for the membership functions to vary the compatibility degrees to the fuzzy sets. For example, a new membership function can be obtained raising the membership value to the power of α, i.e.,

$$\mu'(x) = (\mu(x))^{\alpha}, \ \ 0 < \alpha.$$

By changing the α value we may define different membership function shapes. Figure 4 shows the effect of this approach.

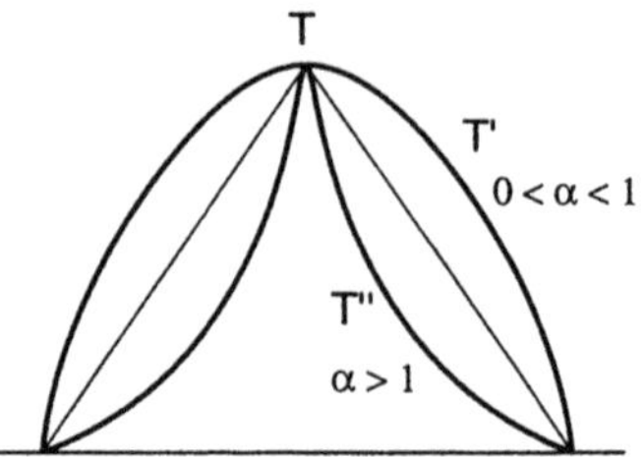

Fig. 4. Definition of the membership function shapes using linguistic modifiers

- *Establishing the context* — A third possibility to change the membership functions is to define the context, i.e, their operation range. It is usually performed by *scaling functions* that map the input and output variables onto the universe of discourse over which the fuzzy sets are defined. From a linguistic point of view, the scaling function can be interpreted as a sort of context information.
  We may distinguish between linear and non-linear scaling functions, which are well known in classical control theory:
  - *Linear context* [34] — A linear scaling function is of the form:

    $$f(x) = \alpha \cdot x + \beta.$$

    The scaling factor α enlarges or reduces the operating range, which in turn decreases or increases the sensitivity of the controller with respect to that input variable, or the corresponding gain in case of an output variable. The parameter β shifts the operating range and plays the role of an offset to the corresponding variable.
  - *Non-linear context* [35,47,55,65] — The main disadvantage of linear scaling is the fixed relative distribution of the membership functions. Non-linear scaling provides a remedy to this problem as it modifies the relative distribution and changes the shape of the membership functions. A common non-linear scaling function for a variable that is symmetric with respect to the origin is of the form:

    $$f(x) = sign(x) \cdot |x|^{\alpha}.$$

    Non-linear scaling increases ($\alpha > 1$) or decreases ($\alpha < 1$) the relative sensitivity in the region around the origin and has the opposite effect at the boundaries of the operating range. Figure 5 depicts some effects caused by the non-linear scaling.

  Derivation methods combining both kinds of context adaptations have been also proposed [23].

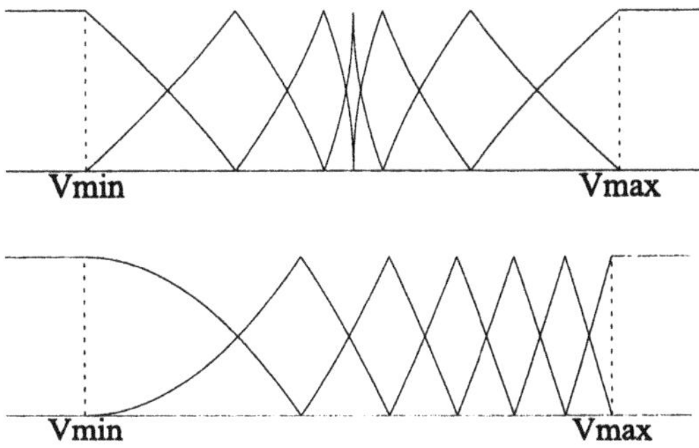

Fig. 5. Non-linear scaling effects

Some other contributions propose methods to define not only the membership function shapes but also the membership function types (such as triangular, trapezoidal, gaussian, and sigmoid) [73].

Interpretability preservation when designing the DB

As said, the DB design may generate intricate semantics that could disturb the expert interpretation, thus losing some legibility degree. Figure 6 illustrates an example where garbled membership functions may involve losing interpretability.

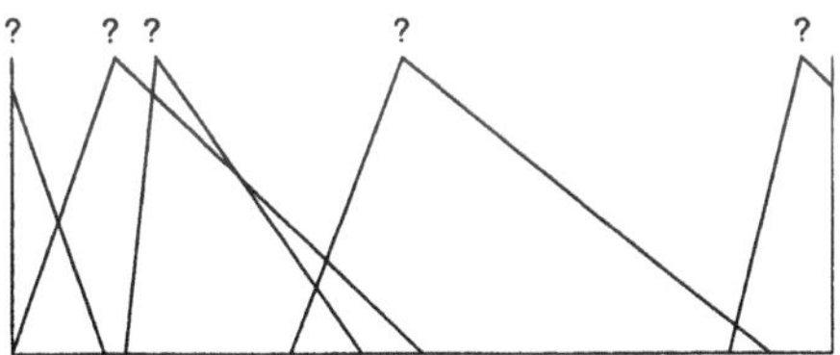

Fig. 6. Certain interpretability may be lost when designing the DB

To decide if a specific DB is interpretable or not is a difficult and subjective task. Nevertheless, some researchers have become aware of this matter and several properties that ensure a good interpretability during the membership function optimization have been proposed [79,80]. Different constrains to the DB design process may be imposed to fulfil all, or some of these properties. A selection of the most important properties and some possible solutions when designing the DB follows:

- *Coverage property* — Every value of the universe of discourse should belong to, at least, a linguistic term. Alternately, a more strict criterion may be considered establishing a minimum level of coverage to be met for the entire universe of discourse.

Possible solutions — This property may be satisfied, e.g., using variation intervals that ensure the overlap between two consecutive fuzzy sets [21] or employing strong fuzzy partitions [17].

- *Normality property* — Every membership function should exhibit full matching with, at least, a value of the universe of discourse. That is, the fuzzy sets should be normal.
  Possible solutions — This property is easily kept forcing the modal points of the extreme membership functions of the variable fuzzy partition to be contained in the universe of discourse.
- *Distinguishability property* — Each linguistic term should have a clear meaning and the associated fuzzy set should clearly define a range of the universe of discourse. In short, the membership functions should be distinct enough from each other.
  Possible solutions — This property may be satisfied, e.g., properly constraining the membership function parameters [54], merging similar membership functions [29], or establishing a semantic order among the linguistic terms [5,6].

Of course, the constrains needed to ensure the semantic integrity make the derivation process less flexible but, moreover of easing the legibility, the risk of overfitting the problem is reduced.

When search techniques are used to design the DB, another possibility to ensure the integrity properties is to include interpretability measures in the objective function, thus guiding the trek to good solutions. Usually, measures of completeness, consistency [45], compactness, or similarity [43] are considered. A more advanced criterion, called conciseness, is proposed in [77] by combining a fuzzy entropy measure, which distinguishes the shapes of the membership functions, with a deviation measure, which evaluates the discrepancy of a membership function from symmetry.

In embedded DB design, another interesting approach is to consider the effect in the RB size caused when defining the number of linguistic terms and their associated membership function shapes. In [68,70], the authors propose to use a simple fuzzy system to select the solution with the desired trade-off between interpretability (number of rules) and accuracy (approximation error) among the different generated KBs.

4.2 Using More Sophisticated RB Learning Methods

These improvements arise as an effort to exploit the accuracy ability of linguistic FRBSs by exclusively focusing on the RB design. In this case, the DB and the model structure keep invariable, thus resulting in the highest interpretability. Usually, all these improvements have the final goal of enhancing the *interpolative reasoning* the FRBS develops. This is one of the most interesting features of FRBSs and plays a key role in their high performance, being a consequence of the cooperative action of the linguistic fuzzy rules. Some specific proposals are explained in the following.

The method COR (cooperative rules) proposed in [12] follows the primary objective of inducing a better cooperation among the linguistic rules. To do that, the RB design is made using global criteria that consider the action of the different rules jointly. It is attained by means of a strong, smart reduction of the search space. The main advantages of the COR methodology are its capability to include heuristic information [14], its flexibility to be used with different metaheuristics [?], and its easy integration within other derivation processes [15].

In [32], a method is proposed to refine a previously obtained RB in disjunctive normal form. The heuristic process is comprised of three sequential steps: firstly, the precision of the model is improved by removing some linguistic terms from the rule antecedents; then, the generalization capability is improved by adding linguistic terms to the antecedents; and finally, the completeness of the RB is attained by the addition of new linguistic fuzzy rules.

4.3 Extending the Model Structure

Another way to improve the LFM accuracy is to extend the usual linguistic model structure to make it more flexible. Some specific possibilities are described in the following.

To use linguistic modifiers

A possibility to relax the rule structure is to include certain operators to slightly change the meaning of the linguistic labels involved in the system when necessary [11,20,33]. As Zadeh highlighted in [87], a way to do so without losing the description to a high degree is to use linguistic hedges or, in a wider sense, *linguistic modifiers*. We must remark that the inclusion of linguistic modifiers in fuzzy rules differs from their use to design the DB [52] (explained in Sect. 4.1).

A linguistic modifier [8,9,86] is an operator that alters the membership functions for the fuzzy sets associated to the linguistic labels, giving a more or less precise definition as a result depending on the case. Thus, a new linguistic rule structure arises as follows:

$$\textbf{IF } X_1 \text{ is } lm_{X_1}\ A_1 \text{ and } \ldots \text{ and } X_n \text{ is } lm_{X_n}\ A_n \textbf{ THEN } Y \text{ is } lm_Y\ B\ ,$$

with lm_{X_i} (lm_Y) being the linguistic modifier to be used (including the identity operator) in the corresponding variable where the membership degree to the linguistic term is given by $\mu_{A_i}^{lm_{X_i}}$ ($\mu_B^{lm_Y}$). An example of a rule with this structure is the following:

$$\textbf{IF } X_1 \text{ is } \textit{very} \text{ high and } X_2 \text{ is low } \textbf{THEN } Y \text{ is } \textit{more-or-less} \text{ large}\ .$$

Actually, the consideration of linguistic modifiers does not define a new meaning to the so-called *primary terms* – *high*, *low*, and *large* in our example – but they are used as generators whose meaning is defined in the context. In other words, thanks to the *attributed-grammar semantic* [88] associated to the linguistic variables, the final membership functions are computed from the knowledge of the membership functions of the primary terms.

Certainly, the fact of using fuzzy rules with linguistic modifiers will have a significative influence in the behavior of the linguistic FRBS because the matching degree of the rule antecedent as well as the output fuzzy set obtained when applied the implication operator in the inference process are changed.

From a linguistic point of view, the linguistic modifiers are usually classified into those that reinforce the characterizations and those that weaken them. From a FM point of view, however, the most interesting effect caused by the modifiers is the alterations in the inference process (support, center of gravity, etc.) that they cause. From this perspective, we can distinguish among three different kinds of linguistic modifiers:

- *Powered modifiers* [86] — The membership degrees of the linguistic terms are modified with non-linear scaling functions by rising them to the power of some factor (see Fig. 7(a)):

  $$\mu'(x) = (\mu(x))^{\alpha}, \quad 0 < \alpha \ ,$$

  For example, the operator "*very*" squares the membership degree of the linguistic term, i.e., $\mu_T^{very}(x) = (\mu_T(x))^2$.
  With $\alpha < 1$ the modifier dilates the fuzzy set, whilst with $\alpha > 1$ the modifier concentrates the fuzzy set. These kinds of linguistic modifiers are known as *linguistic hedges* [86].
- *Expansive/reducted modifiers* [9,59] — These modifiers change the support and core sets of the fuzzy sets enlarging or reducing them, but trying to keep a center of gravity similar to the original (see Fig. 7(b)). The effect over the fuzzy sets, though similar to the previous one (in terms of dilation and concentration), is rather more severe and it is implemented with linear variations.
- *Shifted modifiers* [9,49] — These kinds of modifiers are defined by translations of the membership functions along their domains (see Fig. 7(c)). The effect over the linguistic terms is more severe than the aforementioned ones.

Powered modifiers (with non linear variation) – linguistic hedges – have an important restriction with respect to expansive/reducted modifiers or shifted modifiers (with linear variation): the support and core sets of the fuzzy sets are not altered. Moreover, when a symmetrical fuzzy set is considered, its center of gravity is not changed. On the contrary, the membership degree of a value to the fuzzy set grows in a non-linear way as it gets closer to the core.

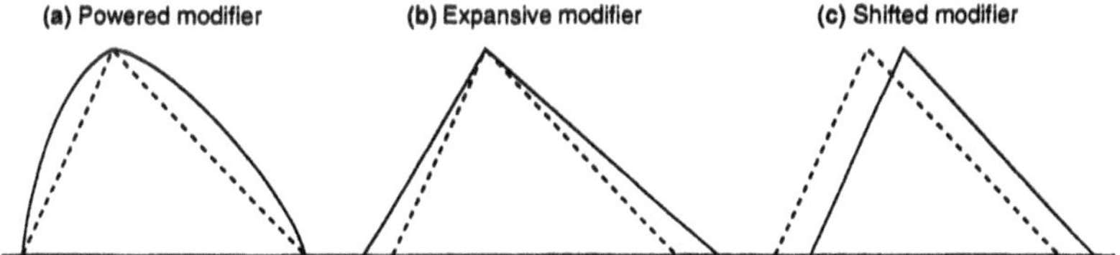

Fig. 7. Types of linguistic modifiers

To use double-consequent rules

This approach involves allowing the RB to present rules where each combination of antecedents may have two consequents associated when it is necessary to improve the model accuracy [22,29,61]. It is clear that this will improve the capability of the model to perform the interpolative reasoning and, thus, its performance. The rule structure obtained will be as follows:

$$\textbf{IF } X_1 \text{ is } A_1 \text{ and } \ldots \text{ and } X_n \text{ is } A_n \textbf{ THEN } Y \text{ is } \{B_1,B_2\} \ .$$

Since each double-consequent fuzzy rule can be decomposed into two different rules with a single consequent, the usual plain fuzzy inference system can be applied. The only restriction imposed is that the defuzzification method must consider the matching degree of the rules fired. For example, the *center of gravity weighted by the matching degree* defuzzification strategy [24] may be used.

When using two consequents per rule, the interpretation of the action performed by every rule may be confusing to some extent. However, we should note this fact does not constitute an inconsistency from the LFM point of view but only a shift of the main labels making the final output of the rule lie in an intermediate zone between both consequents. Indeed, let us suppose that a specific combination of antecedents, "X_1 is A_1 and ... and X_n is A_n," has two different consequents associated, B_1 and B_2. The resulting double-consequent rule may be interpreted as follows [22]:

$$\textbf{IF } X_1 \text{ is } A_1 \text{ and } \ldots \text{ and } X_n \text{ is } A_n \textbf{ THEN } Y \text{ is } \textit{between } B_1 \text{ and } B_2 \ .$$

To use weighted rules

This approach involves using an additional parameter for each rule that indicates its importance degree in the inference process, instead of considering all rules equally important as in the usual case. Thus, more flexibility to improve the interpolative reasoning and, therefore, the model performance, is achieved [1,19,41,62,71,84]. The rule structure will be the following one:

$$\textbf{IF } X_1 \text{ is } A_1 \text{ and } \ldots \text{ and } X_n \text{ is } A_n \textbf{ THEN } Y \text{ is } B \textit{ with } [w] \ ,$$

with w being the real-valued rule weight. In this approach, some changes to the classical inference system must be made to consider the weighted action of each rule.

The operator *with*, which attaches a weight to a rule, may be defined in different ways. One of the most usual options is to multiply the matching degree of the antecedent by the corresponding weight before applying the implication operator, which relates antecedent and consequent. Another possibility is to change the conclusion derived from the implication operator according to the corresponding weight (e.g., changing the support of the obtained fuzzy set).

These weights are usually considered to handle inconsistencies with advanced inference methods [84] or neural networks [19]. Moreover, some proposals make use of them to improve the model accuracy with an automatic learning of weights using different techniques such as heuristic methods [41,71], gradient descent processes [62], or evolutionary algorithms [1].

To use hierarchical knowledge bases

A deeper change in the model structure involves considering hierarchical KBs. In this case, the hierarchical KB is composed of a set of layers where each one contains linguistic partitions with different granularity levels (a layer of the hierarchical DB) and linguistic rules whose linguistic variables take values in these partitions (a layer of the hierarchical RB) [27]. Different learning methods have been proposed to design this extended model structure.

The method proposed in [42] obtains a hierarchical KB by creating several hierarchical linguistic partitions with different granularity levels, generating the complete set of linguistic rules in each of these partitions, taking the union of all of these sets, and finally performing a genetic rule selection process on the whole rule set.

The method introduced in [27] uses an inductive linguistic rule generation method to progressively refine the controversial regions (those covered by linguistic fuzzy rules with a bad performance) by defining new rules in a deeper layer. The obtained hierarchical RB is compacted by a subsequent selection process. Therefore, this latter method follows a *descending* approach refining the regions by increasing the granularity. An *ascending* approach by merging fuzzy sets to progressively reducing the granularity is proposed in [36].

5 Concluding Remarks

The FM research developed in the last two decades was mainly focused on exploiting the flexibility of FM to obtain the maximum accuracy. During this evolution, the derivation methods were improved, the components to be designed were extended, and new model structures were proposed. This search of the accuracy usually set aside the interpretability of the obtained models.

However, we should remember the initial philosophy of fuzzy set theory directed to serve the bridge between the human understanding and the machine processing. In this challenge, the faculty of fuzzy models to express the behavior of the real system in a comprehensible manner acquires a great importance. This is why the current tendency in FM tries to find a better balance between interpretability and accuracy.

This equilibrium is attained from different perspectives. One of the things that attracts the eye is the fact that it is frequently performed by means of previous existing extensions, but used in a more rational and moderate way. Thus, this chapter was aimed to present an introduction to the different tends recently proposed in the specialized literature to improve the accuracy degree of the fuzzy models with the objective of finding the desired trade-off, i.e., preserving their interpretability.

The remaining 14 chapters contained in this volume are excellent works of research in the FM approach studied in this chapter and they properly represent the existing state-of-the-art.

References

1. R. Alcalá, J. Casillas, O. Cordón, and F. Herrera. Cooperative coevolution for linguistic modeling with weighted double-consequent fuzzy rules. In *Proceedings of the 2nd International Conference in Fuzzy Logic and Technology*, pages 237–240, Leicester, UK, 2001.
2. R. Babuška. *Fuzzy modeling for control.* Kluwer Academic, Norwell, MA, USA, 1998.
3. R. Babuška, H. Bersini, D.A. Linkens, D. Nauck, G. Tselentis, and O. Wolkenhauer. Future prospects for fuzzy systems and technology. ERUDIT Newsletter Vol. 6, No. 1. Aachen, Germany, 2000. Available at `http://www.erudit.de/erudit/newsletters/news_61/page5.htm`.
4. A. Bastian. How to handle the flexibility of linguistic variables with applications. *International Journal of Uncertainty, Fuzziness and Knowledge-Based Systems*, 2(4):463–484, 1994.
5. U. Bodenhofer. A similarity-based generalization of fuzzy orderings preserving the classical axioms. *International Journal of Uncertainty, Fuzziness and Knowledge-Based Systems*, 8(5):593–610, 2000.
6. U. Bodenhofer and P. Bauer. Towards an axiomatic approach to interpretability. In *Proceedings of the 6th International Conference on Soft Computing*, pages 334–339, Iizuka, Japan, 2000.
7. P.P. Bonissone, P.S. Khedkar, and Y. Chen. Genetic algorithms for automated tuning of fuzzy controllers: a transportation application. In *Proceedings of the 5th IEEE International Conference on Fuzzy Systems*, pages 674–680, New Orleans, LA, USA, 1996.
8. B. Bouchon-Meunier. Linguistic variables in the knowledge base of an expert system. In J. Rose, editor, *Cybernetics and systems: the way ahead*, pages 745–752. Thales Publications, Lytham St Annes, England, 1987.

9. B. Bouchon-Meunier and Y. Jia. Linguistic modifiers and imprecise categories. *International Journal of Intelligent Systems*, 7:25–36, 1992.
10. B. Carse, T.C. Fogarty, and A. Munro. Evolving fuzzy rule based controllers using genetic algorithms. *Fuzzy Sets and Systems*, 80:273–294, 1996.
11. J. Casillas, O. Cordón, M.J. del Jesus, and F. Herrera. Genetic tuning of fuzzy rule-based systems integrating linguistic hedges. In *Proceedings of the 9th IFSA World Congress and the 20th NAFIPS International Conference*, pages 1570–1574, Vancouver, Canada, 2001.
12. J. Casillas, O. Cordón, and F. Herrera. COR: A methodology to improve ad hoc data-driven linguistic rule learning methods by inducing cooperation among rules. *IEEE Transactions on Systems, Man, and Cybernetics—Part B: Cybernetics*. To appear.
13. J. Casillas, O. Cordón, and F. Herrera. Can linguistic modeling be as accurate as fuzzy modeling without losing its description to a high degree? Technical Report #DECSAI-00-01-20, Department of Computer Science and Artificial Intelligence, University of Granada, Granada, Spain, 2000. Available at `http://decsai.ugr.es/~casillas/`.
14. J. Casillas, O. Cordón, and F. Herrera. Learning fuzzy rules using ants colony algorithms. In *Proceedings of the 2nd International Workshop on Ant Algorithms*, pages 13–21, Brussels, Belgium, 2000.
15. J. Casillas, O. Cordón, F. Herrera, and J.J. Merelo. Cooperative coevolution for learning fuzzy rule-based systems. In *Proceedings of the 5th International Conference on Artificial Evolution*, pages 97–108, Le Creusot, France, 2001.
16. J.L. Castro, C.J. Mantas, and J.M. Benítez. Interpretation of artificial neural networks by means of fuzzy rules. *IEEE Transactions on Neural Networks*, 13(1):101–116, 2002.
17. F. Cheong and R. Lai. Constraining the optimization of a fuzzy logic controller using an enhanced genetic algorithm. *IEEE Transactions on Systems, Man, and Cybernetics—Part B: Cybernetics*, 30(1):31–46, 2000.
18. Z. Chi, H. Yan, and T. Pham. *Fuzzy algorithms with application to image processing and pattern recognition*. World Scientific, Singapore, 1996.
19. J.-S. Cho and D.-J. Park. Novel fuzzy logic control based on weighting of partially inconsistent rules using neurtal network. *Journal of Intelligent and Fuzzy Systems*, 8:99–110, 2000.
20. O. Cordón, M.J. del Jesus, and F. Herrera. Genetic learning of fuzzy rule-based classification systems cooperating with fuzzy reasoning methods. *International Journal of Intelligent Systems*, 13:1025–1053, 1998.
21. O. Cordón and F. Herrera. A three-stage evolutionary process for learning descriptive and approximate fuzzy logic controller knowledge bases from examples. *International Journal of Approximate Reasoning*, 17(4):369–407, 1997.
22. O. Cordón and F. Herrera. A proposal for improving the accuracy of linguistic modeling. *IEEE Transactions on Fuzzy Systems*, 8(3):335–344, 2000.
23. O. Cordón, F. Herrera, L. Magdalena, and P. Villar. A genetic learning process for the scaling factors, granularity and contexts of the fuzzy rule-based system data base. *Information Sciences*, 136(1-4):85–107, 2001.
24. O. Cordón, F. Herrera, and A. Peregrín. Applicability of the fuzzy operators in the design of fuzzy logic controllers. *Fuzzy Sets and Systems*, 86(1):15–41, 1997.

25. O. Cordón, F. Herrera, and P. Villar. Analysis and guidelines to obtain a good fuzzy partition granularity for fuzzy rule-based systems using simulated annealing. *International Journal of Approximate Reasoning*, 25(3):187–215, 2000.
26. O. Cordón, F. Herrera, and P. Villar. Generating the knowledge base of a fuzzy rule-based system by the genetic learning of the data base. *IEEE Transactions on Fuzzy Systems*, 9(4):667–674, 2001.
27. O. Cordón, F. Herrera, and I. Zwir. Linguistic modeling by hierarchical systems of linguistic rules. *IEEE Transactions on Fuzzy Systems*, 10(1):2–20, 2002.
28. D. Driankov, H. Hellendoorn, and M. Reinfrank. *An introduction to fuzzy control.* Springer-Verlag, Heidelberg, Germany, 1993.
29. J. Espinosa and J. Vandewalle. Constructing fuzzy models with linguistic integrity from numerical data-AFRELI algorithm. *IEEE Transactions on Fuzzy Systems*, 8(5):591–600, 2000.
30. P. Glorennec. Constrained optimization of FIS using an evolutionary method. In F. Herrera and J.L. Verdegay, editors, *Genetic algorithms and soft computing*, pages 349–368. Physica-Verlag, Heidelberg, Germany, 1996.
31. A.F. Gómez-Skarmeta and F. Jiménez. Fuzzy modeling with hybrid systems. *Fuzzy Sets and Systems*, 104(2):199–208, 1999.
32. A. González and R. Pérez. A fuzzy theory refinement algorithm. *International Journal of Approximate Reasoning*, 19(3-4):193–220, 1998.
33. A. González and R. Pérez. A study about the inclusion of linguistic hedges in a fuzzy rule learning algorithm. *International Journal of Uncertainty, Fuzziness and Knowledge-Based Systems*, 7(3):257–266, 1999.
34. R.R. Gudwin and F.A.C. Gomide. Context adaptation in fuzzy processing. In *Proceedings of the 1st Brazil-Japan Joint Symposium in Fuzzy Systems*, pages 15–20, Campinas, SP, Brazil, 1994.
35. R.R. Gudwin, F.A.C. Gomide, and W. Pedrycz. Context adaptation in fuzzy processing and genetic algorithms. *International Journal of Intelligent Systems*, 13(10-11):929–948, 1998.
36. S. Guillaume and B. Charnomordic. Generating an interpretable family of fuzzy partitions. Internal report of Cemagref, Montpellier, France.
37. K.M. Hangos, editor. *Special issue on grey box modelling*, volume 9(6) of *International Journal of Adaptive Control and Signal Processing.* John Wiley & Sons, New York, NY, USA, 1995.
38. F. Herrera, M. Lozano, and J.L. Verdegay. Tuning fuzzy controllers by genetic algorithms. *International Journal of Approximate Reasoning*, 12:299–315, 1995.
39. A. Homaifar and E. McCormick. Simultaneous design of membership functions and rule sets for fuzzy controllers using genetic algoritms. *IEEE Transactions on Fuzzy Systems*, 3(2):129–139, 1995.
40. H. Ishibuchi and T. Murata. A genetic-algorithm-based fuzzy partition method for pattern classification problems. In F. Herrera and J.L. Verdegay, editors, *Genetic algorithms and soft computing*, pages 555–578. Physica-Verlag, Heidelberg, Germany, 1996.
41. H. Ishibuchi and T. Nakashima. Effect of rule weights in fuzzy rule-based classification systems. *IEEE Transactions on Fuzzy Systems*, 9(4):506–515, 2001.
42. H. Ishibuchi, K. Nozaki, N. Yamamoto, and H. Tanaka. Selecting fuzzy if-then rules for classification problems using genetic algorithms. *IEEE Transactions on Fuzzy Systems*, 3(3):260–270, 1995.

43. F. Jiménez, A.F. Gómez-Skarmeta, H. Roubos, and R. Babuška. A multi-objective evolutionary algorithm for fuzzy modeling. In *Proceedings of the 9th IFSA World Congress and the 20th NAFIPS International Conference*, pages 1222–1228, Vancouver, Canada, 2001.
44. Y. Jin. Fuzzy modeling of high-dimensional systems: complexity reduction and interpretability improvement. *IEEE Transactions on Fuzzy Systems*, 8(2):212–221, 2000.
45. Y. Jin, W. von Seelen, and B. Sendhoff. On generating FC^3 fuzzy rule systems from data using evolution strategies. *IEEE Transactions on Systems, Man, and Cybernetics—Part B: Cybernetics*, 29(4):829–845, 1999.
46. C.L. Karr. Genetic algorithms for fuzzy controllers. *AI Expert*, 6(2):26–33, 1991.
47. F. Klawonn. Fuzzy sets and vague environments. *Fuzzy Sets and Systems*, 66:207–221, 1994.
48. K. KrishnaKumar and A. Satyadas. GA-optimized fuzzy controller for spacecraft attitude control. In J. Periaux, G. Winter, M. Galán, and P. Cuesta, editors, *Genetic algorithms in engineering and computer science*, pages 305–320. John Wiley & Sons, New York, NY, USA, 1995.
49. G. Lakoff. Hedges: a study in meaning criteria and the logic of fuzzy concepts. *Journal of Philosofical Logic*, 2:458–508, 1973.
50. P. Lindskog. Fuzzy identification from a grey box modeling point of view. In H. Hellendoorn and D. Driankov, editors, *Fuzzy model identification*, pages 3–50. Springer-Verlag, Heidelberg, Germany, 1997.
51. J. Liska and S.S. Melsheimer. Complete design of fuzzy logic systems using genetic algorithms. In *Proceedings of the 3rd IEEE International Conference on Fuzzy Systems*, pages 1377–1382, Orlando, FL, USA, 1994.
52. B.-D. Liu, C.-Y. Chen, and J.-Y. Tsao. Design of adaptive fuzzy logic controller based on linguistic-hedge concepts and genetic algorithms. *IEEE Transactions on Systems, Man, and Cybernetics—Part B: Cybernetics*, 31(1):32–53, 2001.
53. S. López, L. Magdalena, and J.R. Velasco. Genetic fuzzy c-means algorithm for the automatic generation of fuzzy partitions. In B. Bouchon-Meunier, R.R. Yager, and L.A. Zadeh, editors, *Information, Uncertainty, Fusion*, pages 407–418. Kluwer Scientific, Norwell, MA, USA, 1999.
54. A. Lotfi, H.C. Andersen, and A.C. Tsoi. Interpretation preservation of adaptive fuzzy inference systems. *International Journal of Approximate Reasoning*, 15:379–394, 1996.
55. L. Magdalena. Adapting the gain of an FLC with genetic algorithms. *International Journal of Approximate Reasoning*, 17(4):327–349, 1997.
56. L. Magdalena and F. Monasterio-Huelin. A fuzzy logic controller with learning through the evolution of its knowledge base. *International Journal of Approximate Reasoning*, 16(3):335–358, 1997.
57. E.H. Mamdani. Applications of fuzzy algorithms for control a simple dynamic plant. *Proceedings of the IEE 121*, 12:1585–1588, 1974.
58. E.H. Mamdani and S. Assilian. An experiment in linguistic systhesis with fuzzy logic controller. *International Journal of Man-Machine Studies*, 7:1–13, 1975.
59. J.G. Marín-Blázquez, Q. Shen, and A.F. Gómez-Skarmeta. From approximative to descriptive models. In *Proceedings of the 9th IEEE International Conference on Fuzzy Systems*, pages 829–834, San Antonio, TX, USA, 2000.
60. D. Nauck and R. Kruse. Neuro-fuzzy systems for function approximaton. *Fuzzy Sets and Systems*, 101(2):261–271, 1999.

61. K. Nozaki, H. Ishibuchi, and H. Tanaka. A simple but powerful heuristic method for generating fuzzy rules from numerical data. *Fuzzy Sets and Systems*, 86(3):251–270, 1997.
62. N.R. Pal and K. Pal. Handling of inconsistent rules with an extended model of fuzzy reasoning. *Journal of Intelligent and Fuzzy Systems*, 7:55–73, 1999.
63. W. Pedrycz, editor. *Fuzzy modelling: paradigms and practice.* Kluwer Academic, Norwell, MA, USA, 1996.
64. W. Pedrycz. Fuzzy equalization in the construction of fuzzy sets. *Fuzzy Sets and Systems*, 119(2):329–335, 2001.
65. W. Pedrycz, R.R. Gudwin, and F.A.C. Gomide. Nonlinear context adaptation in the calibration of fuzzy sets. *Fuzzy Sets and Systems*, 88(1):91–97, 1997.
66. C.A. Peña-Reyes and M. Sipper. Applying Fuzzy CoCo to breast cancer diagnosis. In *Proceedings of the 2000 Congress on Evolutionary Computation*, pages 1168–1175, Piscataway, NJ, USA, 2000. IEEE Press.
67. C.A. Peña-Reyes and M. Sipper. Fuzzy CoCo: a cooperative-coevolutionary approach to fuzzy modeling. *IEEE Transactions on Fuzzy Systems*, 9(5):727–737, 2001.
68. H. Pomares, I. Rojas, J. Ortega, J. González, and A. Prieto. A systematic approach to a self-generating fuzzy rule-table for function approximation. *IEEE Transactions on Systems, Man, and Cybernetics—Part B: Cybernetics*, 30(3):431–447, 2000.
69. M.A. Potter and K.A. De Jong. Cooperative coevolution: an architecture for evolving coadapted subcomponents. *Evolutionary Computation*, 8(1):1–29, 2000.
70. I. Rojas, H. Pomares, J. Ortega, and A. Prieto. Self-organized fuzzy system generation from training examples. *IEEE Transactions on Fuzzy Systems*, 8(1):23–36, 2000.
71. L. Sánchez, J. Casillas, O. Cordón, and M.J. del Jesus. Some relationships between fuzzy and random set-based classifiers and models. *International Journal of Approximate Reasoning*, 29(2):175–213, 2002.
72. M. Setnes, R. Babuška, and H.B. Verbruggen. Rule-based modeling: precision and transparency. *IEEE Transactions on Systems, Man, and Cybernetics—Part C: Applications and Reviews*, 28(1):165–169, 1998.
73. Y. Shi, R. Eberhart, and Y. Chen. Implementation of evolutionary fuzzy systems. *IEEE Transactions on Fuzzy Systems*, 7(2):109–119, 1999.
74. T. Söderström and P. Stoica. *System identification.* Prentice Hall, Engleweed Cliffs, NJ, USA, 1989.
75. M. Sugeno and T. Yasukawa. A fuzzy-logic-based approach to qualitative modeling. *IEEE Transactions on Fuzzy Systems*, 1(1):7–31, 1993.
76. H. Surmann, A. Kanstein, and K. Goser. Self-organizing and genetic algorithms for an automatic design of fuzzy control and decision systems. In *Proceedings of the 1st European Congresss on Fuzzy and Intelligent Technologies*, pages 1097–1104, Aachen, Germany, 1993.
77. T. Suzuki, T. Kodama, T. Furuhashi, and H. Tsutsui. Fuzzy modeling using genetic algorithms with fuzzy entropy as conciseness measure. *Information Sciences*, 136(1-4):53–67, 2001.
78. P. Thrift. Fuzzy logic synthesis with genetic algorithms. In R.K. Belew and L.B. Booker, editors, *Proceedings of the 4th International Conference on Genetic Algorithms*, pages 509–513, San Mateo, CA, USA, 1991. Morgan Kaufmann Publishers.

79. J. Valente de Oliveira. Semantic constraints for membership function optimization. *IEEE Transactions on Systems, Man, and Cybernetics—Part A: Systems and Humans*, 29(1):128–138, 1999.
80. J. Valente de Oliveira. Towards neuro-linguistic modeling: constraints for optimization of membership functions. *Fuzzy Sets and Systems*, 106(3):357–380, 1999.
81. C.-H. Wang, T.-P. Hong, and S.-S. Tseng. Integrating fuzzy knowledge by genetic algorithms. *IEEE Transactions on Evolutionary Computation*, 2(4):138–149, 1998.
82. L.-X. Wang and J.M. Mendel. Generating fuzzy rules by learning from examples. *IEEE Transactions on Systems, Man, and Cybernetics*, 22(6):1414–1427, 1992.
83. N. Xiong and L. Litz. Fuzzy modeling based on premise optimization. In *Proceedings of the 9th IEEE International Conference on Fuzzy Systems*, pages 859–864, San Antonio, TX, USA, 2000.
84. W. Yu and Z. Bien. Design of fuzzy logic controller with inconsistent rule base. *Journal of Intelligent and Fuzzy Systems*, 2:147–159, 1994.
85. Y. Yuan and H. Zhuang. A genetic algorithm for generating fuzzy classification rules. *Fuzzy Sets and Systems*, 84:1–19, 1996.
86. L.A. Zadeh. A fuzzy-set theoretic interpretation of linguistic hedges. *Journal of Cybernetics*, 2(2):4–34, 1972.
87. L.A. Zadeh. Outline of a new approach to the analysis of complex systems and desision processes. *IEEE Transactions on Systems, Man, and Cybernetics*, 3:28–44, 1973.
88. L.A. Zadeh. The concept of a linguistic variable and its application to approximate reasoning. Parts I, II and III. *Information Science*, 8, 8, 9:199–249, 301–357, 43–80, 1975.

SECTION 2

ACCURACY IMPROVEMENTS CONSTRAINED BY INTERPRETABILITY CRITERIA

COR Methodology: A Simple Way to Obtain Linguistic Fuzzy Models with Good Interpretability and Accuracy

Jorge Casillas, Oscar Cordón, and Francisco Herrera

Department of Computer Science and Artificial Intelligence,
Computer Engineering School, University of Granada,
E-18071 Granada, Spain
e-mail: {casillas,ocordon,herrera}@decsai.ugr.es

Abstract. The chapter introduces a simple learning methodology, the cooperative rules (COR) one, that improves the accuracy of linguistic fuzzy models preserving the highest interpretability. Its operation mode involves a combinatorial search of fuzzy rules performed over a set of previously generated candidate ones.

The accuracy is achieved by developing a smart search space reduction and by inducing the generation of a linguistic fuzzy rule set with good cooperation. COR also ensures a good interpretability by keeping the membership functions and the model structure unaltered, as well as generating a compact rule base.

1 Introduction

Fuzzy modeling, i.e., system modeling with fuzzy rule-based systems (FRBSs), may be considered as an approach used to model a system making use of a descriptive language based on fuzzy logic with fuzzy predicates. In this framework, one of the most important areas is *linguistic* fuzzy modeling, where the interpretability of the obtained model is the main requirement. This task is usually developed by means of linguistic (or Mamdani-type) FRBSs, which use fuzzy rules composed of linguistic variables [29] taking values in a term set with a real-world meaning. Thus, the linguistic fuzzy model consists of a set of linguistic descriptions regarding the behavior of the system being modeled [26].

An interpretable model has no sense if it does not faithfully represent the modeled system, i.e., if it is not accurate enough. Thus, a good trade-off between interpretability and accuracy is needed to perform a useful fuzzy modeling. When the main objective is the interpretability, this trade-off can be attained by using different tools to improve the accuracy.

This approach has been performed by learning/tuning the membership functions by defining their shapes [15,20,18], their types (triangular, trapezoidal, etc.) [25], or their context (defining the whole semantic) [23], learning the granularity (number of linguistic terms) of the fuzzy partitions [11], or extending the model structure by using linguistic modifiers [5,12], weights (importance factors for each rule) [22], or hierarchical architectures (mixing rules with different granularities) [14], among others.

However, all these techniques change the membership functions or extend the original model structure, so that certain interpretability degree is lost. Opposite to it, our aim in this contribution will be to attain the desired balance by increasing the accuracy of linguistic FRBSs keeping the highest interpretability. To do so, a learning procedure that exclusively designs the fuzzy rule set will be introduced: the cooperative rules (COR) methodology. It is based on a combinatorial search of cooperative rules performed over a set of previously generated candidate rule consequents. Additionally, a rule base reduction is developed to improve the interpretability and the accuracy by obtaining compact models without redundancies and inconsistencies.

The chapter is organized as follows. Section 2 describes the COR methodology, Sect. 3 analyzes the behavior and main characteristics of the proposal, Sect. 4 introduces an specific method based on the COR methodology, Sect. 5 performs an experimental study applying our method and other ones to two real-world problems, and, finally, Sect. 6 points out some conclusions.

2 The COR Methodology

A family of efficient and simple methods to derive fuzzy rules guided by covering criteria of the data in the example set, called *ad hoc data-driven methods*, has been proposed in the literature in the last few years [3]. Their high performance, in addition to their quickness and easy understanding, make them very suitable for learning tasks.

However, ad hoc data-driven methods usually look for the fuzzy rules with the best individual performance (e.g. [28]) and therefore the global interaction among the rules of the RB is not considered. This sometimes causes KBs with bad cooperation among the rules to be obtained, thus not being as accurate as desired.

With the aim of addressing these drawbacks keeping the interesting advantages of ad hoc data-driven methods, a new methodology to improve the accuracy obtaining better cooperation among the rules is proposed in [3]: the COR methodology. Instead of selecting the consequent with the highest performance in each subspace like ad hoc data-driven methods usually do, the COR methodology considers the possibility of using another consequent, different from the best one, when it allows the FRBS to be more accurate thanks to having a KB with better cooperation.

COR consists of two stages:

1. *Search space construction* — It obtains a set of candidate consequents for each rule.
2. *Selection of the most cooperative fuzzy rule set* — It performs a combinatorial search among these sets looking for the combination of consequents with the best global accuracy.

A wider description of the COR-based rule generation process is shown in Fig. 1, whilst an example of the operation mode for a simple problem with two input variables and three labels for each linguistic variable is graphically illustrated in Fig. 2.

Since the search space tackled in step 2. is usually large, it is necessary to use approximate search techniques. In [3], accurate linguistic models have been obtained using simulated annealing. Other techniques such as tabu search, genetic algorithms, and ant colony optimization (ACO) have been also applied to COR methodology [4].

3 Analysis of COR Methodology

This section analyzes the behavior and main characteristics of the COR methodology.

3.1 Search Space Reduction

The COR methodology reduces the search space basing on heuristic information. This fact differences COR from other rule base learning methods [27] and allows it to be quicker and to make a better solution exploration.

This search space reduction is performed by two constrains:

1. *Maximum Number of Fuzzy Input Subspaces* — The maximum number of fuzzy input subspaces, and therefore maximum number of fuzzy rules, is limited by the positive example sets. Depending on the approach followed (step 1.1.1. or 1.1.2. in Fig. 1), the selection of the subspaces will be more conservative or more generic.
   Step 1.1.1. divides the input space with a crisp grid bounded by the cross points between labels and, therefore, each example contributes to generate a single rule. On the contrary, step 1.1.2. divides the input space with a fuzzy grid where each example may contribute to generate several rules. Figure 3 graphically shows this fact. In Fig. 3(b), the examples lying in white zones have an influence on the generation of one rule, those lying in light grey zones influence two rules, and the ones lying in dark grey zones influence four rules.
   It is not possible to determine which approach is the best. The crisp grid approach always obtains an equal number or fewer rules than the fuzzy grid one – as in the fuzzy grid approach the examples have an influence on a wider region, thus generating more rules. However, this fact may make the model obtained with the crisp grid approach not to be as accurate as desired sometimes.
2. *Candidate Rule Set in each Subspace* — Once the fuzzy input subspaces are defined, a second search space reduction is made by constraining the set of possible consequents for each antecedent combination, i.e., the candidate rules in each subspace.

Inputs:

- *An input-output data set* – $E = \{e_1, \ldots, e_l, \ldots, e_N\}$, with $e_l = (x_1^l, \ldots, x_n^l, y^l)$, $l \in \{1, \ldots, N\}$, N being the data set size, and n being the number of input variables – *representing the behavior of the problem being solved.*
- *A fuzzy partition of the variable spaces.* In our case, uniformly distributed fuzzy sets are regarded. Let $\mathcal{A}_i$ be the set of linguistic terms of the i-th input variable, with $i \in \{1, \ldots, n\}$, and $\mathcal{B}$ be the set of linguistic terms of the output variable, with $|\mathcal{A}_i|$ ($|\mathcal{B}|$) being the number of labels of the i-th input (output) variable.

Algorithm:

1. *Search space construction*:
   1.1. *Define the fuzzy input subspaces containing positive examples*: To do so, we should define the positive example set ($E^+(S_s)$) for each fuzzy input subspace $S_s = (A_1^s, \ldots, A_i^s, \ldots, A_n^s)$, with $A_i^s \in \mathcal{A}_i$ being a label, $s \in \{1, \ldots, N_S\}$, and $N_S = \prod_{i=1}^n |\mathcal{A}_i|$ being the number of fuzzy input subspaces. Two possibilities could be the following:
      1.1.1. $E^+(S_s) = \{e_l \in E \mid \forall i \in \{1, \ldots, n\}, \forall A_i' \in \mathcal{A}_i,\ \mu_{A_i^s}(x_i^l) \geq \mu_{A_i'}(x_i^l)\}$
      1.1.2. $E^+(S_s) = \{e_l \in E \mid \mu_{A_1^s}(x_1^l) \cdot \ldots \cdot \mu_{A_n^s}(x_n^l) \neq 0\}$,
      with $\mu_{A_i^s}(\cdot)$ being the membership function associated with the label A_i^s.
      Among all the N_S possible fuzzy input subspaces, consider only those containing at least one positive example. To do so, the set of subspaces with positive examples is defined as $S^+ = \{S_j \mid E^+(S_j) \neq \emptyset\}$.
   1.2. *Generate the set of candidate rules in each subspace with positive examples*: Firstly, the candidate consequent set associated with each subspace containing at least an example, $S_j \in S^+$, is defined. Two possibilities follow:
      1.2.1. $C(S_j) = \{B_k \in \mathcal{B} \mid \exists e_l \in E^+(S_j)\ with\ \forall B' \in \mathcal{B},\ \mu_{B_k}(y^l) \geq \mu_{B'}(y^l)\}$.
      1.2.2. $C(S_j) = \{B_k \in \mathcal{B} \mid \exists e_l \in E^+(S_j)\ with\ \mu_{B_k}(y^l) \neq 0\}$.

      Then, the candidate rule set for each subspace is defined as $CR(S_j) = \{R_k =$ [IF X_1 is A_1^j and ... and X_n is A_n^j THEN Y is B_k] $\mid B_k \in C(S_j)\}$.
      In order to allow COR methodology to reduce the initial number for fuzzy rules, the special element $R_\emptyset$ (which means "do not care") is added to each candidate rule set, i.e., $CR(S_j) = CR(S_j) \cup R_\emptyset$. If this element is selected, no rules are used in the corresponding fuzzy input subspace.
2. *Selection of the most cooperative fuzzy rule set* — This stage is performed by running a combinatorial search algorithm to look for the combination $\{R_1 \in CR(S_1), \ldots, R_j \in CR(S_j), \ldots, R_{|S^+|} \in CR(S_{|S^+|})\}$ with the best accuracy.
   An index measuring the cooperation degree of the encoded rule set is considered to evaluate the quality of each solution. In our case, the algorithm uses a global error function called *mean square error* (MSE), which is defined as

$$\text{MSE} = \frac{1}{2 \cdot N} \sum_{l=1}^{N} (F(x_1^l, \ldots, x_n^l) - y^l)^2,$$

   with $F(x_1^l, \ldots, x_n^l)$ being the output obtained from the FRBS when the example e_l is used, and y^l being the known desired output. The closer to zero the measure, the greater the global performance and, thus, the better the rule cooperation.

Fig. 1. COR algorithm

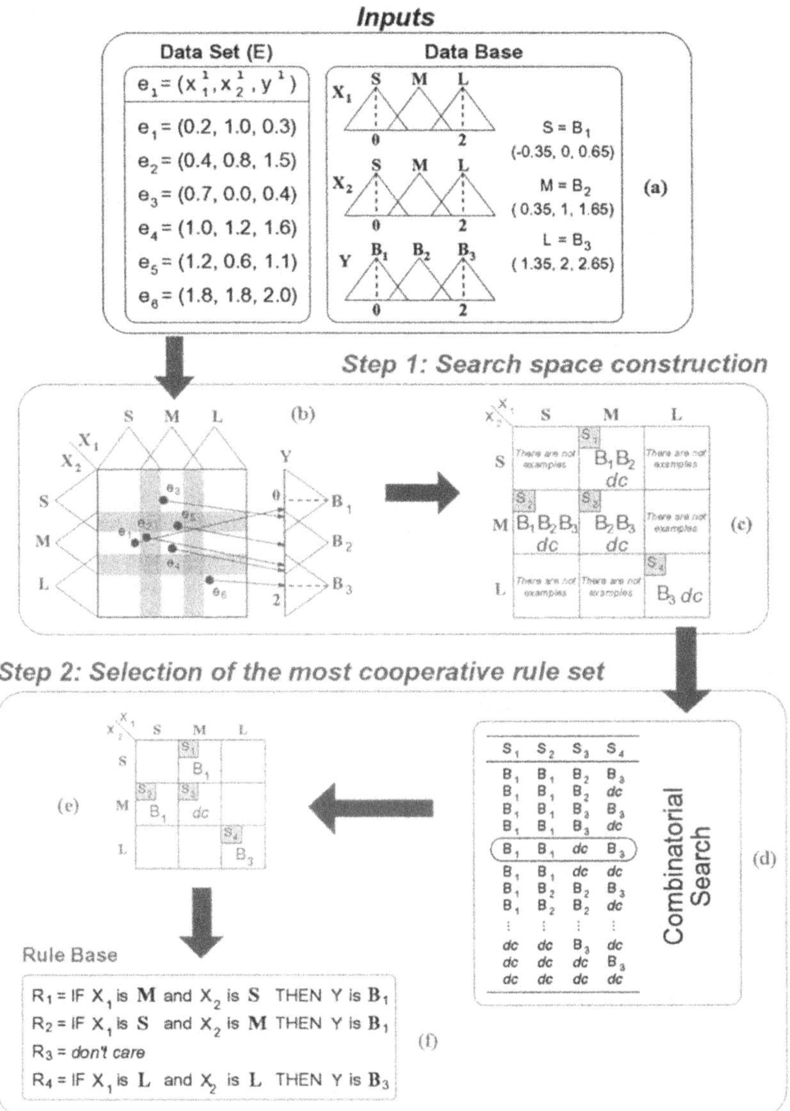

Fig. 2. COR-based learning process for a simple problem with two input variables ($n = 2$) and three labels in the output fuzzy partition ($|\mathcal{B}| = 3$): (a) data set (E) and data base previously defined; (b) the six examples are located in four ($|S^+| = 4$) different subspaces that determine the antecedent combinations and candidate consequents of the rules; (c) set of possible consequents for each subspace, including the special element "don't care" (dc); (d) combinatorial search accomplished within a space composed of 72 different combinations of consequents; (e) rule decision table for the fifth combination; (f) rule base generated from this combination

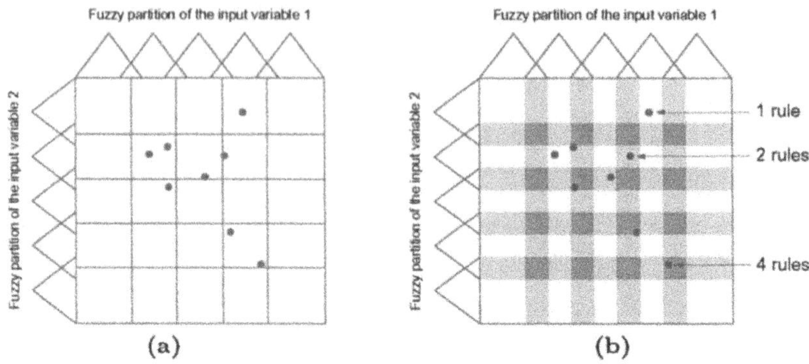

Fig. 3. **(a)** Crisp grid, an example only contributes to the generation of one rule. **(b)** Fuzzy grid, an example may contribute to the generation of several rules

Again, depending on the approach followed (step 1.2.1. or 1.2.2. in Fig. 1), the search space will be different. Step 1.2.1. has a more restrictive condition and generates lesser number of candidate rules than step 1.2.2.. We should remark that the positive example set previously generated will also influence in the candidate rule sets.

This search space reduction is graphically depicted in Fig. 4 when solving a real-world modeling problem (see Sect. 5.1 for a description of it) with two input variables and five linguistic terms for each fuzzy partition. The approaches 1.1.1+1.2.1 (COR-1) and 1.1.2+1.2.2 (COR-2) in the proposed methodology are compared with a well-known rule base learning method, the one proposed by Thrift in [27].

From the example distribution tackled – Fig. 4(a) –, the Thrift method considers all the available fuzzy input subspaces and all the possible consequents (including the "don't care" symbol) in each of them – Fig. 4(c) –. However, the COR methodology only considers a subset of subspaces and candidate consequents for each – Fig. 4(d) –. This significative reduction decreases the solution space size – Fig. 4(b) – and ease the search process.

3.2 Interpretability and Accuracy Issues

This section analyzes some aspects included in the COR methodology that allow it to obtain linguistic fuzzy models with a good interpretability and accuracy degree.

- *Cooperation Among Rules to Improve Accuracy* — The cooperation induced by the COR methodology allows us to obtain linguistic fuzzy models with a better accuracy. This is due to the interpolative reasoning developed by FRBSs, which is one of the most interesting features of these kinds of systems and plays a key role in their high performance, being

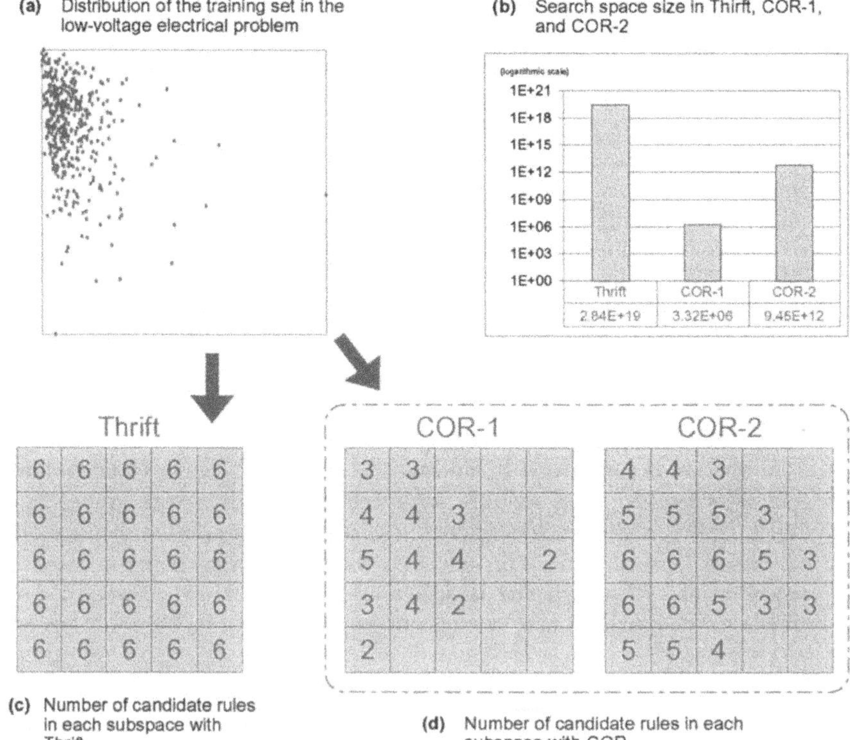

Fig. 4. Comparison between the search space tackled by the Thrift method and the COR methodology in a real-world problem with five linguistic terms for each variable

a consequence of the cooperative action among the linguistic rules. On the other hand, the fact of globally processing these rules makes COR be more robust against noise.

- *Model Structure and Membership Functions Invariable for an Excellent Interpretability* — The COR methodology is an effort to exploit the accuracy ability of linguistic FRBSs by exclusively focusing on the rule base design. In this case, the membership functions and the model structure keep invariable, thus resulting in the highest interpretability. Indeed, instead of improve the accuracy by deriving the shape of the membership functions or by extending the model structure (weighted rules, linguistic hedges, hierarchical knowledge bases, etc.), COR methodology improves the accuracy inducing cooperation among the linguistic fuzzy rules.
- *Rule Base Reduction to Improve Interpretability and Accuracy* — A problem when defining a rule base is that one can not be sure whether the rules are correctly defined, i.e., without redundant rules or rules that generate

conflicts with others in certain situations. To face this problem, a *rule reduction process* can be developed by combining rules and/or selecting a subset of rules from a given rule base to achieve the goal of minimizing the number of rules used while maintaining – or even improving – the FRBS performance. The badly defined and conflicting rules are eliminated by the method because their existence degrades the system performance. Some methods have been proposed to search for an optimized subset of rules, usually achieved by genetic algorithms [6,13,14,16]. These proposals generally perform the reduction with a postprocessing stage, once the rule base has been derived.

The COR methodology, however, achieves the reduction process at the same time as the learning one with the aim of improving the accuracy (the cooperation among rules and thus the system performance can be improved by removing rules) and interpretability (a model with less rules is more interpretable) of the learned model.

This process is performed by adding the null rule ($R_\emptyset$) to the candidate rule set corresponding to each subspace, as shown in the step 1.2. of Fig. 1. In this way, if such a element is selected for a specific subspace, this will mean that no rules will take part for the corresponding antecedent combination. Although the addition of $R_\emptyset$ in each candidate rule set increases the search space, more accurate and interpretable solutions can be obtained.

4 Application of Ant Colony Optimization to COR Methodology

COR is characterized by its flexibility to be used with different techniques of optimization and search. In [3], successful linguistic models have been obtained using simulated annealing. Nevertheless, these results could be improved incorporating heuristic information to the learning process. This consideration would guide the algorithm in the search, making it quicker on finding good solutions. ACO [9] is a good support for such intention thanks to the inherent use of heuristic information. Therefore, this section describes the use of ACO in the COR methodology. The following subsection briefly introduces ACO algorithms and, after that, the following five subsections present the different components of the algorithm.

4.1 Introduction to Ant Colony Optimization

ACO algorithms [8] constitute a new family of global search bio-inspired algorithms that has recently appeared. Since the first proposal, the Ant System algorithm [10] – applied to the Traveling Salesman Problem –, numerous models have been developed to solve a wide set of optimization problems (refer to [8] for a review of models and applications).

ACO algorithms draw inspiration from the social behavior of ants to provide food to the colony. In the food search process, consisting of the food find and the return to the nest, the ants deposit a substance called *pheromone*. The ants have the ability of sniffing the pheromone and pheromone trails guide the colony during the search. When an ant is located at a branch, it decides to take the path according to the probability defined by the pheromone existing in each trail. In this way, the depositions of pheromone terminate in constructing a path between the nest and the food that can be followed by new ants. The progressive action of the colony members involves the length of the path is progressively reduced. The shortest paths are finally the more frequently visited ones and, therefore, the pheromone concentration is higher on them. On the contrary, the longest paths are less visited and the associated pheromone trails are evaporated.

The basic operation mode of ACO algorithms is as follows [10]: at each iteration, a population of a specific number of ants progressively construct different tracks on the graph (i.e., solutions to the problem) according to a *probabilistic transition rule* that depends on the available information (heuristic and pheromone trails). After that, the pheromone trails are updated. This is done by first decreasing them by some constant factor (corresponding to the evaporation of the pheromone) and then reinforcing the solution attributes of the constructed solutions considering their quality. This task is developed by the *global pheromone trail update rule*.

Several extensions to this basic operation mode have been proposed. Their improvements mainly consist of using different transition and update rules, introducing new components, or adding a local search phase [2,9,24].

To apply ACO algorithms to a specific problem, the five steps shown in Fig. 5 have to be performed. The following sections describe these aspects particularized to the COR methodology.

1. *Problem representation*: Interpret the problem to be solved as a graph or a similar structure easily traveled by ants.
2. *Heuristic information*: Define the way of assigning a heuristic preference to each choice that the ant has to take in each step to generate the solution.
3. *Pheromone initialization*: Establish an appropriate way of initializing the pheromone.
4. *Fitness function*: Define a fitness function to be optimized.
5. *ACO algorithm*: Select an ACO algorithm and apply it to the problem.

Fig. 5. Steps followed to apply ACO algorithms to a specific problem

4.2 Problem Representation

For applying ACO in the COR methodology, it is convenient to see it as a combinatorial optimization problem with the capability of being represented on a graph. In this way, we can face the problem considering a fixed number of subspaces and interpreting the learning process as the way of assigning consequents – i.e., labels of the output fuzzy partition – to these subspaces with respect to an optimality criterion (i.e., following the COR methodology).

Hence, we are in fact dealing with an assignment problem and the problem representation can be similar to the one used to solve the *quadratic assignment problem* (QAP) [1], but with some peculiarities. We may draw an analogy between *subspaces* and locations and between *consequents* and facilities. However, unlike the QAP, the *set of possible consequents for each subspace may be different* and *it is possible to assign a consequent to more than one subspace* (two rules may have the same consequent). We can deduce from these characteristics that the order of selecting each subspace to be assigned a consequent is not determinant since one assignment does not restrict the remaining ones, i.e., the assignment order is irrelevant.

Therefore, according to Fig. 1, each node $S_j \in S^+$ is assigned to each candidate consequent $B_k \in C(S_j)$ and the especial symbol "don't care" that stands for absence of rules in such a subspace.

Figure 6 depicts the graph corresponding to the example represented in Fig. 2.

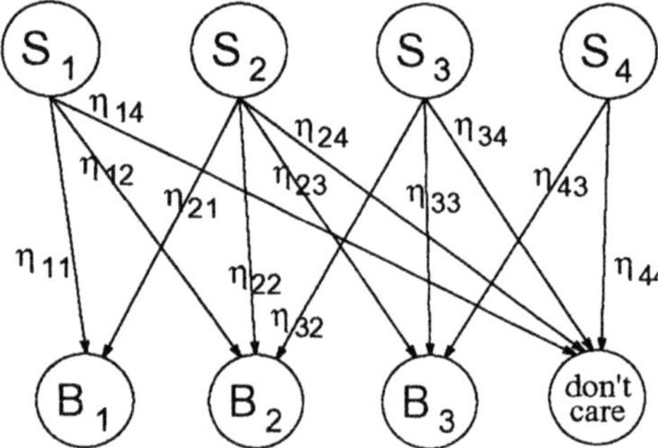

Fig. 6. Graph used in the COR-based ACO algorithm for the example shown in Fig. 2

4.3 Heuristic Information

The heuristic information on the potential preference of selecting a specific consequent, B_k, in each antecedent combination (subspace) is determined as described in Fig. 7.

For each subspace $S_j \in S^+$ do:

1. Build the sets $E^+(S_j)$ and $C(S_j)$ as shown in Fig. 1.
2. For each $B_k \in C(S_j)$, make use of an initialization function based on covering criteria to give a heuristic preference degree to each choice. Many different possibilities may be considered. Three of them are the following:
   (a) $\eta_{jk} = H_1(S_j, B_k) = \max_{e_l \in E^+(S_j)} Min \ \ \mu_{A^j}(x^l), \mu_{B_k}(y^l)$.
   (b) $\eta_{jk} = H_2(S_j, B_k) = \frac{1}{|E^+(S_j)|} \sum_{e_l \in E^+(S_j)} Min \ \ \mu_{A^j}(x^l), \mu_{B_k}(y^l)$.
   (c) $\eta_{jk} = H_3(S_j, B_k) = H_1(S_j, B_k) \cdot H_2(S_j, B_k)$.
   with $\mu_{A^j}(x^l) = Min \ \ \mu_{A_1^j}(x_1^l), \ldots, \mu_{A_n^j}(x_n^l)$.
3. For each $B_k \notin C(S_j)$, make $\eta_{jk} = 0$.
4. Finally, for the "don't care" symbol, make the following:

$$\eta_{j,|\mathcal{B}|+1} = \frac{1}{|\mathcal{B}|} \sum_{B_k \in \mathcal{B}} \eta_{jk}.$$

Fig. 7. Heuristic assignment process

4.4 Pheromone Initialization

The initial pheromone value of each assignment is obtained as follows:

$$\tau_0 = \frac{1}{|S^+|} \sum_{S_j \in S^+} \max_{B_k \in \mathcal{B}} \eta_{jk}.$$

In this way, the initial pheromone will be the mean value of the path constructed taking the best consequent in each rule according to the heuristic information (a greedy assignment).

4.5 Fitness Function

The fitness function will be the said MSE, defined in Fig. 1.

4.6 Ant Colony System with Local Search Algorithm

Once the previous components have been defined, an ACO algorithm has to be given to solve the problem. In this contribution, the well-known ant colony system [9] is considered. A local search procedure is added to improve the behavior. The components adapted to our problem is introduced in the following.

Solution Construction. The algorithm introduces a transition rule that establishes a balance between biased exploration and exploitation of the available information. The node k (i.e., the consequent B_k) is selected for the subspace S_j as follows:

$$k = \begin{cases} arg \max_{R_u \in CR(S_j)} \{(\tau_{ju})^\alpha \cdot (\eta_{ju})^\beta\}, & \text{if } q < q_0 \\ T, & \text{otherwise} \end{cases},$$

with τ_{jk} being the pheromone of the trail (j,k); η_{jk} being the heuristic information; α and β being parameters which determine the relative influence of the pheromone strength and the heuristic information; q being a random variable uniformly distributed over $[0,1]$; $q_0 \in [0,1]$ being a threshold defining the probability of selecting the more hopeful coupling (exploitation); and with T being a random node selected according to the following transition rule (biased exploration):

$$p(j,k) = \begin{cases} \dfrac{(\tau_{jk})^\alpha \cdot (\eta_{jk})^\beta}{\sum_{R_u \in CR(S_j)} (\tau_{ju})^\alpha \cdot (\eta_{ju})^\beta}, & \text{if } R_k \in CR(S_j) \\ 0, & \text{otherwise} \end{cases},$$

We should note that, as in the QAP, the transition rule becomes an assignment rule but, contrary to that problem, there is not a need for the ant to keep a tabu list with the previous assignments made, since the same consequent can be assigned to different rules.

Pheromone Trail Update Rule. The pheromone trail update rule is performed in two stages, global and local:

- *Global pheromone trail update rule*: Only an ant – the one who generated the best solution (T_{best}) till now – releases pheromone on a coupling. The formula is the following:

$$\tau_{jk} \leftarrow (1-\rho) \cdot \tau_{jk} + \rho \cdot \Delta\tau_{jk} ,$$

  with $\rho \in [0,1]$ being the pheromone evaporation parameter, m being the number of ants, T_a being the solution constructed by the ant a, and with

$$\Delta\tau_{jk} = \begin{cases} \dfrac{1}{\text{MSE}(RB_{best})}, & \text{if } (j,k) \in T_{best} \\ 0, & \text{otherwise} \end{cases}.$$

- *Local pheromone trail update rule*: Each time an ant covers a coupling, a local pheromone update is done as follows:

$$\tau_{jk} \leftarrow (1-\rho) \cdot \tau_{jk} + \rho \cdot \Delta\tau_{jk} .$$

  In this paper, we will consider $\Delta\tau_{jk} = \tau_0$ [9].

Local Search. One of the most usual ways to improve the performance of ACO algorithms is the use of local search techniques [19,2,24]. This approach entails employing a local optimization technique to refine the solutions obtained after one or several iterations.

In spite of using local search procedures usually improve the efficacy of the ACO algorithm, it increases the number of evaluations at each iteration and therefore the runtime of the learning method, thus losing efficiency. Moreover, we must consider that in our problem, opposite to other applications, the time needed to evaluate a neighboring solution is greater than the one needed to construct a new solution.

The local search technique usually is applied to the solution generated by each ant. However, in order to accelerate the process, in our case *the local search process will be only applied to the best solution generated at each iteration.* After this process, the global pheromone trail is updated in the usual way.

The proposed local search will consist of the simple hill-climbing algorithm described in Fig. 8.

Let $C_i = \{B_1^i, \ldots, B_{c_i}^i\}$ – with $c_i \in \{1, \ldots, N_c\}$ – be the candidate consequent set of the i-th rule. Let LSi and LSn be two values previously given by the learning method designer to respectively define the maximum number of iterations and the number of neighbors to create at each iteration. Do the following:

Let $\mathcal{S}_{best} = \{R_1, \ldots, R_j, \ldots, R_{|S^+|}\}$ be the solution corresponding to the best track found in the current ACO algorithm iteration.

Set $\mathcal{S}_{cur} \leftarrow \mathcal{S}_{best}$.

For $h = 1, \ldots, LSi$ do:

For $q = 1, \ldots, LSn$ do:

- Obtain the solution $\mathcal{S}'_q$ applying a *neighbor generation mechanism* to $\mathcal{S}_{cur}$, $\mathcal{S}'_q \leftarrow N(\mathcal{S}_{cur})$. This operator randomly selects a specific $j \in \{1, \ldots, |S^+|\}$ and changes R_j by $R'_j \in CR(S_j) - \{R_j\}$. Therefore, $\mathcal{S}'_q = \{R_1, \ldots, R'_j, \ldots, R_{S^+}\}$.
- If $q = 1$, set $\mathcal{S} \leftarrow \mathcal{S}'_1$. Else, if $\mathcal{S}'_q$ is better than $\mathcal{S}$, set $\mathcal{S} \leftarrow \mathcal{S}'_q$.

- If $\mathcal{S}$ is better than $\mathcal{S}_{cur}$, set $\mathcal{S}_{cur} \leftarrow \mathcal{S}$ and continue. Otherwise, break the loop.

Set the best track to the optimized solution, $\mathcal{S}_{best} \leftarrow \mathcal{S}_{cur}$.

Fig. 8. Local search process used in the ant colony system algorithm

5 Experimental Study

This section shows and analyzes some experimental results obtained by the COR-based ACO algorithm previously presented.

Two different approaches within the COR methodology are used. The former one (COR-1), makes a very strong search space reduction that directly affects to the number of rules generated and the solution space explored (see Sect. 3.1 for more details). This version uses the approaches 1.1.1 and 1.2.1 in the algorithm shown in Fig. 1. The latter one (COR-2), however, tackles a wider search space that, although generates higher number of rules, leads the method to obtain more accurate solutions. This second version uses the approaches 1.1.2 and 1.2.2 in the algorithm of Fig. 1.

Moreover, four methods have been selected to compare their performance with our proposals. The first one, proposed by Wang and Mendel in [28], is a simple algorithm that, although does not obtain good accuracy results, is a traditional reference in the area. The second one, proposed by Nozaki *et al.* in [21], uses linguistic fuzzy rules with double consequents and weights associated to them, moreover of considering an additional membership function parameter α to perform a non-linear scaling over the membership functions. The third one, proposed by Thrift in [27], is a basic GA-based learning method that only defines the fuzzy rule set. Finally, the fourth process, proposed by Liska and Melsheimer in [17], is a sophisticated learning method based on two stages that firstly designs the fuzzy rule set and the corresponding membership functions with a genetic algorithm-based process and then performs a final tuning process to refines the model.

On the other hand, the analyzed methods have been applied to two different real-world problems[1]. Table 1 collects their main characteristics.

Table 1. Summary on the two applications considered and their main characteristics

Application	#V	#Tra	#Tst	#LT
Electrical line length	2	396	99	7
Electrical maintenance costs	4	847	212	5

#V = number of input variables, #Tra = training data set size, #Tst = test data set size, #LT = number of linguistic terms considered for each fuzzy partition

These applications are briefly described in the two following subsections. After that, the obtained results and an analysis of them are introduced in Sect. 5.3.

[1] The data sets used in this experiment can be downloaded at the web page `http://decsai.ugr.es/~casillas/FMLib/`

5.1 The Electrical Low Voltage Line Length Problem

This problem involves finding a model that relates the total length of low voltage line installed in Spanish rural towns [7]. This model will be used to estimate the total length of line being maintained by an electrical company. We were provided with a sample of 495 towns in which the length of line was actually measured and the company used the model to extrapolate this length over more than 10,000 towns with these properties. We will limit ourselves to the estimation of the *total length of low voltage line installed in a town*, given the inputs *number of inhabitants of the town* and *distance from the center of the town to the three furthest clients*. To develop the different experiments in this contribution, the sample has been randomly divided in two subsets, the training and test ones, with an 80%-20% of the original size respectively. Thus, the training set contains 396 elements, whilst the test one is composed by 99 elements. Seven labels are considered.

5.2 The Electrical Network Maintenance Costs Problem

Estimating the maintenance costs of an electrical network in a town [7] is a complex but interesting problem. Since an actual measure is very difficult to obtain when medium or low voltage lines are used, the consideration of models becomes useful. These estimations allow electrical companies to justify their expenses. Moreover, the model must be able to explain how a specific value is computed for a certain town. Our objective will be to relate the *maintenance costs of medium voltage line* with the following four variables: *sum of the lengths of all streets in the town*, *total area of the town*, *area that is occupied by buildings*, and *energy supply to the town*. We will deal with estimations of minimum maintenance costs based on a model of the optimal electrical network for a town. We were provided with a sample of 1,059 simulated towns.

The sample has been randomly divided into two subsets, the training one with 847 elements and the test one with 212 elements (80%-20%). Five linguistic terms for each variable are considered.

5.3 Results and Analysis

Table 2 collects the results obtained by the analyzed learning methods, where *#R* stands for the number of rules, MSE_{tra} and MSE_{tst} for the values of the MSE over the training and test data sets respectively, and *EBS* for the number of evaluations needed to obtain the best solution. The best results for both applications are shown in boldface.

As regards the values of parameters used in the Nozaki *et al.*'s method, the best results were obtained with $\alpha = 5$ in both problems, which are the results shown in the table. In the Thrift's method, a population size of 61

Table 2. Results obtained by the analyzed methods

	Electrical line length				*Electrical maintenance costs*			
Method	**#R**	**MSE$_{tra}$**	**MSE$_{tst}$**	**EBS**	**#R**	**MSE$_{tra}$**	**MSE$_{tst}$**	**EBS**
Wang [28]	24	222,654	239,962	—	66	71,294	80,934	—
Nozaki [21]	64	185,395	170,489	—	536	35,859	42,218	—
Thrift [27]	49	169,077	175,739	26,706	534	34,063	42,116	40,464
Liska [17]	49	**167,014**	167,383	47,672	625	57,911	69,277	43,409
COR-1	**19**	171,492	188,966	2,785	**52**	47,096	51,649	**7,987**
COR-2	29	173,823	**160,753**	**417**	234	**30,348**	**39,435**	12,837

individuals, 1,000 generations, 0.6 as crossover probability, and 0.2 as mutation probability per chromosome were used. In the Liska and Melsheimer's method, 61 individuals, 1,000 generations, 49 and 625 as maximum number of rules for the electrical line length and electrical maintenance costs problems respectively, 0.6 as crossover probability, 0.1 as mutation probability per chromosome, and 0.1 as creep probability were used.

In the ACO algorithm of the COR method, the values of parameters were: $\rho = \{0.2, 0.4, 0.6, 0.8\}$ (any of these values obtained the same results), $\alpha = \{1, 2\}$ (the same results with both values), $\beta = 2$ and $\beta = 1$ for the electrical line length and the electrical maintenance costs problems respectively, $q_0 = 0.4$ and $q_0 = 0.2$ respectively for each problem, $LSi = 10$, and $LSn = 32$ and $LSn = 20$ respectively for each problem. With respect to the heuristic information considered, the H_3 and H_1 functions (see Sect. 4.3) was used for the electrical line length and the electrical maintenance costs problems, respectively. The number of iterations were 50 for all the cases.

From the obtained results, we can verify the good behavior of the COR methodology. The proposal COR-2 obtains the best generalization degrees (MSE$_{tst}$) in both problems. Moreover, this good accuracy is attained using lesser number of rules than the rest of analyzed methods, which improves the interpretability of the model.

Focusing on the two COR-based proposals shown, we can see the different interpretability and accuracy degrees obtained by them. The COR-1 method performs a very strong search space reduction that allows it to obtain fuzzy models with very compact rule sets, i.e., with small number of rules. Therefore, this approach is very suitable when the interpretability of the obtained model is an important issue. On the contrary, the COR-2 method generates fuzzy models with higher number of rules, thus obtaining better accuracy degrees. These results lead us to see the COR methodology as a useful tool to regulate the trade-off between both requirements.

Compared with the Nozaki *et al.*'s method, the COR-2 method obtains more accurate and interpretable models. While the former method needs using weighted double-consequents rules and a non-linear scaling factor to attain the shown accuracy, our method obtains more accurate method with simple, but cooperative linguistic fuzzy rules.

Opposite to the Thrift's method, COR-2 also obtains more accurate models in the second problem and better generalization degree in the former, with a significantly lesser number of rules. We must remark the difference of accuracy in the electrical maintenance costs, where in spite of being the search space of our method included into the one of the Thrift's method, a best solutions is found with the COR-based method. This fact relates with the good exploration of the search space performed by our method by reducing the possible solution set and using heuristic information during the search.

Compared with the Liska and Melsheimer's method, one or both COR-based methods again obtain better accuracy degrees and significantly less number of rules. It is remarkable the fact that COR-1 method uses a fuzzy rule base 92% smaller than the one used by Liska and Melsheimer's method in the second problem, obtaining moreover better accuracy degrees. This turns out to be more surprising if we realize the latter method also derives the shape of the membership functions. The results seem to be related with the search spaces tackled by both approaches, which in the Liska and Melsheimer's method case is excessively large to be properly explored.

Finally, we may analyze the quickness of each learning process by comparing the number of evaluations of the fitness function needed to find the solution finally returned by the optimization algorithms. Thus, comparing the Thrift's method, the Liska and Melsheimer's one, and our proposals, we can verify that our methods not only obtain accurate and interpretable models, but they also generates them a far more quick than the other two methods. This aspect is interesting when several learning methods are hybridized to perform a more sophisticated modeling process and makes COR methodology very suitable for such purposes.

6 Concluding Remarks

This chapter has introduced a learning methodology to quickly generate accurate and simple linguistic fuzzy models, the COR methodology. It is based on a combinatorial search of fuzzy rules performed over a set of candidate ones to find those best cooperating. Therefore, instead of selecting the consequent with the highest performance in each fuzzy input subspace as other methods usually do, COR considers the possibility of using another consequent, different from the best one, when it allows the fuzzy model to be more accurate thanks to having a rule set with best cooperation.

The obtained experimental results lead us to think that the simple learning procedure performed by the COR methodology obtains linguistic fuzzy models with an excellent interpretability and good accuracy. The interpretability is ensured by keeping the membership functions and the model structure unaltered, as well as generating a compact rule base. The accuracy is achieved by developing a smart search space reduction and by inducing the generation of a linguistic fuzzy rule set with good cooperation.

References

1. E. Bonabeau, M. Dorigo, and G. Theraulaz. *Swarm intelligence. From natural to artificial systems.* Oxford University Press, Oxford, UK, 1999.
2. B. Bullnheimer, R.F. Hartl, and C. Strauss. A new rank based version of the ant system: a computational study. *Central European Journal for Operations Research and Economics*, 7(1):25–38, 1999.
3. J. Casillas, O. Cordón, and F. Herrera. COR: A methodology to improve ad hoc data-driven linguistic rule learning methods by inducing cooperation among rules. *IEEE Transactions on Systems, Man, and Cybernetics—Part B: Cybernetics.* To appear.
4. J. Casillas, O. Cordón, and F. Herrera. Different approaches to induce cooperation in fuzzy linguistic models under the COR methodology. In B. Bouchon-Meunier, J. Gutiérrez-Ríos, L. Magdalena, and R.R. Yager, editors, *Techniques for constructing intelligent systems*, volume 1, pages 321–334. Physica-Verlag, Heidelberg, Germany, 2002.
5. O. Cordón, M.J. del Jesus, and F. Herrera. Genetic learning of fuzzy rule-based classification systems cooperating with fuzzy reasoning methods. *International Journal of Intelligent Systems*, 13:1025–1053, 1998.
6. O. Cordón and F. Herrera. A proposal for improving the accuracy of linguistic modeling. *IEEE Transactions on Fuzzy Systems*, 8(3):335–344, 2000.
7. O. Cordón, F. Herrera, and L. Sánchez. Solving electrical distribution problems using hybrid evolutionary data analysis techniques. *Applied Intelligence*, 10(1):5–24, 1999.
8. M. Dorigo and G. Di Caro. The ant colony optimization meta-heuristic. In D. Corne, M. Dorigo, and F. Glover, editors, *New ideas in optimization*, pages 11–32. McGraw-Hill, New York, NY, USA, 1999.
9. M. Dorigo and L.M. Gambardella. Ant colony system: a cooperative learning approach to the traveling salesman problem. *IEEE Transactions on Evolutionary Computation*, 1(1):53–66, 1997.
10. M. Dorigo, V. Maniezzo, and A. Colorni. The ant system: optimization by a colony of cooperating agents. *IEEE Transactions on Systems, Man, and Cybernetics—Part B: Cybernetics*, 26(1):29–41, 1996.
11. J. Espinosa and J. Vandewalle. Constructing fuzzy models with linguistic integrity from numerical data-AFRELI algorithm. *IEEE Transactions on Fuzzy Systems*, 8(5):591–600, 2000.
12. A. González and R. Pérez. A study about the inclusion of linguistic hedges in a fuzzy rule learning algorithm. *International Journal of Uncertainty, Fuzziness and Knowledge-Based Systems*, 7(3):257–266, 1999.
13. F. Herrera, M. Lozano, and J.L. Verdegay. A learning process for fuzzy control rules using genetic algorithms. *Fuzzy Sets and Systems*, 100:143–158, 1998.
14. H. Ishibuchi, K. Nozaki, N. Yamamoto, and H. Tanaka. Selecting fuzzy if-then rules for classification problems using genetic algorithms. *IEEE Transactions on Fuzzy Systems*, 3(3):260–270, 1995.
15. Y. Jin, W. von Seelen, and B. Sendhoff. On generating FC^3 fuzzy rule systems from data using evolution strategies. *IEEE Transactions on Systems, Man, and Cybernetics—Part B: Cybernetics*, 29(4):829–845, 1999.
16. A. Krone, P. Krause, and T. Slawinski. A new rule reduction method for finding interpretable and small rule bases in high dimensional search spaces. In

Proceedings of the 9th IEEE International Conference on Fuzzy Systems, pages 693–699, San Antonio, TX, USA, 2000.
17. J. Liska and S.S. Melsheimer. Complete design of fuzzy logic systems using genetic algorithms. In *Proceedings of the 3rd IEEE International Conference on Fuzzy Systems*, pages 1377–1382, Orlando, FL, USA, 1994.
18. B.-D. Liu, C.-Y. Chen, and J.-Y. Tsao. Design of adaptive fuzzy logic controller based on linguistic-hedge concepts and genetic algorithms. *IEEE Transactions on Systems, Man, and Cybernetics—Part B: Cybernetics*, 31(1):32–53, 2001.
19. V. Maniezzo and A. Colorni. The ant system applied to the quadratic assignment problem. *IEEE Transactions on Knowledge and Data Engineering*, 11(5):769–778, 1999.
20. D. Nauck and R. Kruse. Neuro-fuzzy systems for function approximaton. *Fuzzy Sets and Systems*, 101(2):261–271, 1999.
21. K. Nozaki, H. Ishibuchi, and H. Tanaka. A simple but powerful heuristic method for generating fuzzy rules from numerical data. *Fuzzy Sets and Systems*, 86(3):251–270, 1997.
22. N.R. Pal and K. Pal. Handling of inconsistent rules with an extended model of fuzzy reasoning. *Journal of Intelligent and Fuzzy Systems*, 7:55–73, 1999.
23. W. Pedrycz, R.R. Gudwin, and F.A.C. Gomide. Nonlinear context adaptation in the calibration of fuzzy sets. *Fuzzy Sets and Systems*, 88(1):91–97, 1997.
24. M. Setnes and H. Hellendoorn. Orthogonal transforms for ordering and reduction of fuzzy rules. In *Proceedings of the 9th IEEE International Conference on Fuzzy Systems*, pages 700–705, San Antonio, TX, USA, 2000.
25. Y. Shi, R. Eberhart, and Y. Chen. Implementation of evolutionary fuzzy systems. *IEEE Transactions on Fuzzy Systems*, 7(2):109–119, 1999.
26. M. Sugeno and T. Yasukawa. A fuzzy-logic-based approach to qualitative modeling. *IEEE Transactions on Fuzzy Systems*, 1(1):7–31, 1993.
27. P. Thrift. Fuzzy logic synthesis with genetic algorithms. In R.K. Belew and L.B. Booker, editors, *Proceedings of the 4th International Conference on Genetic Algorithms*, pages 509–513, San Mateo, CA, USA, 1991. Morgan Kaufmann Publishers.
28. L.-X. Wang and J.M. Mendel. Generating fuzzy rules by learning from examples. *IEEE Transactions on Systems, Man, and Cybernetics*, 22(6):1414–1427, 1992.
29. L.A. Zadeh. The concept of a linguistic variable and its application to approximate reasoning. Parts I, II and III. *Information Science*, 8, 8, 9:199–249, 301–357, 43–80, 1975.

Constrained optimization of genetic fuzzy systems

France Cheong[1] and Richard Lai[2]

[1] School of Business IT,
RMIT University, Melbourne, Victoria 3000, Melbourne
[2] Department of Computer Science and Computer Engineering,
La Trobe University, Bundoora, Victoria 3083, Australia

Abstract. Although fuzzy systems and artificial neural networks are both universal function approximators, fuzzy systems have the advantage of using a form of knowledge representation which in general is interpretable by human beings. However, unlike neural networks, fuzzy systems have no built-in learning mechanism. To compensate for this deficiency, genetic algorithms (GAs) have often been used to automate the design of fuzzy systems. Unfortunately, the inconsiderate use of GAs and the temptation to automate every conceivable aspect of the design can lead to the design of fuzzy systems with obscure structures that are not easily interpretable. In this chapter, we describe our efforts for constraining the use of genetic algorithms and evolutionary programming in order to design fuzzy logic controllers (fuzzy systems used in control applications) which are both accurate and interpretable.

1 Introduction

Fuzzy systems are systems that tolerate a reasonable amount of imprecision, vagueness and uncertainty with the aim of reducing complexity. They use fuzzy logic (or continuous-valued logic) which is based on Zadeh's theory of fuzzy sets [1–4]. Fuzzy systems are similar to artificial neural networks (optimization and learning algorithms based loosely on concepts inspired by research into the nature of the brain) [5–10] in the sense that they are both model-free input-output function approximators. Unlike statistical approximators, they approximate a function without a mathematical model of how the outputs depend on the inputs. They are dynamical systems that are trained to approximate input-output functions from sample data.

Fuzzy systems are often preferred to neural networks as function approximators because they use a form of knowledge representation that is interpretable by human beings. In fuzzy systems, knowledge is structured as production rules which ressemble the rules of expert systems except that they are of a fuzzy nature. Unlike fuzzy systems, the knowledge of neural networks is generally not interpretable by human beings as they are encoded as weights in the network.

The knowledge of these function approximators is normally acquired using an automated training technique which approximates the functions from sample data. It is not possible to configure the weights of a neural network manually. An automated method must be used and neural networks have built-in learning mechanisms such as the back-propagation algorithm for performing this function. Fuzzy systems do not possess a built-in learning mechanism and since the knowledge representation is interpretable by human beings, it is possible to manually configure a fuzzy system. It is often very tempting to manually configure a fuzzy system, claiming that expert (or naive!) knowledge about the application domain was used, especially in situations where it is hard to obtain sample data. However, there is no guarantee that such a system will perform satisfactorily over a wide range of operating conditions. We believe that any available expert knowledge about the application domain is better used for generating, filtering or augmenting sample data rather than direct tuning of the fuzzy system. Once sample data is available, the learning of the function which maps the inputs to the output(s) is best left to an automated technique.

To compensate for the learning inability of fuzzy systems, artificial neural networks and genetic algorithms (GAs) have often been used as learning mechanisms in view of automating the design of optimized fuzzy systems. GAs are exploratory search and optimization techniques based on the principles of natural evolution and population genetics. Although the creation of GAs is often attributed to Holland [11], in fact Fraser [12] and Bremermann *et al.* [13] proposed similar algorithms before Holland. Fuzzy systems utilizing neural networks as learning mechanisms are known as neuro-fuzzy systems (NFSs) [14–16] while those using GAs are known as genetic fuzzy systems (GFSs) [17–26] .

We believe that in many cases, GAs have been used inconsiderately in the design of fuzzy systems. Furthermore, since GAs are fairly simple to use, researchers have been tempted to automate every conceivable aspect of the design process. We believe that all these factors have led to the design of fuzzy systems with obscure structures that are not easily interpretable. In this chapter, we describe how we constrain evolutionary algorithms in the design of fuzzy systems in order to preserve the interpretability of the designed fuzzy systems while maintaining their accuracy. We also demonstrate how we leverage expert knowledge about the application domain to design and configure optimized genetic fuzzy systems.

2 Fuzzy logic control

The biggest success of fuzzy systems in industrial and commercial applications has been achieved with fuzzy logic controllers (FLCs). FLCs are fuzzy systems used in control applications and were first introduced by Assilian and Mamdani [27]. FLCs are gaining in popularity across a broad array of disciplines because they allow a more human approach to control. Mathematical modelling knowledge as used in conventional control design methods is not required, and this is a great advantage since with the increasing complexity of systems, the ability to describe them mathematically becomes difficult.

Figure 1 shows the main components of a fuzzy logic control system.

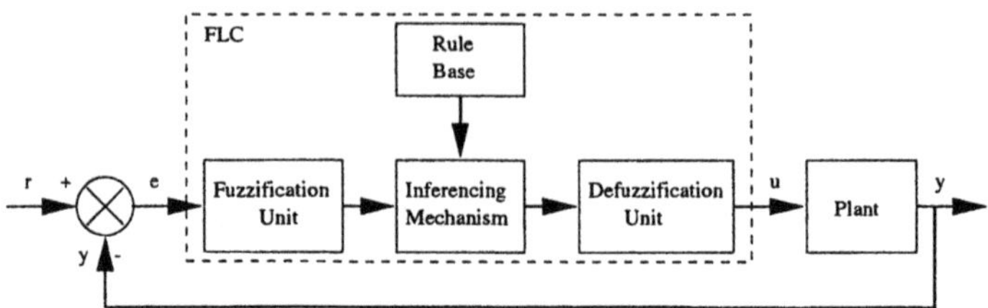

Fig. 1. Components of a fuzzy logic control system

The fuzzification unit converts the *crisp* error values (usually the error and its derivative for a Proportional-Derivative-like controller) into *fuzzy* values. Figure 2a shows the partitioning of the universe of discourse of the error into three fuzzy sets described by triangular membership functions. It shows that an error value of 0.125 belongs to the fuzzy sets *Zero* and *Positive* with membership values of 0.875 and 0.125 respectively. The membership value represents the degree to which a crisp value belongs to a fuzzy set. Similarly, the change in error is fuzzified using the membership functions of its reference fuzzy sets (i.e the number and shape of the fuzzy sets).

The inferencing mechanism uses the input fuzzy sets and their membership values to look up the appropriate rules in a rule base to compute a fuzzy output value. Figure 2b depicts a rule base represented as a 3 by 3 fuzzy associative matrix (FAM) [14].

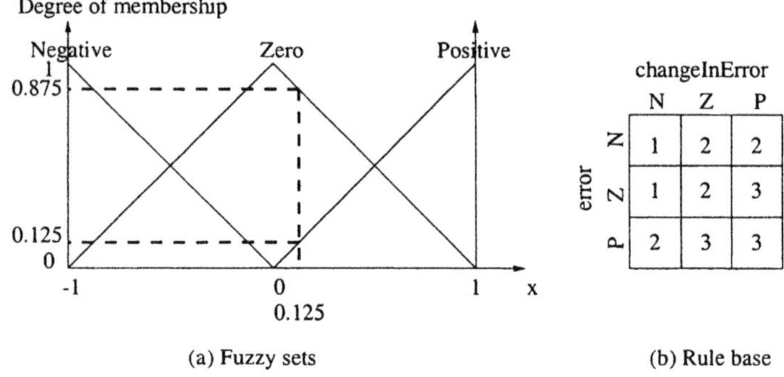

(a) Fuzzy sets (b) Rule base

Fig. 2. Fuzzy sets and rule base

The rows of the matrix represent the three fuzzy sets of the *error*, the columns represent the three fuzzy sets of the *changeInError* and the cells of the matrix represent the fuzzy sets of the *output* which are denoted as 1,2 and 3 for *Negative*, *Zero* and *Positive* respectively to facilitate representation and manipulation. The rules in the rules base are interpreted as follows (in rowwise order):

```
Rule 1: IF error=Negative AND changeInError=Negative
        THEN output=Negative
Rule 2: IF error=Negative AND changeInError=Zero
        THEN output=Negative
                  .
                  .
                  .
Rule 9: IF error=Positive AND changeInError=Positive
        THEN output=Positive
```

The defuzzification unit aggregates the outputs of all the rules that have fired for the particular input fuzzy sets to produce a crisp output.

Currently, there are two main types of FLCs, the Mamdani FLC and the Takagi-Sugeno-Kang FLC [28]. The Mamdani FLC was described in the preceding paragraphs. The inputs to the Proportional-Derivative (PD) version are the error and the change in error and the output is the change in output (not the value of the output itself).

When designing PD-like Mamdani FLCs, expert knowledge is available in the form of the MacVicar-Whelan meta rules [29]. MacVicar-Whelan derived these meta rules by combining the experience gained with the first FLCs developed by Assilian and Mamdani [27] and King and Mamdani [30] with common engineering sense. Assuming the input variables to the FLC are the error and change in error, then a standard template rule base can be built from the following MacVicar-Whelan meta rules:

1. If both the error and change in error are zero, then change in output is zero.
2. If the error is tending to zero at a satisfactory rate, then change in output is zero.
3. If the error is not self-correcting, then change in output is not zero and depends on the sign and magnitude of the error and change in error.

This standard template rule base can be adjusted by excluding, modifying or adding new control rules based on the specificity of the control problem. A standard template 7 by 7 rule base is shown in Figure 3 (see [29,31] for a detailed formulation of the rule base).

Error \ Change in Error	NB	NM	NS	ZE	PS	PM	PB
NB	1	1	1	1	2	3	4
NM	1	2	2	2	3	4	5
NS	1	2	3	3	4	5	6
ZE	1	2	3	4	5	6	7
PS	2	3	4	5	5	6	7
PM	3	4	5	6	6	6	7
PB	4	5	6	7	7	7	7

Fig. 3. 7 by 7 MacVicar-Whelan rule base

3 Problems with genetic fuzzy logic controllers

We have identified two problems with genetic fuzzy logic controllers (GFLCs), more particularly with the uncontrolled use of GAs for optimally configuring the FLCs and we have labelled them *messy fuzzy sets* and *scrambled rule base*.

Figure 4 shows an input variable fuzzified into seven fuzzy sets, Figure 4a showing *well-formed fuzzy sets* and Figure 4b showing *messy fuzzy sets* produced by the uncontrolled use of GAs.

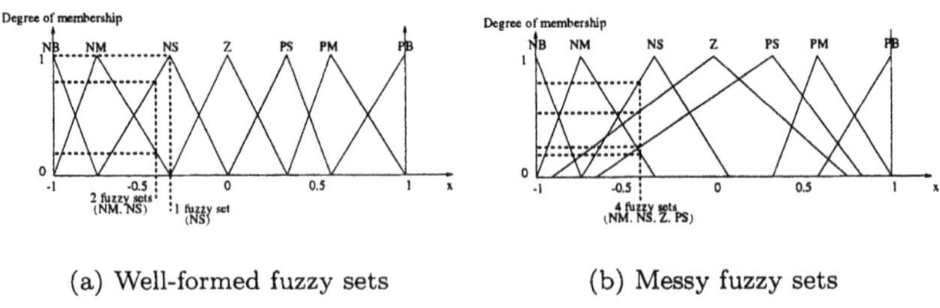

(a) Well-formed fuzzy sets

(b) Messy fuzzy sets

Fig. 4. Fuzzy sets

In well-formed fuzzy sets, a maximum of two fuzzy sets overlap (i.e. a degree of ovelapping of two) at any point in the universe of discourse; this is not the case with messy fuzzy sets. From Figure 4a, it can be seen that fuzzification of the input variable can produce membership of either 1 or 2 fuzzy sets. Membership of 1 fuzzy set is obtained when the input variable assumes a value that is directly under the apex of one of the triangles. In such a case, the membership value is 1.0 for a certain fuzzy set and 0.0 for the other two overlapping fuzzy sets. If the input value assumes any other value, then membership of two fuzzy sets will result. If we are considering an FLC with 2 inputs, then we will have 1, 2 or 4 rules firing at any one time, with 4 rules firing most of the time.

From Figure 4b, it can be seen that if the input variable assumes the value shown in the figure, this will result in membership of 4 fuzzy sets. If we have an FLC with two inputs and if the other input variable results in membership of, say 3 fuzzy sets, this will cause 12 rules to fire. Thus, messy fuzzy sets cause a high number of rules to fire at any one time because of the high number of fuzzy sets overlapping each other. In addition to this high number of rules firing, there is also great variability in the number of rules firing as opposed to the more or less fixed number of 4 (sometimes 1 or 2) for well-formed fuzzy sets. The high and variable number of rules firing at any one time have a negative impact on computation speed and hence this presents a strong case for the elimination of messy fuzzy sets.

Figure 5a shows the rules for a well-formed FLC rule base according to the MacVicar-Whelan meta-rules [29]. The rule base has been modified for 7 fuzzy sets (instead of 8) and the *error* defined as *set point* minus *output* (instead of the other way round) according to Yager and Filev [31]. The *error*, *changeInError* and *output* have been fuzzified into 7 fuzzy sets each. The fuzzy sets are *NegativeBig*, *NegativeMedium*, *NegativeSmall*, *Zero*, *PositiveSmall*, *PositiveMedium* and *PositiveBig*; they are referred to as 1, 2, 3, 4, 5, 6 and 7 respectively. The cells of the matrix contain the fuzzy outputs and their possible range of variation is indicated by the arrows and the values at the bottom left hand corner. The values shown by the arrows indicate the permissible values applicable to the cells of the matrix that lie in the path parallel to the diagonal from bottom left to top right.

In many cases, the unconstrained use of GAs for optimizing the FLC rule base will result into a *scrambled rule base* containing rules which do not make sense. Such a rule base is shown in Figure 5b. The shaded entries are fuzzy outputs which are in contradiction with common sense and the MacVicar-Whelan meta-rules. For example, the first element of the rule base matrix should not contain 7 (*PositiveBig*) since this would request the actuator controlling the system to open much more and this will result in increased error instead of decreasing it. The same reasoning applies to the other elements of the matrix; if they do not contain fuzzy outputs of the correct magnitude and direction, the FLC will not perform correctly. Since the FLC does not produce an output based on a single individual rule (an output is produced by a combination of several rules firing at the same time), incorrect individual rules may not affect the combined output; correct overall output may still be produced. If the scrambled rule base still produces the correct output, then what is the motivation for producing well-formed rule bases? The motivation is knowledge acquisition; well-formed rule bases can be interpreted and analysed for correctness.

Many studies have reported ways for minimizing the number of rules in the FLC rule base [32,33,20,34,35] to decrease computation costs. For example, the number of rules in a 7 by 7 rule base might be minimized from 49 to say 11. We believe that this is not necessary with well-formed fuzzy sets since the

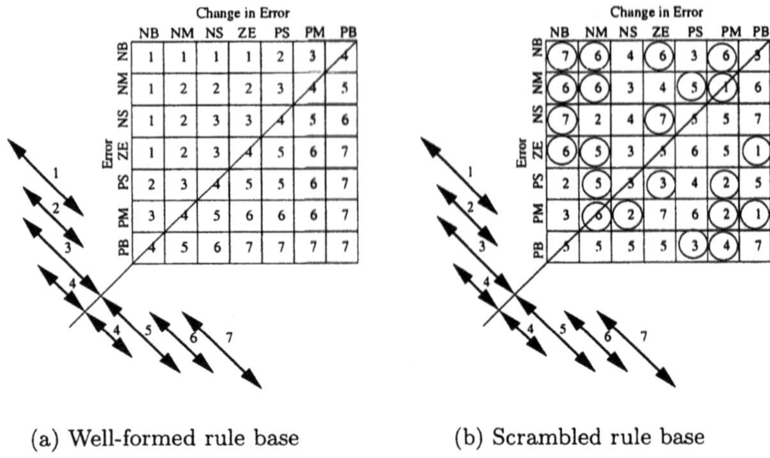

(a) Well-formed rule base

(b) Scrambled rule base

Fig. 5. Rule bases

number of rules firing at any one time will be 4 whatever the size of the rule base (3 by 3, 5 by 5 or 7 by 7). It seems more logical to minimize the number of rules firing at any one time than to minimize the total number of rules in the rule base. Keeping all the possible rules also ensures the robustness of the FLC, since in the event of failure of a rule (if implemented in hardware), one or more other rules will still fire. Had the rule base been minimized, failure of a rule might result in no output at all. With larger rule bases, only the look-up table representing the rule base matrix will be larger; the computation speed will be the same as a smaller rule base.

It can be argued that in fuzzy systems with more than two input variables, keeping all the rules impacts on the interpretability of the rules and storage requirements. This is not the case if hierarchical FLCs (HFLCs) are adopted. In HFLCs, the number of rules do not increase exponentially as the number of inputs increases; instead the number of rules is a linear function of the number of input variables [36].

4 A solution

After having explained the problems associated with unconstrained GFLCs, we now propose a solution to eliminate them and make GFLCs more interpretable while maintaining a high level of accuracy. We performed two sets of experiments to achieve this objective. First of all, we devised an enhanced GA and used it to optimally configure both the rule base and the membership functions of an FLC. We then simplified our solution by using another evolutionary algorithm named evolutionary programming (which is better suited for real-valued optimization than GAs) to configure only the membership functions of the FLC; the rule base used being an unmodified standard MacVicar-Whelan rule base.

Before we describe the experiments we performed for designing interpretable GFLCs, we next present a brief overview of evolutionary algorithms, genetic algorithms and evolutionary programming.

5 Evolutionary algorithms

Evolutionary algorithms (EAs) are a class of algorithms that use some of the known mechanics of evolution, more specifically the processes of selection, reproduction and mutation to search for the best solution to a problem. The interest in EAs is mainly due their flexibility, adaptibility and robustness in solving difficult optimization problems. Unlike many classical optimizing techniques, EAs do not require the computing of local derivatives to guide the search process; only an objective function needs to be computed. Furthermore, EAs are more likely to arrive at the global optimum because they work on a population of points instead of a point by point approach as used by conventional optimization techniques.

The algorithm for a typical EA is shown in Figure 6. Given an optimization problem, the parameters concerned are grouped into a structure (an individual) and a collection of such individuals (a population) is created by either randomly generating the parameters or using expert knowledge about the problem. The EA runs iteratively on the population of individuals using the genetic operators in a random way but based on the fitness of the structures to perform such tasks as selecting, copying, exchanging and perturbing portions of individuals to create new generations of individuals and eventually find the best individual representing the solution to the problem. Currently, there are several EAs: genetic algorithms (GAs), genetic programming (GP), evolution strategies (ESs), evolutionary programming (EP) and differential evolution (DE). They differ in the representation of the problem and the use of genetic operations (selection, reproduction and mutation).

5.1 Genetic algorithms

Genetic algorithms (GAs) were first invented by Fraser [12] and Bremermann *et al.* [13]. They were further developed by Holland [11] and Goldberg [37]. Figure 7 shows the basic algorithm for a simple GA.

In general, a GA consists of three fundamental operators: reproduction (selection), crossover and mutation. Given an optimization problem, the parameters concerned are encoded into a population of chromosomes. The GA then runs iteratively using the three operators in a random way but based on the fitness of the chromosomes to perform the basic tasks of copying and exchanging portions of chromosomes, and finally find and decode the

```
Algorithm EA
1. Start with a randomly initialized population
2. Evaluate fitness of individuals
3. While not done do
        3.1 Increment generation counter
        3.2 Select parents for reproduction
        3.3 Recombine genes of selected parents
        3.4 Mutate population stochastically
        3.5 Evaluate fitness of individuals
        3.6 Select survivors
   End while
```

Fig. 6. Generic evolutionary algorithm

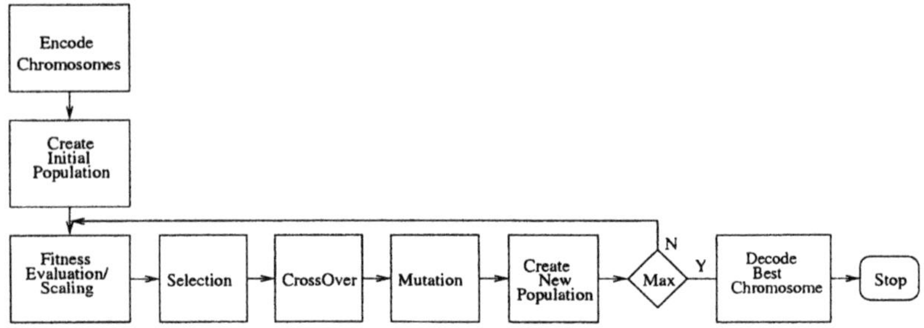

Fig. 7. Algorithm for a simple GA

best chromosomes representing the solution to the problem. Comprehensive overviews and introductions to GAs can be found in [37–39].

5.2 Evolutionary programming

Evolutionary programming (EP) is an EA that was conceived by Fogel et al. [40]. EP has traditionally used representations that are tailored to the problem domain such as real-valued vectors for real-valued optimization problems, ordered lists for travelling salesman problems and graphs for finite state machine problems. EP is an abstraction of the evolution process at the species level and not at the individual level like GAs and ESs. Thus, it does not use the recombination mechanism at all since recombination does not occur within species.

An offspring is created by mutating a parent and this can be expressed mathematically as:

$$\sigma^{t+1} = \sqrt{\alpha + \beta * fitness(X^t)} \quad \text{and} \tag{1}$$

$$X^{t+1} = X^t + N(0, \sigma^{t+1}) \tag{2}$$

where α and β are system parameters. Since it is sometimes hard to set these parameters, there has been several studies [41–43] on the use of adaptive system parameters to optimize them. When adaptive system parameters are used, the system parameters are evolved together with the parameters of the problem to be solved.

Selection of survivors is done by stochastic methods via a tournament based on fitness. If there are n individuals in the population, each individual is mutated to produce an offspring. Each individual from the $2n$ (parents plus offsprings) population is entered into a competition against a pre-selected number of opponents and receives a *win* if its fitness is equal to or better than its opponent. n individuals with the highest *wins* are then selected as the survivors for the next generation. More details on EP can be found in [40,44,45].

6 Constrained optimization of rule base and membership functions

Our goal in this section is to produce FLCs that are optimized as well as interpretable. To this end we propose an enhanced GA which generates optimized FLCs by searching for well-formed fuzzy sets and rule bases and we compare the performance of these FLCs with conventional PD controllers.

We devised specialized FLC genetic operators to maintain the feasibility of the chromosomes by preventing the formation of messy fuzzy sets and scrambled rule bases during the optimizing process. We also experimented with a penalty function defined for the purpose. The FLC-specific methods for coding, initialization, mutation and migration of chromosomes are described below.

Our design of FLCs is based on the use of an odd number of fuzzy sets as opposed to an even number in order to eliminate *positive zero* and *negative zero* fuzzy sets in the universe of discourse and we limited the possible number of fuzzy sets to 3, 5 and 7. The role of the enhanced GA is to generate three optimized FLCs (3-fuzzy-set, 5-fuzzy-set and 7-fuzzy-set controllers) by searching for the optimal configuration of membership functions and rule bases. The systems designer can then choose one of the three FLCs depending on the requirements of a particular situation.

6.1 Constraints

There are several strategies that can be used by GAs for constrained optimization problems and the existing strategies can be roughly classified into [46]:

- *Rejecting strategy:* All infeasible (or undesirable) solutions are discarded during the evolution process.
- *Repairing strategy:* Infeasible solutions are converted to feasible ones.
- *Modified genetic operators strategy:* The feasibility of chromosomes are maintained by specialized problem-specific genetic operators. The advantage of this strategy is that it never generates infeasible solutions; however, it does not consider points outside the feasible region and movement through infeasible regions of the search space might lead to faster optimization and better solutions than limiting search trajectories only to feasible regions [47].
- *Penalty strategy:* This strategy essentially converts the constrained optimization problem into an unconstrained problem by penalizing infeasible solutions. The main issue with this strategy is the formulation of the penalty function for effective search of the optimum solution. Several techniques have been proposed, but there are no general guidelines for its formulation as it is often problem-dependent [46].

Our proposed constrained GA for optimising FLCs is based on the third strategy. Using the modified genetic operators strategy, we maintained the feasibility of chromosomes by preventing the occurrence of messy fuzzy sets and scrambled rule bases at any time during the evolution process. Messy fuzzy sets are prevented from occuring by coinciding of the apex of a triangle representing a fuzzy set with the ends of two other overlapping triangles. Furthermore, we constrained these points to occur within certain ranges in the normalized universe of discourse of [-1, 1]. The apices of the first and last triangles were fixed at -1 and 1 respectively and the apices of the other triangles constrained between range [*x1, x2*] which is defined as follows:

$$[x1, x2] = (2/n * (s-1)) \pm a \tag{3}$$

where $n = 3, 5, 7$, i.e. the number of fuzzy sets in the universe of discourse, $1 < s < n$, i.e. the fuzzy set being considered, and a is a constant whose value has to be determined experimentally. Good values of a were found to be 0.75, 0.37 and 0.25 for 3-, 5- and 7-fuzzy-set universes of discourse. Figure 8 shows the range [*x1, x2*] for the apex of fuzzy set 2 (*Zero*) in a 3-fuzzy-set universe of discourse.

Scrambled rule bases are prevented from occuring by allowing only rules that conform with the MacVicar-Whelan meta rules and tolerating some degree of variation within the range shown in Figure 5a.

6.2 Coding of chromosomes

We used real values for coding the membership functions and integers for coding the rule base instead of the classical *concatenated binary mapping* [37] to code the chromosomes. In the concatenated binary mapping, each parameter is coded as a binary value and the codes of all the parameters

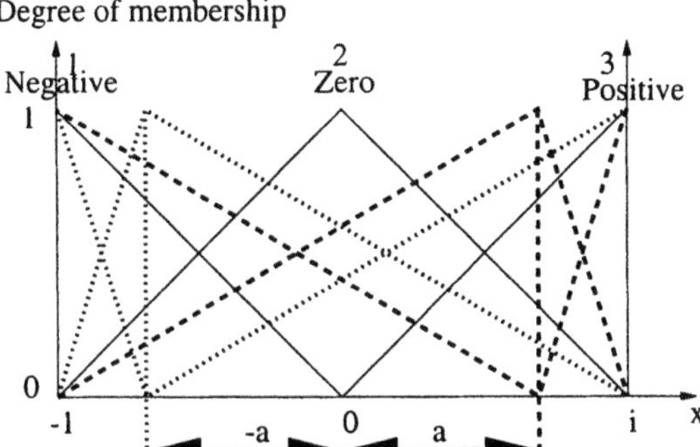

Fig. 8. Range [$x1$, $x2$] for apex of fuzzy set 2 in a 3-fuzzy-set universe of discourse

are joined together into one composite string. We used real/integer values because they are conceptually closer to the problem space than the binary representation and furthermore, they allow easy and efficient implementation of the genetic operators [39].

Figure 8 shows that only one parameter is required to describe the membership functions of a 3-fuzzy-set universe of discourse and this is the location of the apex of the triangle describing the *Zero* fuzzy set; the apices of the other two triangles are fixed at -1 and 1. In several studies, the triangles representing the membership functions of fuzzy sets are described by three parameters each: left, top and right corners of the triangle. With well-formed fuzzy sets this is not necessary because the left and right corners are fixed at points described by the apices of other triangles. In fact, it is the use of three parameters for describing the triangles that cause the occurrence of messy sets, since when the left and right points are loose, they can go anywhere.

The number of parameters required for representing 3-, 5- and 7-fuzzy-set universes of discourse are 1, 3 and 5 respectively, and since we have three variables (error, change in error and output) we need 3, 9 and 15 parameters respectively to code the membership functions. We used a vector of real numbers as opposed to the conventional binary strings to code the membership functions and a matrix of integers of size 3 by 3, 5 by 5 and 7 by 7 for 3-, 5- and 7-fuzzy-set systems respectively. The choice of representation for the membership functions and the rule base was also guided by our strategy for implementing the crossover and mutation operators which are described later.

6.3 Initialization of chromosomes

Chromosomes are normally initialized with random values or with values derived from expert knowledge of the system to be implemented with the latter facilitating earlier convergence to the optimal solution. We used a combination of these two methods since we used random values constrained within certain permissible ranges as determined by our knowledge of the FLCs.

We initialized the parameters of the membership functions with random values but constrained within the permissible ranges as defined by (3). Concerning the initialization of the rule base, we filled it with random values representing the codes assigned to the linguistic labels of the output. In an attempt to conform to the MacVicar-Whelan meta rules, we constrained the random values to occur between the ranges indicated in Figure 9. Furthermore, we fixed the values at the extremities and the centres of the rule bases. We believe that they should be fixed just as we believe that the apices of the first and last triangles describing the fuzzy sets should be fixed at the extremities of the universe of discourse.

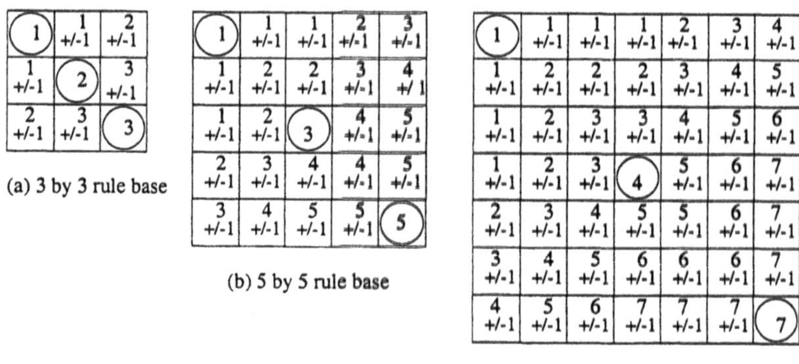

Fig. 9. Initialization of rule bases

6.4 Crossover

Crossover is a genetic operator that aims to combine the better qualities among the preferred chromosomes. It is usually performed by exchanging segments of binary code from parent chromosomes at a randomly selected location (one-point crossover) in order to produce offsprings. In addition to one-point crossover, there are also multiple-point crossovers whereby several segments of code are exchanged instead of one segment only. Figure 10a shows one-point crossover for binary coded chromosomes.

To ensure that the membership functions of the offsprings generated by the crossover operator are always valid (i.e. ordered values within the permissible ranges in the universe of discourse) our crossover operator generates random values in the intervals represented by the values of the parents. For

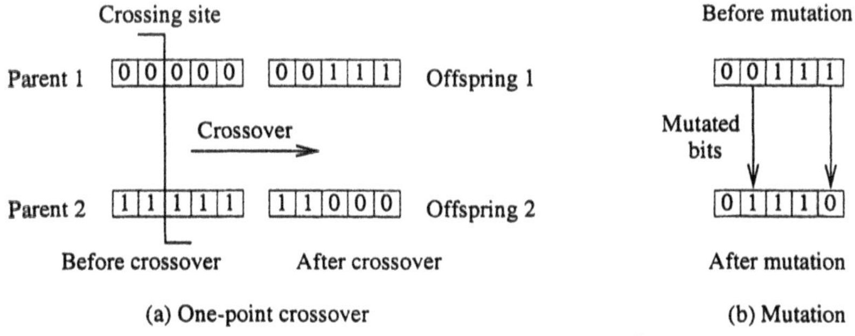

Fig. 10. Conventional genetic operators

example, if the apex of fuzzy set 4 is 0.5 for one parent and 0.6 for another parent, then the apex of that fuzzy set will be a random value between 0.5 and 0.6 for each offspring. Since this method can only produce one offspring, we need to generate another random number within the same interval to obtain a second offspring. The crossover of the membership functions is shown in Figure 11a.

For the reproduction of the rule base, since we are dealing with squares, we exchanged square sections of random sizes and locations between the parents as shown in Figure 11b. Although the generation of one random square is sufficient to produce two offsprings, we chose to generate two random squares to be consistent with the behaviour of our crossover operator for membership functions. This is the reason, why two squares of random sizes and locations are shown in Figure 11b.

6.5 Mutation

Mutation is usually accomplished by arbitrarily altering one or more bits from 1 to 0 and vice-versa at selected locations of the binary coded chromosome. Mutation of binary coded chromosomes is shown in Figure 10b. The idea behind mutation is to introduce some extra genetic variability into the population. We accomplished this by making mutation similar to initialization. Mutation of the membership functions involved generating a random real value within the permissible ranges as defined by (3). Mutation of the rule base involved generating a random integer value within the ranges indicated in Figure 9.

6.6 A Parallel GA

Since we are optimizing three types of FLCs (3-, 5- and 7-fuzzy-set controllers) coded as chromosomes of different sizes we cannot evolve them as

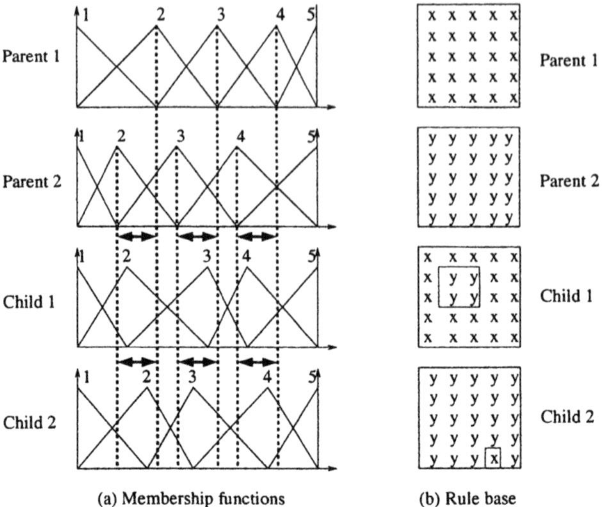

Fig. 11. Crossover

one population because this difference in size would prevent the use of genetic operators such as crossover. Even if we could devise a crossover operator for this situation, this would be a meaningless operation as it is not logical to cross individuals of different species. Thus, we had to evolve three populations of different sized chromosomes representing 3-, 5- and 7-fuzzy-set FLCs. Each population was evolved using a steady-state GA [48] which replaces only a percentage of the population at each new generation. We used the steady-state GA because of its superior performance [49].

However, instead of evolving the three populations sequentially, we evolved them in parallel with migration of individuals (chromosomes) between the populations. Migration is useful because it introduces new genetic material into the populations and this can help avoid premature convergence to local minimas [50]. At the end of each generation we selected 5 best individuals from each population and migrated them to the other populations as shown in Figure 12. Migration of individuals between populations caused a problem because we cannot directly introduce the migrants into the receiving populations due to the difference in chromosome size and structure. We solved the problem by devising a scheme to create migrants from one type of chromosome to another as explained in the next section.

6.7 Migration scheme

Figure 13 shows how the membership functions are converted from one type to another. This is essentially generating random numbers between certain

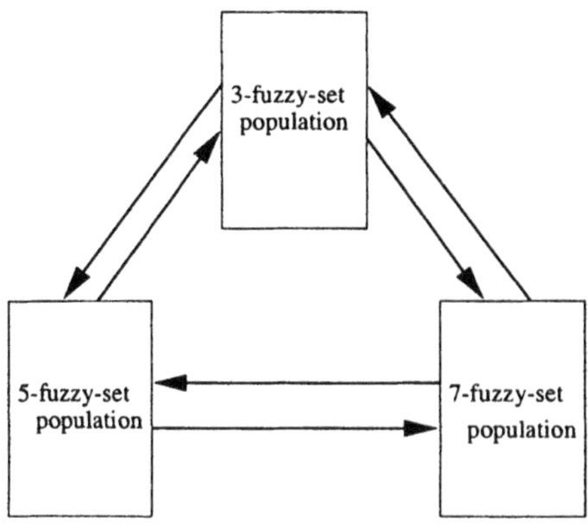

Fig. 12. Migration model

intervals derived from the original chromosome. Figure 13 shows how to convert 3- to 5-, 5- to 7-fuzzy-set membership functions and vice-versa. To convert from 3- to 7-fuzzy-set membership functions and vice-versa, 5-fuzzy-set membership functions have to be used as intermediary.

The conversion of the rule base proved to be more challenging, but since in this study we are aiming to produce rule bases which agree with the MacVicar-Whelan meta-rules, we used the meta-rules to devise the scheme.

When expanding a rule base to a larger one, there are two problems: (a) the codes representing the output have to be mapped to a greater range, and (b) new output codes have to be generated for the extra cells in the rule base. For example, when expanding a 3 by 3 to a 5 by 5 rule base, output codes are mapped from [1,3] to [1,5]. When one output code can be mapped to several ones in the bigger range, one code is selected at random from the possible ones. For example an output of 1 in a 3-fuzzy-set universe of discourse can correspond to 1 or 2 in a 5-fuzzy-set universe of discourse. To generate output values for the extra cells, adjoining cells are copied along the diagonal of the rule base as shown by the arrows in Figure 14a and Figure 14b. Some fixed output values are derived from the meta-rules and inserted at the upper left, bottom right extremities and center of the rule base as shown in the diagram.

Collapsing a rule base to a smaller one involves copying cells at the same locations in the bigger rule base, except for the extremities and the center of the rule base which will contain fixed values from derived from the meta-rules as before. However, in an attempt to use more information from the larger rule base, the values of adjoining cells along the diagonal are also taken into consideration. This is shown by the arrows in Figure 14b and Figure 14d. The value to be copied and mapped to a lower range is randomly selected from either the current cell or one or more adjoining cells.

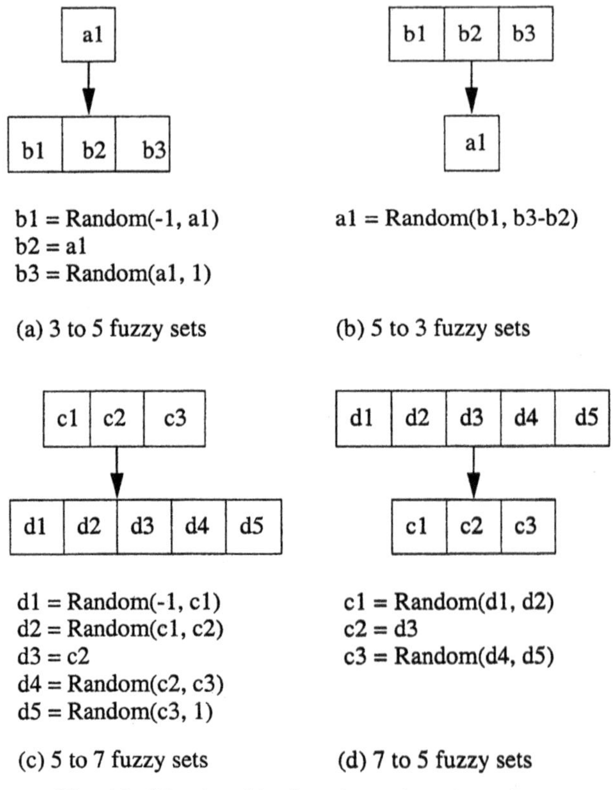

Fig. 13. Membership function migration scheme

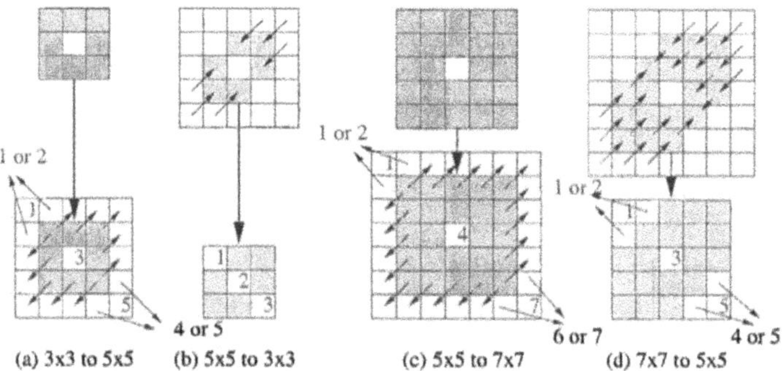

Fig. 14. Rule base migration scheme

6.8 Testing the enhanced GA

We tested the viability of our enhanced GA by generating FLCs for controlling three plant processes with second order transfer functions, labelled plant A, plant B and plant C as used in [51] and defined as follows: $G_A(s) = \frac{2}{s(s+1.4)+2}$, $G_B(s) = \frac{2}{(s+1)(s+2)}$, $G_C(s) = \frac{1}{s(1+0.1s)}$. Figure 15 shows the closed loop step responses of the three plants without and with a conventional PD controller tuned with the help of MATLAB [1].

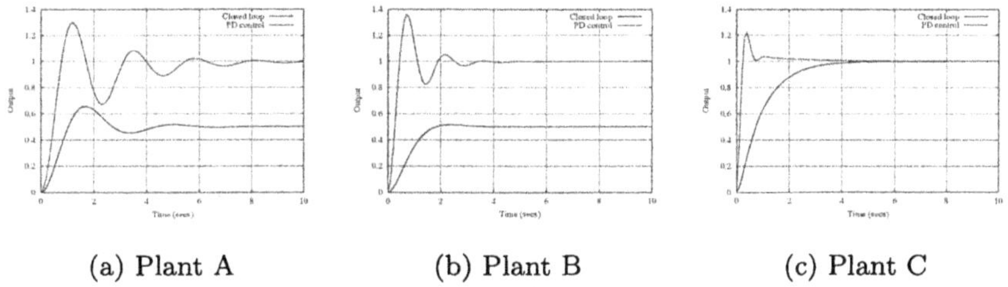

(a) Plant A (b) Plant B (c) Plant C

Fig. 15. Closed loop and PD controller step responses

The performance of the FLCs was judged by unit step responses and we used the Integral-of-Time-multiplied Absolute-Error (ITAE) [2] criterion as objective function. ITAE is defined as $\int t\,|e(t)|\,dt$. Since ITAE is a cost function and since it is common practice to use GAs for maximization problems rather than minimization, we defined the performance index as $1/ITAE$.

6.9 Results and analysis

The step response of the best FLCs obtained is shown in Figures 16, 17 and 18. It can be seen from the figures that these FLCs outperform their conventional counterparts.

Figure 18 also shows that our enhanced GA did find optimum FLC5 and FLC7 but not FLC3 for plant C. Controlling plant C seems to be a hard problem for FLC3 which is handicapped by the small size of its rule base and the constraints imposed by our methodology i.e. only one point can be varied in the 3 reference fuzzy sets and the use of singleton defuzzifier instead of centre of gravity defuzzifier. In most cases, FLC5 or FLC7 would be chosen for a control application although FLC3 could do the job equally well (as was the case for plants A and B). We have included FLC3 in our design methodology to show that it is possible to have a good FLC with 3 fuzzy sets since people tend to automatically favour 5 and 7 fuzzy sets. Furthermore,

[1] MATLAB is a trademark of The Math Works Inc.

[2] It should be noted that the ITAE cost function does not optimally tune the FLCs.

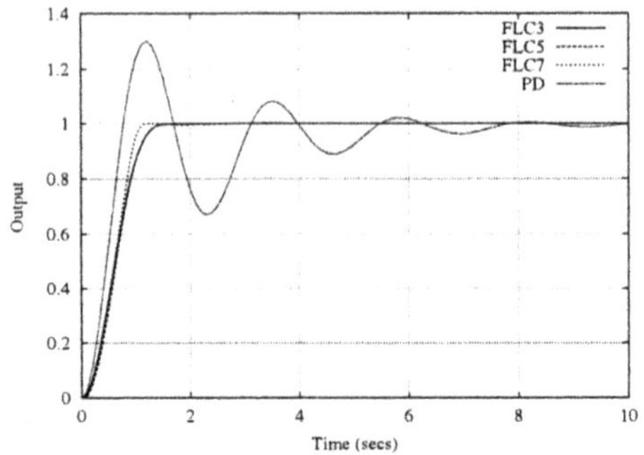

Fig. 16. Step responses for plant A

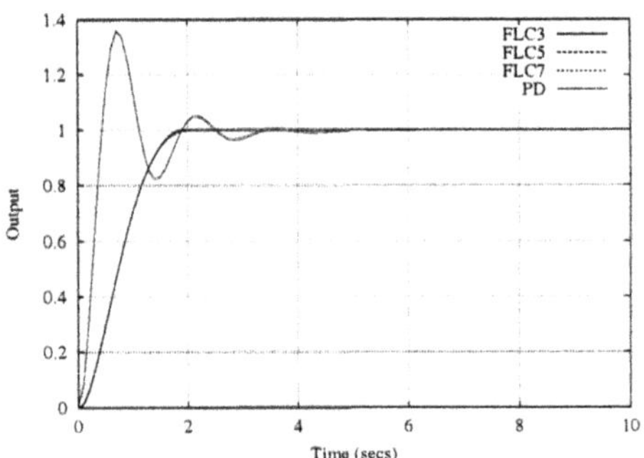

Fig. 17. Step responses for plant B

FLC3 forms part of the design environment particularly for providing new genetic material for the good operation of the migration strategy.

7 Constrained optimization of membership functions only

In the previous section, we proposed an enhanced GA for generating FLCs with optimized, well-formed fuzzy sets and rule bases. We believe that in certain cases, optimized FLCs can be built using standard McVicar-Whelan rule bases without any modification to the rules. In such situations, the design of an FLC can be simplified by optimizing the membership functions only. We chose evolutionary programming (EP) for optimizing the membership functions because EP is designed to work with real numbers as opposed to GAs which traditionally operate on binary representations. In order to test the feasibility of our simplified design method, we used it to configure FLCs

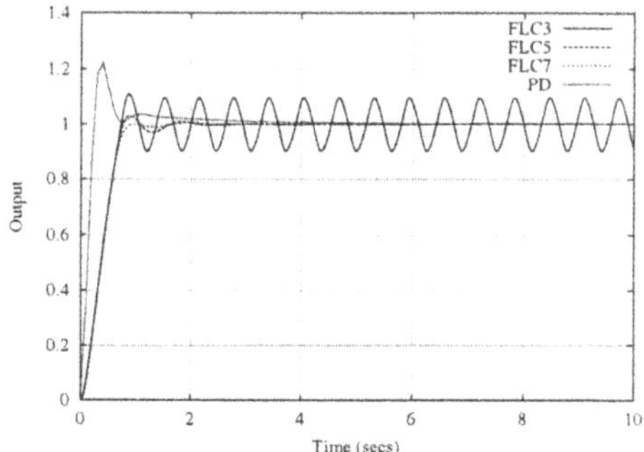

Fig. 18. Step responses for plant C

for plants A, B and C and we compared the performances of the FLCs with those obtained in the previous section where both the rule base and the membership functions were optimized.

The successful use of EAs requires the choice of the appropriate problem representation, objective function, selection mechanisms, genetic operators and system parameters. Some EAs are fairly straightforward to configure since their operating mechanisms are fixed and only a small number of parameters have to be set. However, others require the selection of mechanisms from a wide available range and have a large number of parameters to be set. EP was chosen to optimize the design of the FLCs because of its ease of configuration and its ability to work directly on the real-valued representation of the problem domain. To keep the experiments simple, we used the standard plain original version of EP, no strategy was used for self-adapting the system parameters.

The problem to be solved by EP is the finding of the membership functions of the three variables *error*, *change in error* and *change in output* of an FLC whose rule base has been determined according to the MacVicar-Whelan meta-rules such that the FLC controls an unknown plant process optimally according to some performance measure. The performance of the FLCs is assessed by means of a unit step response and compared with the performance of FLCs obtained in the previous section.

We used triangular membership functions for the inputs and singletons [52] for the outputs and we used seven fuzzy sets per input/output variable of the FLCs. To ensure the formation of well-formed fuzzy sets, we used fuzzy sets with a degree of overlapping of two as shown in Figure 4a. This results in the use of only one parameter to describe a fuzzy set (the two other parameters required for triangular fuzzy sets are defined by the neighbouring fuzzy sets). Furthermore, since we use a universe of discourse normalized to the range [-1.0, 1.0] and we avoid the use of trapezoidal fuzzy sets for the first and last fuzzy set, we fix the apices of the first and last fuzzy sets to -1.0 and

1.0 respectively. This further reduces the number of parameters to represent the membership functions by two. Thus, only 5 parameters are required per input variable such that a total of 15 parameters is required to define the membership functions of the two inputs and one output of the FLC.

Each potential solution to the problem (i.e an individual) is represented as a set of 15 real-valued parameters. For an individual to be valid, the parameter set should consist of three subsets of 5 parameters sorted in ascending order, in the range [-1.0, 1.0] and all values within the subsets should be unique (otherwise there will be less than seven fuzzy sets). No other constraints were used. We used a population of 100 individuals and initialized the parameters of each individual with random numbers in the range [-1.0, 1.0]. To ensure the validity of the individuals, we grouped their parameters into three subsets (each one representing the membership function of an FLC variable) and sorted them in ascending order.

At the beginning of each EA cycle, individuals are selected to be parents for creating offsprings; in EP all individuals are selected to be parents. At the end of each EA cycle, another selection mechanism is required to select survivors from the population of parents and offsprings to form the next generation. In EP, survivors are selected using a probabilistic function (tournament) based on fitness (see section 5.2). We used a tournament size of 10.

EP uses mutation only, it does not use recombination. We used the standard mutation operator as described in section 5.2 and experimented with several values of α and β and found good values to be 0.0 and 0.1 respectively.

Each solution FLC for the three plant processes was obtained by evolving EP for 100 runs and each run consisted of 300 generations.

7.1 Results and analysis

The performance of the FLCs was judged by comparing the steps responses of the FLCs with and without optimized rule bases. Step responses for plants A, B and C are shown in Figures 19, 20 and 21 respectively. It can be seen that the performance of the FLCs designed by our simplified method is equal to that of the FLCs with both optimized rule bases and membership functions except for plant C. Slightly better control of plant C is achieved with an FLC using an optimized rule base. This tends to suggest that in some cases, the generic MacVicar-Whelan rule base may not perform adequately and in these situations rule base optimization might be required. Unlike many rule base optimization methods, our method does not compromise the interpretability of the rules.

8 Conclusions

In this chapter, we have raised the issue of the unconstrained use of GAs in genetic fuzzy logic controllers which compromise the interpretability of

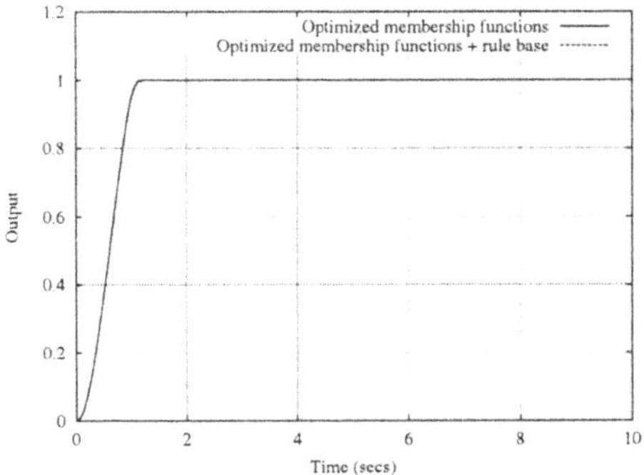

Fig. 19. Step responses for plant A

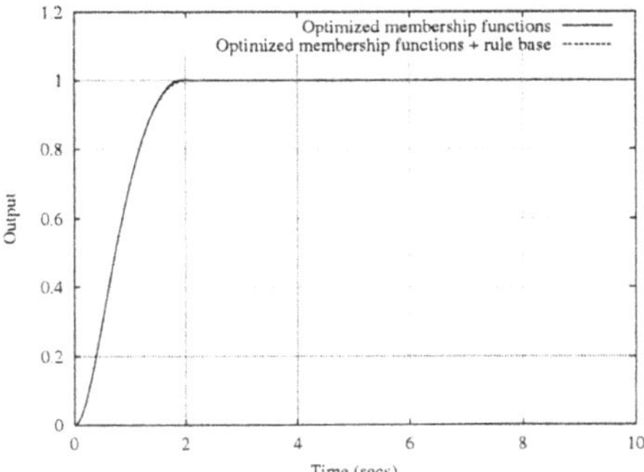

Fig. 20. Step responses for plant B

these fuzzy systems. In particular, this leads to the formation of messy fuzzy sets and scrambled rule bases. We proposed a solution to this problem by developing an enhanced GA which constrains the optimization of the FLCs in order to produce controllers with well-formed fuzzy sets and rule bases. Well-formed fuzzy sets are obtained by constraining the GA to produce fuzzy sets with a degree of overlapping of two. Well-formed rule bases are produced by constraining the GA to generate rule bases which conform to the MacVicar-Whelan meta rules within some reasonable limits. The MacVicar-Whelan meta rules embody expert knowledge in fuzzy control as they are derived from the experience gained with FLCs and common engineering sense. We used our enhanced GA to generate FLCs for controlling unknown plant processes and we found out that the performance of these FLCs were better than their conventional counterparts.

We further simplified our FLC design procedure by using evolutionary programming to optimize only the membership functions; the rule base used

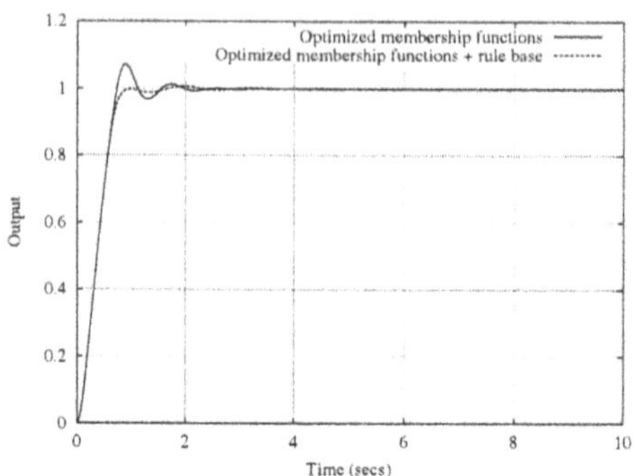

Fig. 21. Step responses for plant C

being a standard unmodified MacVicar-Whelan rule base. This is based on our belief that using a standard rule base and adjusting only the membership functions can provide enough latitude to meet the requirements of the control problem. We used evolutionary programming because we believe it is better suited for the real-valued optimization of the membership functions. The simplified design procedure was used to generate FLCs for the same plant processes used before. The performance of FLCs with generic MacVicar-Whelan rule bases was as good as the performance of FLCs with optimized rule base and membership functions in two cases out of three. This confirms our belief that a generic MacVicar-Whelan rule base can be used for some control applications and that optimization of the rule base is not required for each and every control application.

Although the FLCs optimized in this chapter seem to be fairly restrictive since they have only two inputs, our aim in this chapter was to develop a simple and reliable technique to optimize such FLCs which can be used as building blocks for more complex systems. We successfully built a hierarchical FLC with four inputs in [53] using the techniques described in this chapter to optimize the lower level FLCs.

References

1. L. A. Zadeh, "Fuzzy Sets," *Information and Control*, vol. 21, pp. 338–353, 1965.
2. L. A. Zadeh, "Fuzzy Set and Systems," in *Proc. Symp. Syst. Theory*, pp. 29–37, Polytech. Inst. Brooklyn, 1965.
3. L. A. Zadeh, "Outline of a New Approach to the Analysis of Complex Systems and Decision Processes," *IEEE Transactions on Systems, Man & Cybernetics*, vol. 1, pp. 28–44, 1973.
4. L. A. Zadeh, "The Concept of a Linguistic Variable and its Application to Approximate Reasoning," *Information Sciences*, vol. 8, pp. 43–80, 1975.

5. W. S. McCulloch and W. A. Pitts, "A logical calculus of the ideas immanent in nervous activity," *Bulletin of Mathematics and Biophysics*, vol. 5, pp. 115–133, 1943.
6. B. G. Farley and W. A. Clark, "Simulation of self-organizing systems by digital computers," *Institute of Radio Engineers – Transactions of Professional Group of Information Theory*, vol. PGIT-4, pp. 76–84, 1954.
7. F. Rosenblatt, "The perceptron: A probabilistic model for information storage and organization in the brain," *Psychoanalytic Review*, vol. 65, pp. 386–408, 1958.
8. B. Widrow and M. E. Hoff, "Adaptive switching circuits," in *WESCON Convention, Record Part IV*, pp. 96–104, 1960.
9. E. R. Caianiello, "Outline of a theory of thought-processes and thinking machines," *Journal of Theoretical Biology*, vol. 2, pp. 204–235, 1961.
10. K. Steinbuch, "Die Lernmatrix," *Kybernetik*, vol. 1, pp. 36–45, 1961.
11. J. H. Holland, *Adaptation in natural and artificial systems*. Ann Arbor, Michigan: The University of Michigan, 1975.
12. A. S. Fraser, "Simulation of Genetic Systems by Automatic Digital Computers," *Australian Journal of Biological Sciences*, vol. 10, pp. 484–491, 1957.
13. H. J. Bremermann, M. Rogson, and S. Salaff, "Global Properties of Evolution Processes," in *Natural Automata and Useful Simulations* (H. H. Patee, E. A. Edlsack, L. Fein, and A. B. Callahan, eds.), pp. 3–41, H. J. Bremermann and Spartan Books, 1966.
14. B. Kosko, *Neural Networks and Fuzzy Systems*. Englewood Cliffs, NJ: Prentice Hall, 1992.
15. R. Jang, "Self-learning Fuzzy Controllers Based on Temporal Back Propagation," *IEEE Transactions on Neural Networks*, vol. 3, no. 5, pp. 714–723, 1992.
16. R. Jang, "Fuzzy Controller Design Without Domain Experts," in *Proc. IEEE International Conference on Fuzzy Systems*, pp. 289–296, 1992.
17. C. Karr, "Genetic Algorithms for Fuzzy Controllers ," *AI Expert*, vol. 6, pp. 26–33, February 1991.
18. D. T. Pham and D. Karaboga, "Optimum Design of Fuzzy Logic Controllers Using Genetic Algorithms," *Journal of Systems Engineering*, vol. 1, no. 2, pp. 114–118, 1991.
19. A. Homaifar and E. McCormick, "Full design of fuzzy controllers using genetic algorithms," in *SPIE Vol. 1766 - Neural and Stochastic Methods in Image and Signal Processing*, pp. 393–404, 1992.
20. M. A. Lee and H. Takagi, "Integrating Design Stages of Fuzzy Systems using Genetic Algorithms," in *Proc. 2nd IEEE International Conference on Fuzzy systems*, pp. 612–617, 1993.
21. K. C. Ng and Y. Li, "Design of Sophisticated Fuzzy Logic Controllers Using Genetic Algorithms," in *Proc. 3rd IEEE International Conference on Fuzzy systems*, pp. 1708–1712, 1994.
22. M. G. Cooper, "Evolving a Rule-Based Fuzzy Controller," *Simulation*, vol. 65, no. 1, pp. 67–73, 1995.
23. D. A. Linkens and H. O. Nyonggesa, "Genetic Algorithms for fuzzy control Part 1: Offline system development and application," *IEE Proc.-Control Theory Appl.*, vol. 142, pp. 161–176, May 1995.

24. D. A. Linkens and H. O. Nyonggesa, "Genetic Algorithms for fuzzy control Part 2: Online system development and application," *IEE Proc.-Control Theory Appl.*, vol. 142, pp. 177–185, May 1995.
25. M. Setnes and H. Roubos, "GA-Fuzzy Modeling and Classification: Complexity and Performance," *IEEE Transactions on Fuzzy Systems*, vol. 8, no. 5, pp. 509–522, 2000.
26. O. Cordón, F. Herrera, F. Hoffman and L. Magdalena (eds.), "Genetic Fuzzy Systems - Evolutionary Tuning and Learning of Fuzzy Knowledge Bases," in *Advances in Fuzzy Systems - Applications and Theory*, vol. 19, World Scientific, 2001.
27. S. Assilian and E. H. Mamdani, "An Experiment in Linguistic Synthesis with a Fuzzy Logic Controller," *International Journal of Man-Machine Studies*, vol. 7, no. 1, pp. 1–13, 1974.
28. T. Takagi and M. Sugeno, "Fuzzy Identification of Systems and Its Applications to Modeling and Control," *IEEE Transactions on Systems, Man, and Cybernetics*, vol. SMC-15, pp. 116–132, January/February 1985.
29. P. J. MacVicar-Whelan, "Fuzzy sets for man-machine interaction," *International Journal of Man-Machine Studies*, vol. 8, pp. 687–697, 1976.
30. P. J. King and E. H. Mamdani, "The Application of Fuzzy Control Systems to Industrial Processes," *Automatica*, vol. 13, pp. 235–242, 1977.
31. R. R. Yager and D. P. Filev, *Essentials of fuzzy modeling and control.* New York, NY: John Wiley and Sons, Inc, 1994.
32. D. S. Feldman, "Fuzzy Logic Synthesis with Genetic Algorithms," in *Proc. 5th International Conference on Genetic Algorithms*, pp. 312–17, 1993.
33. H. Ishibuchi and T. Murata, "Minimizing the Fuzzy Rule Base and Maximizing Its Performance by a Multi-Objective Genetic Algorithms," in *Proc. 6th IEEE International Conference on Fuzzy systems*, pp. 259–264, 1997.
34. R. Rovatti, R. Guerrieri, and G. Baccarani, "An Enhanced Two-Level Boolean Synthesis Methodology for Fuzzy Rules Minimization," *IEEE Transactions on Fuzzy Systems*, vol. 3, pp. 288–299, August 1995.
35. P. Thrift, "Fuzzy Logic Synthesis with Genetic Algorithms," in *Proc. 4th International Conference on Genetic Algorithms*, pp. 509–513, 1992.
36. G. Raju, J. Zhou, and R. A. Kisner, "Hierarchical Fuzzy Control," *International Journal of Control*, vol. 54, no. 5, pp. 1201–1216, 1991.
37. D. E. Goldberg, *Genetic Algorithms in Search, Optimization and Machine-Learning.* Ann Arbor, Michigan: Addison-Wesley, 1989.
38. L. Davis, *A handbook of genetic algorithms.* New York: Van Nostrand Reinhold, 1990.
39. Z. Michalewicz, *Genetic Algorithms + Data Structures = Evolution Programs.* New York: Springer-Verlag, 1992.
40. L. J. Fogel, A. J. Owens, and M. J. Walsh, *Artificial Intelligence Through Simulated Evolution.* New York: Wiley Publishing, 1966.
41. D. B. Fogel, L. J. Fogel, and J. W. Atmar, "Meta-Evolutionary Programming," in *Proc. 25th Asilomar Conference on Signals, Systems & Computers* (R. R. Chen, ed.), (Pacific Grove, CA), pp. 540–545, 1991.
42. D. B. Fogel, L. J. Fogel, J. W. Atmar, and G. B. Fogel, "Hierarchic methods of Evolutionary Programming," in *Proc. 1st Annual Conference on Evolutionary Programming*, (San Diego, CA), pp. 175–182, Evolutionary Programming Society, 1992.

43. N. Saravanan, D. B. Fogel, and K. M. Nelson, "A Comparison of Methods for Self-Adaptation in Evolutionary Algorithms," *BioSystems*, vol. 36, pp. 157–166, 1995.
44. D. B. Fogel, *Evolutionary Computation: Toward a New Philosophy of Machine Intelligence.* Piscataway, NJ: IEEE Press, 1995.
45. T. Bäck, *Evolutionary Algorithms in Theory and Practice.* New York: Oxford University Press, 1996.
46. M. Gen and R. Cheng, "A survey of Penalty Techniques in Genetic Algorithms," in *Proc. 1996 IEEE International Conference on Evolutionary Computation*, (Japan), pp. 804–809, 1996.
47. F. Glover and H. Greenberg, "New approaches for heuristic search: a bilateral linkage with artificial intelligence," *European Journal of Operational Research*, vol. 39, pp. 119–130, 1989.
48. G. Syswerda, "Uniform crossover in genetic algorithms," in *Proc. 3rd IEEE International Conference on Genetic Algorithms*, pp. 2–9, 1989.
49. G. Syswerda, "A study of reproduction in generational and steady state genetic algorithms," in *Proc. of a workshop on foundations of genetic algorithms and classifier systems* (G. J. E. Rawlins, ed.), (Hillsdale, NJ), pp. 94–101, Lawrence Eribaum Publishers, 1991.
50. M. Rebaudengo and M. S. Reorda, "An experimental analysis of the effects of Migration in Parallel Genetic Algorithms," in *Proc. EUROMICRO Workshop on Parallel and Distributed Processing*, pp. 232–238, 1992.
51. T. C. Chin, H. C. Hua, K. W. Yeo, and W. L. Yeo, "Genetic Algorithm-based Optimal Fuzzy Controller," in *Proc. 2nd International Conference on Intelligent Systems*, pp. B135–B140, 1994.
52. D. I. Brubaker, "Fuzzy-logic basics: intuitive rules replace complex math," *EDNASIA Magazine*, vol. 37, no. 13, pp. 111–116, 1992.
53. F. Cheong and R. lai, "Designing a hierarchical fuzzy logic controller using differential evolution," in *Proc. IEEE International Conference on Fuzzy Systems (FUZZ-IEEE'99)*, pp. I–277–I–282, 1999.

Trade-off between the Number of Fuzzy Rules and Their Classification Performance

Hisao Ishibuchi and Takashi Yamamoto

Department of Industrial Engineering, Osaka Prefecture University
1-1 Gakuen-cho, Sakai, Osaka 599-8531, Japan

Summary: In this chapter, we examine the trade-off between the size of fuzzy rule-based systems and their classification performance through computer simulations on commonly used data sets in the literature. The size of fuzzy systems (i.e., the number of fuzzy rules) depends on several factors such as input selection, fuzzy partition, and rule selection. First we illustrate the effect of the fuzzy partition of each input variable on the performance of fuzzy systems. Next we examine the effect of the input selection on the performance of fuzzy systems through computer simulations using sequential feedforward input selection. The effect of the fuzzy partition is also examined in the same computer simulations. Finally, we directly examine a trade-off between the number of fuzzy rules and their classification performance using a genetic algorithm-based fuzzy rule selection method with two objectives: to minimize the number of fuzzy rules and to maximize the classification performance.

Keywords: Pattern classification, fuzzy rule-based systems, input selection, fuzzy partition, rule selection, multi-objective genetic algorithms.

1 Introduction

Fuzzy rule-based systems have been successfully applied to various fields such as control [1-3] and classification [4,5]. One advantage of fuzzy systems over other nonlinear models is their interpretability. That is, human users can easily understand each fuzzy rule because its antecedent and consequent are specified by linguistic values such as *small* and *large*. Another advantage is the high approximation ability of fuzzy systems. It is well-known that they are universal approximators of nonlinear functions as multi-layer feedforward neural networks. This means that high accuracy can be realized by fuzzy systems. Many approaches have been proposed for designing fuzzy systems using fuzzy neural techniques and fuzzy genetic techniques. In many fuzzy neural techniques, fuzzy systems are represented in multi-layer network structures so that the back-propagation learning algorithm [6] can be naturally applied. Roughly speaking, any continuous parameters in fuzzy systems can be adjusted by fuzzy neural techniques. Fuzzy genetic techniques are more flexible than fuzzy neural techniques because genetic algorithms can be applied to combinatorial optimization as well as continuous optimization. For example, fuzzy genetic techniques can perform input selection and fuzzy rule selection as well as membership function tuning. Hierarchical structures of fuzzy systems can be also determined by genetic fuzzy techniques. It should be noted that there exists a trade-off between the two advantages of fuzzy systems: Interpretability and accuracy. This trade-off is illustrated in Fig. 1. In many applications, simple fuzzy systems with high interpretability involve large errors. On the contrary, complicated fuzzy systems with high accuracy do not have high interpretability.

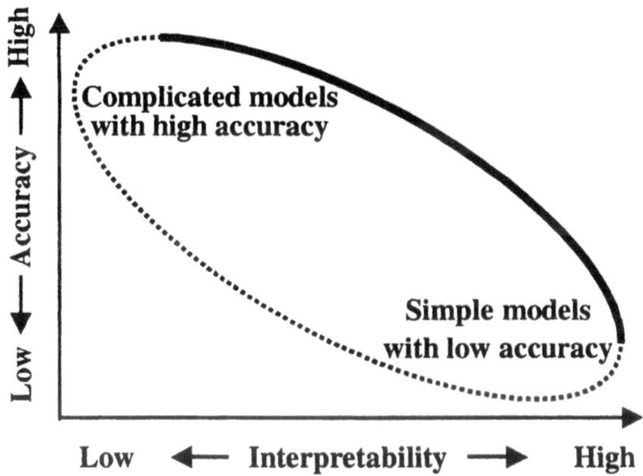

Fig. 1. Trade-off between interpretability and accuracy. The bold curve corresponds to non-dominated fuzzy systems with respect to the two criteria: interpretability and accuracy.

While the performance maximization has been the most commonly used criterion in modeling and classification, recently emphasis is placed on the interpretability of fuzzy systems in some studies [7-16]. When fuzzy modeling and fuzzy classification are handled as single-objective problems with the performance maximization, complicated fuzzy systems are obtained in many applications. This is because the improvement in the accuracy of fuzzy systems often leads to the decrease in their interpretability as shown in Fig. 1. Thus the point is to find a good trade-off between interpretability and accuracy. Each alternative fuzzy system can be viewed as a point in the two-dimensional objective space in Fig. 1. In many applications, there are no fuzzy systems with both high interpretability and high accuracy. The ellipse in Fig. 1 is depicted so that it includes all fuzzy systems for a particular problem. The bold curve in this figure (i.e., the right-top boundary of the ellipse) shows the Pareto front. In general, the Pareto front corresponds to the set of Pareto-optimal solutions of a multi-objective optimization problem. In Fig. 1, the Pareto front corresponds to fuzzy systems that are not dominated by any other ones with respect to interpretability and accuracy. When fuzzy modeling and fuzzy classification are handled as two-objective problems, our task is to find fuzzy systems on the Pareto front. The final fuzzy system should be chosen from them according to the decision maker's preference.

The aim of this chapter is to examine the shape of the Pareto front with respect to the number of fuzzy rules and their classification performance. Through computer simulations on commonly used pattern classification problems in the literature, we examine the classification performance of fuzzy systems of various sizes. Some fuzzy systems consist of only a few fuzzy rules. Others have many fuzzy rules. In our computer simulations, we use a simple heuristic method [17] for generating fuzzy rules from numerical data, in which a pattern space is homogeneously partitioned into fuzzy subspaces by subdividing each axis into linguistic values. The number of fuzzy rules depends on several factors such as input selection, fuzzy partition, and rule selection.

In this chapter, we first illustrate the effect of the fuzzy partition of each input variable on the performance of fuzzy systems. We use four fuzzy partitions in Fig. 2 where K denotes the number of antecedent fuzzy sets on each input variable. Next we examine the effect of input selection on the performance of fuzzy systems. We use a sequential feedforward input selection method [22] where a single input variable is sequentially added to the set of selected ones. The optimal combination of two input variables is found by examining all combinations before the sequential feedforward input selection method is applied. The effect of the fuzzy partition is also examined through computer simulations using the sequential feedforward input selection method. Then we directly examine a trade-off between the number of fuzzy rules and their classification performance using a genetic algorithm-based fuzzy rule selection method with two objectives: to minimize the number of fuzzy rules and to maximize the classification performance [8]. The main advantage of our two-objective rule selection method is that a number of non-dominated rule sets can be simultaneously obtained from its single run. This is because we use a two-objective genetic algorithm. The obtained non-dominated rule sets approximately correspond to fuzzy systems on the Pareto front in Fig. 1. Typical relations between the complexity of fuzzy systems (e.g., the number of fuzzy rules) and error rates are shown in Fig. 3. Error rates on training data usually decrease monotonically as the complexity increases. On the other hand, the relation between the complexity of fuzzy systems and error rates on test data is not monotonic. Error rates on test data first decrease until the complexity reaches its appropriate specification. After that, error rates on test data monotonically increase. Simulation results on some data sets show these typical relations between the number of fuzzy rules and their classification performance.

Other simulation results show that the existence of unnecessary input variables and/or unnecessary fuzzy rules decreases not only error rates on test data but also those on training data. This suggests the importance of input selection and rule selection in the design of fuzzy systems.

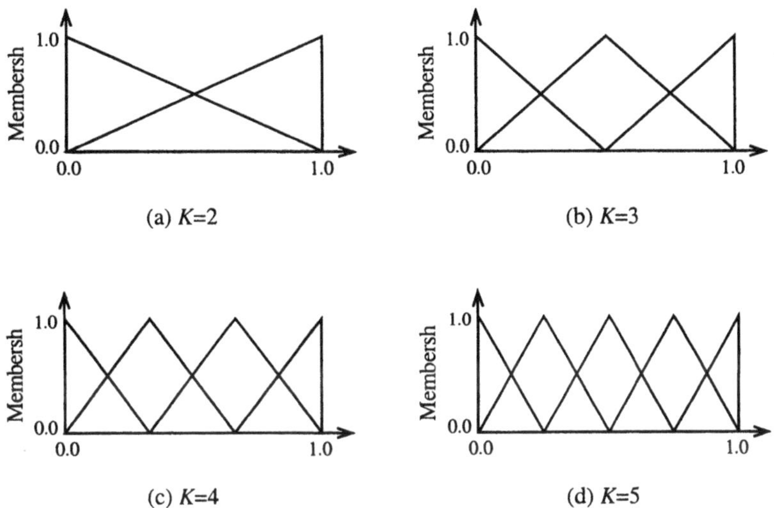

Fig. 2. Four fuzzy partitions examined in this chapter.

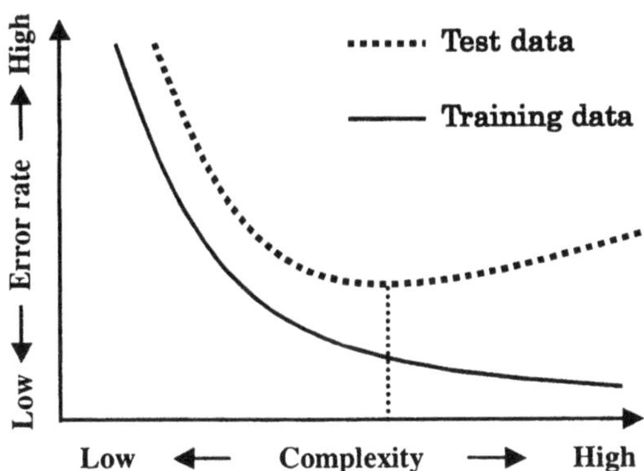

Fig. 3. Typical relations between the complexity of fuzzy systems and error rates.

2 Fuzzy Rule-based Classification

In this section, we briefly explain a fuzzy rule generation method from training patterns and a classification method of new patterns, which were proposed in [17] and used in fuzzy rule selection methods [8,16]. Cordon *et al.* [18] examined alternative formulations of fuzzy rules and fuzzy reasoning.

2.1 Fuzzy Rule Generation

Let us consider an n-dimensional pattern classification problem with m training patterns $\boldsymbol{x}_p = (x_{p1}, ..., x_{pn})$, $p = 1, 2, ..., m$ from M classes. For simplicity of explanation, we assume that each attribute value is normalized into a real number in the unit interval $[0, 1]$. Thus the pattern space of our pattern classification problem is the n-dimensional unit hypercube $[0, 1]^n$. For our pattern classification problem, we use fuzzy rules of the following form:

$$\text{Rule } R_j\text{: If } x_1 \text{ is } A_{j1} \text{ and ... and } x_n \text{ is } A_{jn} \text{ then Class } C_j \text{ with } CF = CF_j, \tag{1}$$

where R_j is the label of the j-th fuzzy rule, A_{ji} is an antecedent fuzzy set on the unit interval $[0,1]$, C_j is the consequent class (*i.e.*, one of the given M classes), and CF_j is the certainty grade of the fuzzy rule R_j. As the antecedent fuzzy set A_{ji}, we use a triangular fuzzy set generated by homogeneously partitioning each axis of the pattern space (see Fig. 2). The antecedent fuzzy set A_{ji} can be a special fuzzy set "*don't care*" whose membership values are always unity over the domain of the corresponding input variable. The effect of the certainty grade on the performance of fuzzy systems was examined in [19].

The fuzzy rule R_j in (1) is viewed as a kind of fuzzy association rule $\boldsymbol{A}_j \Rightarrow \text{Class } C_j$ where $\boldsymbol{A}_j = (A_{j1}, ..., A_{jn})$. The calculation of the confidence of standard association rules is naturally extended to the case of fuzzy association rules [20]. The confidence of the fuzzy association rule $\boldsymbol{A}_j \Rightarrow \text{Class } h$ is calculated as

$$c(\boldsymbol{A}_j \Rightarrow \text{Class } h) = \frac{\sum_{p \in \text{Class } h} \mu_{\boldsymbol{A}_j}(\boldsymbol{x}_p)}{\sum_{p=1}^{m} \mu_{\boldsymbol{A}_j}(\boldsymbol{x}_p)}, \quad h = 1, 2, ..., M, \tag{2}$$

where the compatibility of each pattern $\boldsymbol{x}_p$ with the antecedent part $\boldsymbol{A}_j$ is defined by the product operation as

$$\mu_{\boldsymbol{A}_j}(\boldsymbol{x}_p) = \mu_{A_{j1}}(x_{p1}) \times \cdots \times \mu_{A_{jn}}(x_{pn}). \tag{3}$$

As we can see from (2), the confidence of $\boldsymbol{A}_j \Rightarrow \text{Class } h$ is the ratio of patterns from Class h among compatible patterns with the antecedent part $\boldsymbol{A}_j$. The consequent class C_j of the fuzzy rule R_j in (1) is specified as

$$c(\mathbf{A}_j \Rightarrow \text{Class } C_j) = \max\{c(\mathbf{A}_j \Rightarrow \text{Class } 1), \ldots, c(\mathbf{A}_j \Rightarrow \text{Class } M)\}. \quad (4)$$

Its certainty grade CF_j is specified as

$$CF_j = c(\mathbf{A}_j \Rightarrow \text{Class } C_j) - \bar{c}. \quad (5)$$

where $\bar{c}$ is the average confidence over the $(M-1)$ classes except for the consequent class C_j:

$$\bar{c} = \frac{\sum_{h \neq C_j} c(\mathbf{A}_j \Rightarrow \text{Class } h)}{M-1}. \quad (6)$$

The specification of CF_j in (5) can be easily understood if we consider the case of two-class pattern classification problems (i.e., $M = 2$). In this case, (5) is rewritten as

$$CF_j = |c(\mathbf{A}_j \Rightarrow \text{Class } 2) - c(\mathbf{A}_j \Rightarrow \text{Class } 1)|. \quad (7)$$

As shown in (5), the discounted confidence is used as the certainty grade CF_j. It is also possible to directly use the confidence as the certainty grade as in [18]:

$$CF_j = c(\mathbf{A}_j \Rightarrow \text{Class } C_j). \quad (8)$$

It was shown in [20] that the use of the discounted confidence in (5) leads to better classification rates than the direct use of the confidence in (8). Thus we use the formulation in (5) for calculating the certainty grade of each fuzzy rule in this chapter.

In the case of low-dimensional pattern classification problems, we can generate fuzzy rules by examining all combinations of antecedent fuzzy sets. The consequent class and the certainty grade for each combination of antecedent fuzzy sets are specified by the above-mentioned procedure. When our pattern classification problem involves many input variables (i.e., many features), such an exhaustive examination is impractical due to the exponential increase in the number of combinations of antecedent fuzzy sets. Thus we only examine short fuzzy rules with a few antecedent conditions. The length of a fuzzy rule is defined by the number of its antecedent conditions. Short fuzzy rules can be viewed as having many *don't care* conditions. Even when the total number of fuzzy rules (i.e., combinations of antecedent fuzzy sets) is huge, the number of short fuzzy rules is not so large. Thus we can generate a tractable number of fuzzy rules by examining only short fuzzy rules.

2.2 Fuzzy Reasoning

Let us denote the set of generated fuzzy rules by *S*. The rule set *S* can be viewed as a fuzzy rule-based classification system. We use a single winner method [21] for classifying a new pattern $\boldsymbol{x}_p = (x_{p1}, ..., x_{pn})$ by the fuzzy system S. The single winner rule R_w is determined for the new pattern $\boldsymbol{x}_p$ as

$$\mu_{A_w}(\boldsymbol{x}_p) \cdot CF_w = \text{Max}\{ \mu_{A_j}(\boldsymbol{x}_p) \cdot CF_j \mid R_j \in S \}. \tag{9}$$

That is, the winner rule has the maximum product of the compatibility grade and the certainty grade. If multiple fuzzy rules have the same maximum product but different consequent classes for the new pattern $\boldsymbol{x}_p$, the classification of $\boldsymbol{x}_p$ is rejected. The classification is also rejected if no fuzzy rule is compatible with the new pattern $\boldsymbol{x}_p$.

While the fuzzy reasoning method in (9) is very simple, it can generate complicated non-linear classification boundaries even if we use simple grid-type fuzzy partitions. This is because the certainty grade of each fuzzy rule is taken into account in the fuzzy reasoning [19]. For example, let us consider the following four fuzzy rules in Fig. 4:

If x_1 is *small* and x_2 is *small* then Class 1 with CF_1, (10)

If x_1 is *small* and x_2 is *large* then Class 2 with CF_2, (11)

If x_1 is *large* and x_2 is *small* then Class 3 with CF_3, (12)

If x_1 is *large* and x_2 is *large* then Class 4 with CF_4. (13)

If we use the same certainty grade (*e.g.*, $CF_1 = CF_2 = CF_3 = CF_4 = 1.0$), the classification boundary generated by these four fuzzy rules is very simple as shown in Fig. 4. That is, the two-dimensional pattern space $[0,1] \times [0,1]$ is uniformly divided into four decision regions of the same size by the four fuzzy rules.

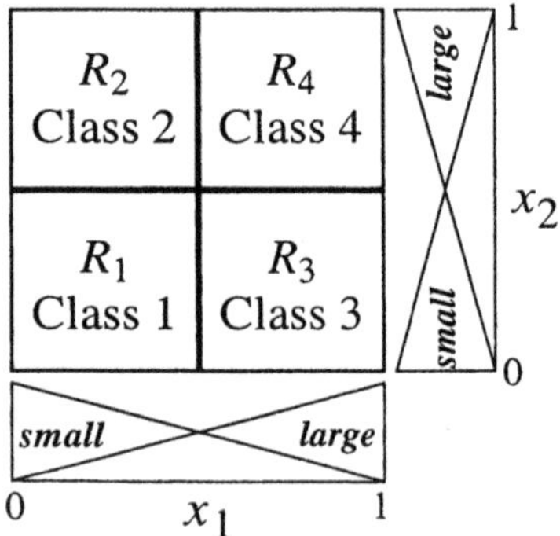

Fig. 4. Four fuzzy rules and the corresponding classification boundary when the same certainty grade is assigned.

Non-linear classification boundaries can be generated by assigning a different certainty grade to each fuzzy rule. In Fig. 5, we show some examples of classification boundaries generated by the four fuzzy rules in Fig. 4. The value of the certainty grade of each fuzzy rule is also shown in Fig. 5. As shown in this figure, the larger the certainty grade is, the larger the corresponding decision region is.

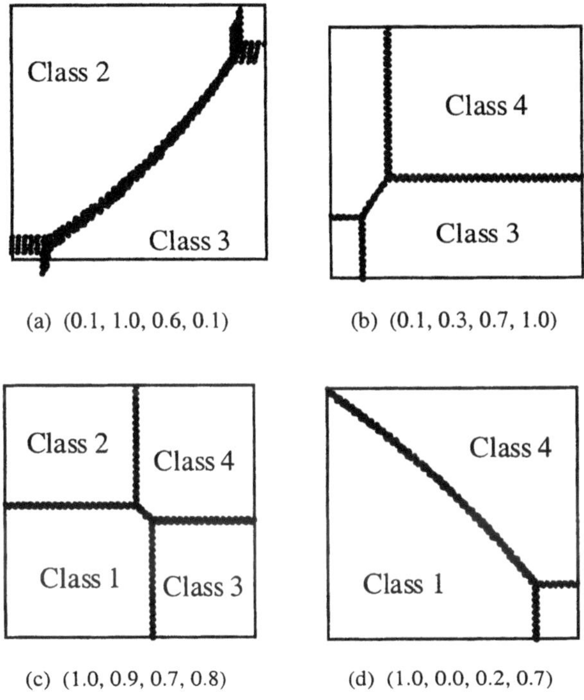

(a) (0.1, 1.0, 0.6, 0.1) (b) (0.1, 0.3, 0.7, 1.0)

(c) (1.0, 0.9, 0.7, 0.8) (d) (1.0, 0.0, 0.2, 0.7)

Fig. 5. Classification boundaries generated by the four fuzzy rules with different certainty grades. The certainty grades (CF_1, CF_2, CF_3, CF_4) are shown for each figure.

3 Fuzzy Partition and Input Selection

In this section, we first illustrate the effect of the fuzzy partition of each input variable on the performance of fuzzy rule-based classification systems. Next we describe a sequential feedforward input selection method. Then we examine the effect of the fuzzy partition and the input selection through computer simulations on commonly used pattern classification problems in the literature: Appendicitis data, Ljubljana cancer data, and wine data.

3.1 Fuzzy Partition

As an example of a pattern classification problem, let us consider 20 training patterns from two classes in the two-dimensional pattern space $[0,1]\times[0,1]$ in Fig. 6. We generated fuzzy rules from these training patterns for illustrating the effect of the fuzzy partition of each axis on the classification performance of fuzzy systems. First we used the 2×2 fuzzy grid in Fig. 4 for generating four fuzzy rules. That is, each axis was divided into two antecedent fuzzy sets as shown in Fig. 4. The following four fuzzy rules were generated from the 20 training patterns using the heuristic rule generation method in the previous section:

If x_1 is *small* and x_2 is *small* then Class 1 with $CF_1 = 0.34$, (14)

If x_1 is *small* and x_2 is *large* then Class 1 with $CF_2 = 0.10$, (15)

If x_1 is *large* and x_2 is *small* then Class 1 with $CF_3 = 0.05$, (16)

If x_1 is *large* and x_2 is *large* then Class 2 with $CF_4 = 0.55$. (17)

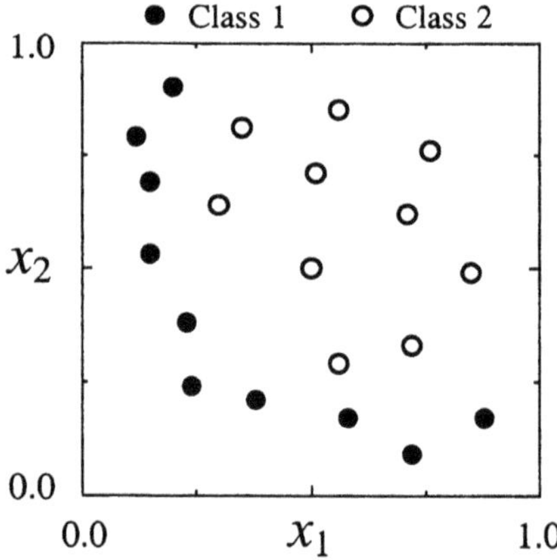

Fig. 6. Training patterns in a two-class pattern classification problem.

We also generated fuzzy rules using other fuzzy partitions in Fig. 2. That is, we used 3×3, 4×4, and 5×5 fuzzy grids for generating fuzzy rules. As a result, we have four fuzzy systems, each of which was generated from a different fuzzy grid. Using each of the four fuzzy systems, we calculated the classification boundary between two classes. Simulation results are shown in Fig. 7. From Fig. 7, we can see that all training patterns are correctly classified by the fuzzy system with $K = 4$. Usually, the finer the fuzzy partition is, the higher the classification rate on training data is. This does not always mean that high classification rates on test data (i.e., new input patterns) are also obtained from fine fuzzy partitions.

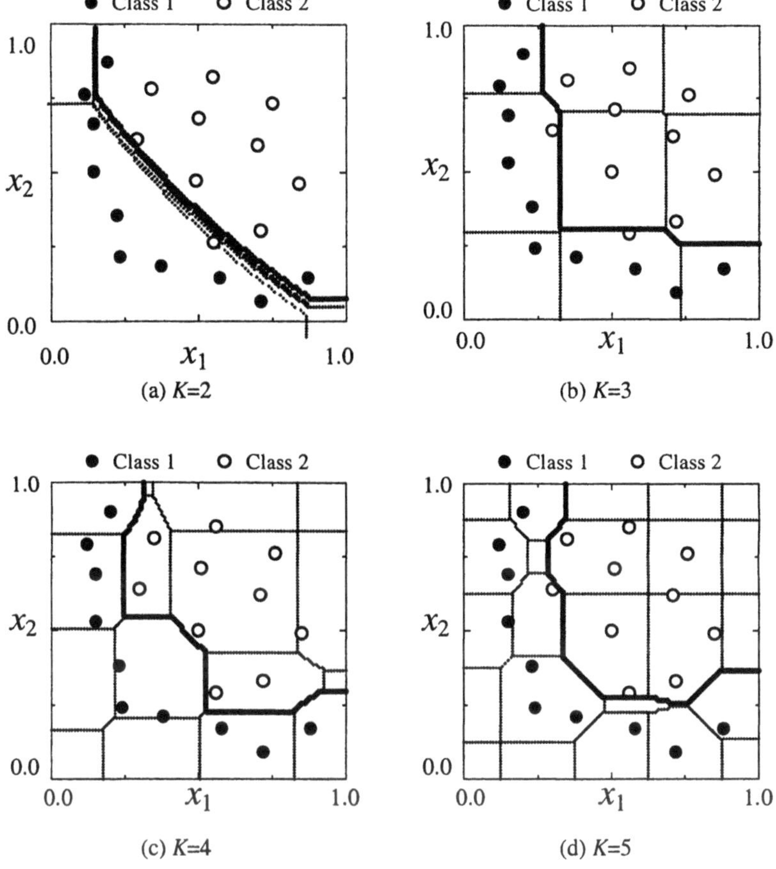

Fig. 7. Classification boundaries by fuzzy systems that are generated from 20 training patterns. Each cell shows the decision region of the corresponding fuzzy rule. The value of K denotes the number of antecedent fuzzy sets on each axis as in Fig. 2.

In the above computer simulations, we use the same fuzzy partition for each axis as shown in Fig. 4. This is for demonstrating the effect of the fuzzy partition of each axis on the classification performance of fuzzy systems in a simple manner. Of course, we can use a different fuzzy partition for each axis. The use of different fuzzy partitions in a single fuzzy system, however, makes the design of computer simulations and the interpretation of simulation results difficult (e.g., consider the comparison between 2×6 and 4×3 fuzzy grids). Thus we use the same fuzzy partition for each axis in computer simulations of this section. Different fuzzy partitions are used in the next section where the relation between the number of fuzzy rules and their classification performance is discussed.

3.2 Input Selection

For examining the effect of input selection on the classification performance of fuzzy systems, we use a sequential feedforward input selection method for fuzzy rule-based classification systems [22]. Input selection is a very active research area in the fields of machine learning and data mining [23-25].

In our sequential feedforward input selection method, first we specify the fuzzy partition for each input variable. Next we examine all the ${}_nC_2$ combinations of two input variables out of the given n variables (i.e., n features). For each pair of input variables, we construct a fuzzy rule-based system and examine its classification rate on training data. The two input variables with the highest classification rate are selected. When multiple pairs have the same highest classification rate, all those pairs are examined in the succeeding input selection procedure. Then we examine $(n-2)$ combinations of three input variables. Each combination is generated by adding a single input variable to the best combination of two input variables. We choose the set of three input variables with the highest classification rate among the examined $(n-2)$ combinations. If multiple sets of three input variables have the same highest classification rate, all the tie combinations are examined in the succeeding procedure. In this sequential feedforward manner, we can choose an arbitrary number of input variables. The outline of our sequential feedforward input selection method is written as follows (tie situations are not taken into account in the following outline for simplicity of explanation):

Step 1: Selection of two input variables.

Let Ψ be the given n input variables. Examine all the ${}_nC_2$ combinations of two input variables from Ψ by designing a fuzzy system for each combination. Let Ω be the two input variables with the highest classification rate on training patterns among the ${}_nC_2$ combinations.

Step 2: Sequential selection of a single input variable.

Examine $(n-|\Omega|)$ combinations of $(|\Omega|+1)$ attributes, each of which is constructed by adding a single attribute in $(\Psi-\Omega)$ to Ω. Replace Ω with the $(|\Omega|+1)$ attributes that have the highest classification rate on training patterns among the $(n-|\Omega|)$ combinations.

Step 3: Termination test.

If the stopping condition is not satisfied, return to Step 2.

While there may be some cases where a single input variable has high classification ability, this input selection method starts with the search for the best combination of two input variables. When a fuzzy system has only a single input variable, the decision region of each fuzzy rule is an interval. This means that each fuzzy rule can be replaced with the corresponding interval rule when a fuzzy system has only a single input variable. See Holte [31] for finding a single input variable for generating interval rules for classification problems.

We use two versions of this sequential feedforward input selection. In one version, the length of each fuzzy rule is the same as the number of selected input variables. That is, we do not use *don't care* conditions. The other version uses *don't care* conditions for avoiding the exponential increase in the number of fuzzy rules with the number of selected input variables. In our computer simulations with the second version, we generate fuzzy rules of the length two. That is, the number of antecedent conditions in each fuzzy rule is always two in our computer simulations with the second version. This means that fuzzy rules are generated from two-dimensional rule tables. For example, when three input variables x_i, x_j and x_k are selected, fuzzy rules are generated from three (i.e., ${}_3C_2$) two-dimensional rule tables, each of which has a pair of input variables (x_i, x_j), (x_j, x_k) or (x_i, x_k). On the other hand, the number of antecedent conditions is the same as the number of selected input variables in the first version of our sequential feedforward input selection.

3.3 Computer Simulations

In our computer simulations in this chapter, we use the following three data sets that have been frequently used in the literature:

Appendicitis data: This data set is a two-class problem involving 106 samples with seven attributes. Weiss & Kulikowski [25] examined the performance of ten non-fuzzy classification methods using the appendicitis data. Grabisch & Nicolas [26] examined six fuzzy classification methods.

Ljubljana cancer data: This data set is a two-class problem involving 286 samples with nine attributes. Weiss & Kulikowski [25] examined the performance of ten non-fuzzy classification methods using this data set. Grabisch [27] examined three fuzzy integral classifiers.

Wine data: This data set is a three-class problem involving 178 samples with 13 attributes. The wine data set has been used by many authors for examining the performance of fuzzy rule-based classification systems such as Setnes & Roubos [15], Ishibuchi *et al.* [16,29], and Castillo *et al.* [30]. This data set is available from the UCI database.

In our computer simulations, we used four fuzzy partitions in Fig. 2 (i.e., four specifications of K: $K = 2,3,4,5$). We also used two versions of the sequential feedforward input selection method. In the first version with no *don't care* conditions, we continued input selection until five input variables were selected. On the other hand, it was continued until all the input variables were selected in the second version where the rule length was two. When the classification performance on training data was examined, all samples in each data set were used as training data. When the classification performance on test data was examined, we used the leaving-one-out (LV1) procedure for the appendicitis data set as in Weiss & Kulikowski [25], the random sub-sampling procedure with 30% test samples for the Ljubljana cancer data as in Weiss & Kulikowski [25], and the 10-fold cross-validation (10-CV) for the wine data. Simulation results are summarized in Figs. 8-13.

In the case of the first version of the sequential feedforward input selection method with no *don't care* conditions, higher classification rates on training data were obtained from more input variables and finer fuzzy partitions on all the three data sets (see Fig. 8 (a), Fig. 10 (a) and Fig. 12 (a)). This is because the increase in the number of input variables means the decrease in the size of the decision region of each fuzzy rule. The use of finer fuzzy partitions also decreases the size of the decision region. As a result, the pattern space of each data set is divided into smaller fuzzy subspaces. This leads to higher classification rates on training patterns. Discretization of the pattern space into smaller fuzzy subspaces, however, does not always mean higher classification rates on test data. When the number of input variables increased, classification rates on test data increased in Fig. 12 (b) but decreased in Fig. 10 (b). The use of the finer fuzzy partitions leads to high classification rates on test data in Fig. 12 (b) but low classification rates in Fig. 10 (b).

When we used the second version of the sequential feedforward input selection method based on fuzzy rules of the length two, simulation results were totally different from the case of the first version. The use of more input variables does not always lead to the increase in classification rates on training data (see Fig. 9 (a), Fig. 11 (a) and Fig. 13 (a)). Since the rule length is always two in the second version, the increase in the number of selected input variables does not mean the decrease in the size of the decision region of each fuzzy rule. On the contrary, the pattern space is always divided into $K \times K$ fuzzy subspaces independent of the number of selected input variables. This is because each fuzzy rule has only two antecedent conditions. The increase in the number of selected input variables only means the increase in the number of fuzzy rules (i.e., the number of two-dimensional $K \times K$ fuzzy rule tables). Thus many fuzzy rules from multiple fuzzy rule tables overlaps with each other in the pattern space. This may lead to the generation of unnecessary fuzzy rules that have bad effects on the classification performance of fuzzy systems on training data as well as test data.

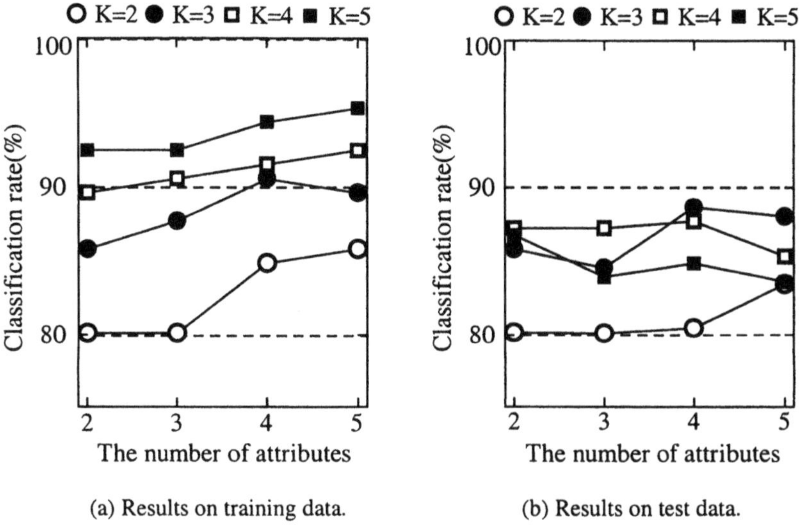

(a) Results on training data. (b) Results on test data.

Fig. 8. Simulation results on the appendicitis data using the first version of the feedforward input selection method.

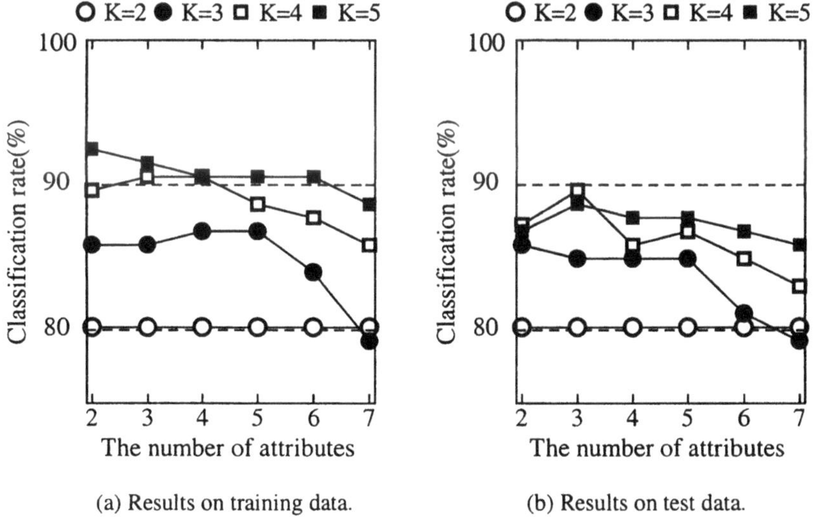

(a) Results on training data. (b) Results on test data.

Fig. 9. Simulation results on the appendicitis data using the second version of the feedforward input selection method.

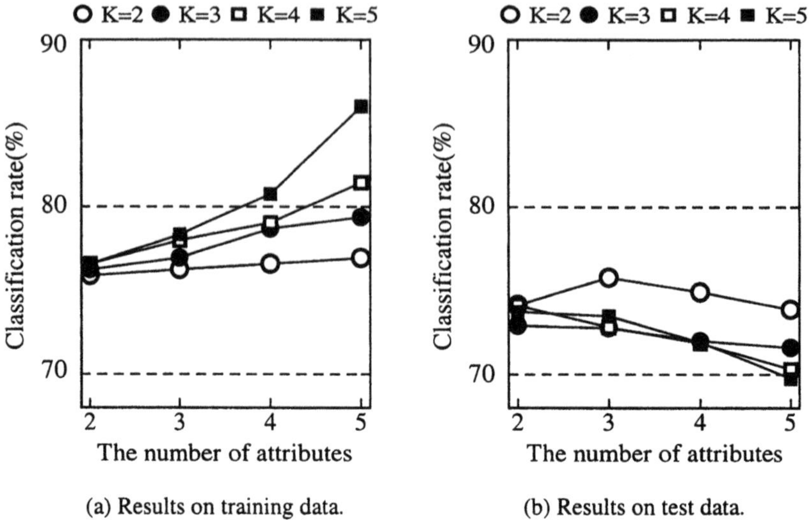

(a) Results on training data.

(b) Results on test data.

Fig. 10. Simulation results on the cancer data using the first version of the feedforward input selection method.

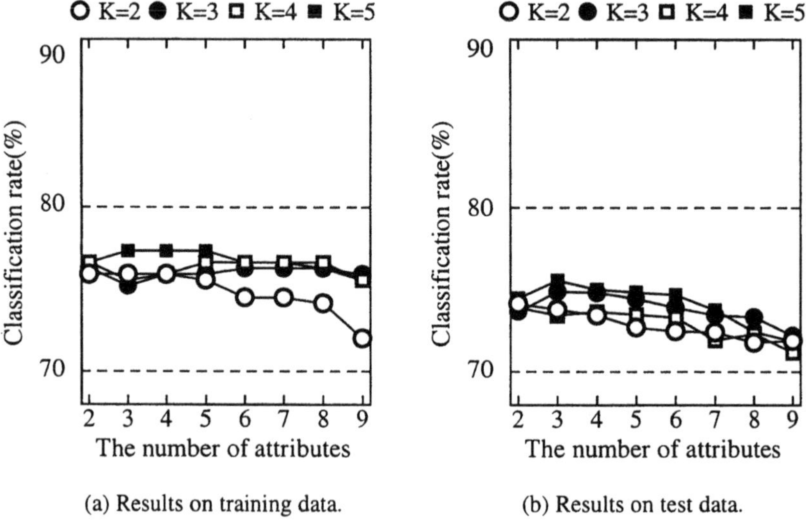

(a) Results on training data.

(b) Results on test data.

Fig. 11. Simulation results on the cancer data using the second version of the feedforward input selection method.

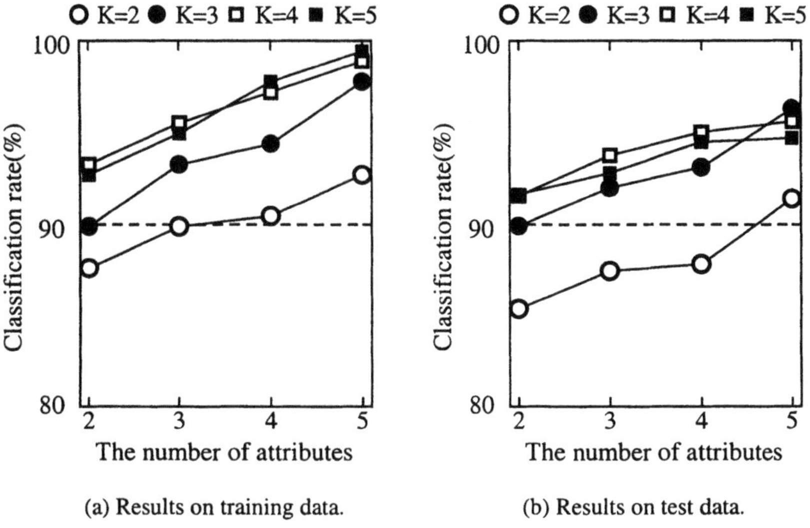

(a) Results on training data. (b) Results on test data.

Fig. 12. Simulation results on the wine data using the first version of the feedforward input selection method.

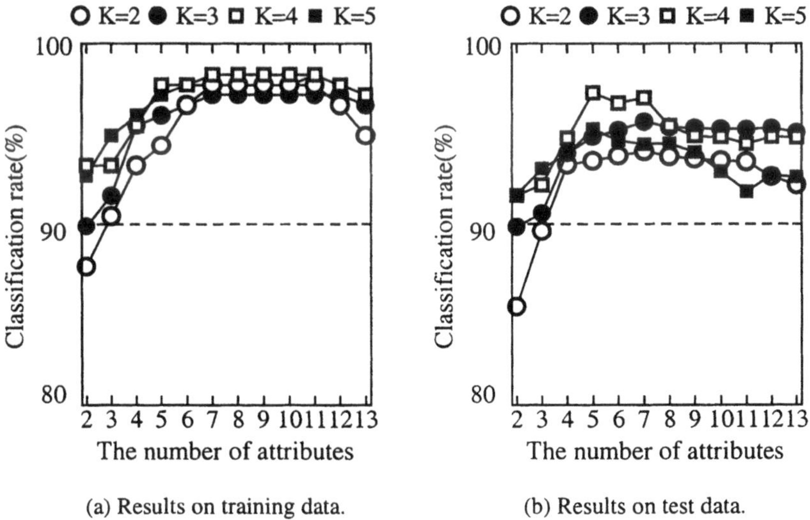

(a) Results on training data. (b) Results on test data.

Fig. 13. Simulation results on the wine data using the second version of the feedforward input selection method.

3.4 Comparison with Other Classifiers

The performance of ten non-fuzzy classification methods was examined in Weiss & Kulikowski [26] using the appendicitis data. Reported results in [26] are summarized in Table 1. Using the same data, Grabisch & Nicolas examined the performance of six fuzzy classification methods. Reported results in [27] are also shown in Table 1. For comparison, the best result (i.e., highest classification rate on test data) by each version of the sequential feedforward input selection method is shown in Table 1. From Table 1, we can see that very good results were obtained by our fuzzy systems for the appendicitis data. Especially, the 89.6% classification rate by the second version is the best among 18 methods in Table 1. In this case, three input variables were selected and four antecedent fuzzy sets were used for each input variable. This means that fuzzy rules were generated from three (i.e., ${}_3C_2$) rule tables, each of which had a 4×4 fuzzy partition. On the other hand, fuzzy rules were generated from a single $3\times3\times3\times3$ fuzzy partition in the case of the first version with the 88.7% classification rate in Table 1.

Table 1. Performance comparison (Appendicitis). n^* shows the number of selected input variables with the highest classification rate on test data.

Non-fuzzy methods in Weiss & Kulikowski [26]	Linear	86.8%
	Quadratic	73.6%
	Nearest neighbor	82.1%
	Bayes independence	83.0%
	Bayes 2nd order	81.1%
	Neural net (BP)	85.8%
	Neural net (ODE)	86.8%
	PVM rule	89.6%
	Optimal rule size 2	89.6%
	CART tree	84.9%
Fuzzy methods in Grabisch & Nicolas [27]	F*k*NN	86.8%
	MIN	86.8%
	MEAN	80.2%
	PERC	79.2%
	QUAD	86.8%
	CLMS	87.7%
Fuzzy systems (1st version)	$n^* = 4$ and $K = 3$	88.7%
Fuzzy systems (2nd version)	$n^* = 3$ and $K = 4$	89.6%

Weiss & Kulikowski [26] also used the cancer data for examining the performance of ten non-fuzzy classification methods. Reported results in [26] are summarized in Table 2. Grabisch [28] examined the performance of fuzzy integral classifiers using the same data. Reported results in [28] are also shown in Table 2 together with simulation results by the two versions of the sequential feedforward input selection method. In the case of the first version with the 75.8% classification rate, fuzzy rules were generated from a 2×2×2 fuzzy partition. On the other hand, three (i.e., ${}_3C_2$) 5×5 fuzzy partitions were used for generating fuzzy rules in the case of the second version with the 75.5% classification rate.

Table 2. Performance comparison (Cancer).

Non-fuzzy methods in Weiss & Kulikowski [26]	Linear	70.6%
	Quadratic	65.6%
	Nearest neighbor	65.3%
	Bayes independence	71.8%
	Bayes 2nd order	65.6%
	Neural net (BP)	71.5%
	Neural net (ODE)	72.4%
	PVM rule	77.1%
	Optimal rule size 2	72.0%
	CART tree	77.1%
Fuzzy integral classifiers in Grabisch [28]	QUAD	68.5%
	CLMS	72.9%
	HLMS	77.4%
Fuzzy systems (1st version)	$n^* = 3$ and $K = 2$	75.8%
Fuzzy systems (2nd version)	$n^* = 3$ and $K = 5$	75.5%

Highest classification rates in Fig. 12 (b) and Fig. 13 (b) on test data of the wine data are as follows:

96.3% by 5 input variables with 3 fuzzy sets (1st version in Fig. 12 (b)),
97.3% by 5 input variables with 4 fuzzy sets (2nd version in Fig. 13(b)).

In the case of the first version of the sequential feedforward input selection method, fuzzy rules were generated from a 3×3×3×3×3 fuzzy partition. Note that fuzzy rules cannot be generated when there are no compatible training patterns. Thus the number of fuzzy rules is not always the same as the number of fuzzy subspaces. On the other hand, ten (i.e., ${}_5C_2$) rule tables with 4×4 fuzzy partitions were used for generating fuzzy rules in the case of the second version. The above results by simple fuzzy grids are comparable to reported results on the wine data in the literature. For example, Castillo et al. [30] reported a 96.76% classification rate on test data (30% of the wine data) where the average number of fuzzy rules was 5.2 over five independent trials. In their SLAVE algorithm, fixed membership functions were used as in our approach. Since their approach uses a union (i.e., disjunction) of multiple linguistic values as a single antecedent fuzzy set, the number of fuzzy rules can be decreased. For example, the fuzzy rule "If x_1 is *small* or *medium* and x_2 is *medium* or *large* then Class 2" is handled as a single rule in [30] while it is handled as four rules in this paper. On the other hand, Setnes & Roubos [15] reported a 98.3% classification rate on training data (178 patterns) by three fuzzy rules. Classification rates higher than 98.3% were obtained in Fig. 12 (a) on training data with five input variables and fine fuzzy partitions.

In our computer simulations in this section, we used all fuzzy rules generated from simple grid-type fuzzy partitions. Thus much more fuzzy rules were included in our fuzzy systems than those in Castillo et al. [30] and Setnes & Roubos [15]. In the next section, we directly examine the relation between the number of fuzzy rules and their classification performance using a genetic algorithm-based rule selection method with two objectives: to minimize the number of fuzzy rules and to maximize the classification ability on training data.

4 Fuzzy Rule Selection

Our rule selection method consists of two phases: Candidate rule generation and genetic algorithm-based rule selection. We first generate a large number of candidate rules from training data. Then we search for non-dominated rule sets (i.e., subsets of candidate rules) with respect to the number of fuzzy rules and their classification performance by a two-objective genetic algorithm.

4.1 Problem Formulation

Let us denote a set of fuzzy rules by S. In our rule selection method, S is a subset of candidate rules. For examining the trade-off between the number of fuzzy rules and their classification performance, we formulated the following two-objective rule selection problem [8]:

$$\text{Maximize } NCP(S) \text{ and minimize } |S|, \tag{18}$$

where $NCP(S)$ is the number of correctly classified training patterns by the rule set S, and $|S|$ is the number of fuzzy rules in S.

4.2 Candidate Rule Generation

In our rule selection method, the rule set S in (18) is selected from a large number of candidate rules. Thus the first phase of our rule selection method is the candidate rule generation from training data. For generating candidate rules, we use all the 14 fuzzy sets in the four fuzzy partitions in Fig. 2 and *don't care* as antecedent fuzzy sets because we do not know an appropriate fuzzy partition for each input variable. This is the main difference between our former studies [8,16] and this work. While the same fuzzy partition was used for all input variables in our former studies, each input variable can use different fuzzy partitions in this work. Some input variables may need fine fuzzy partitions, and coarse fuzzy partitions may be appropriate for others. Since the number of antecedent fuzzy sets for each input variable is 15 (i.e., 14 fuzzy sets in Fig. 2 and *don't care*), the total number of combinations of antecedent fuzzy sets is 15^n for an n-dimensional pattern classification problem. It is impractical to examine all the 15^n combinations for generating candidate rules for high-dimensional pattern classification problems. Thus we only examined fuzzy rules of the length two or less for generating candidate rules in our computer simulations.

4.3 Two-Objective Genetic Algorithm for Rule Selection

Let us assume that N fuzzy rules were generated as candidate rules from training data. A subset S of those candidate rules is represented by a binary string of the length N as $S = s_1 s_2 \cdots s_N$ where $s_j = 1$ and $s_j = 0$ mean the inclusion of the j-th candidate rule in S and the exclusion from S, respectively.

Our task is not to find a single rule set but to find multiple non-dominated rule sets with respect to the two objectives of our rule selection problem in (18). For this task, we use a two-objective genetic algorithm with some heuristic operations [8]. Characteristic features of our two-objective genetic algorithm are as follows: Variable weights in a scalar fitness function, a secondary population that stores non-dominated solutions (i.e., non-dominated rule sets), the use of stored non-dominated solutions as elite solutions, biased mutation probabilities, and the elimination of unnecessary fuzzy rules from each solution.

In our two-objective genetic algorithm, first we randomly generate a pre-specified number of binary strings of the length N to construct an initial population. Next the two objectives of each string are evaluated. The first objective $NCP(S)$ of each string S is calculated by classifying all the given training patterns by S. Since we use the single winner method for fuzzy reasoning (i.e., for classification), the classification is performed by finding a single winner rule for each training pattern. Thus there are many cases where some rules are not chosen as winner rules for any training patterns. We can remove those rules without degrading the first objective $NCP(S)$. At the same time, the second objective $|S|$ is improved by removing unnecessary rules. Thus we remove all fuzzy rules that are not selected as winner rules of any patterns from the rule set S. The removal of unnecessary rules is performed for each rule set (i.e., for each binary string) in the current population by changing the corresponding 1's to 0's. From the combinatorial nature of our rule selection problem, some fuzzy rules with no contribution in one rule set may have large contribution in another rule set. Thus we cannot remove any fuzzy rules from all rule sets in the current population without examining each rule set. The second objective $|S|$ is calculated for each rule set after unnecessary rules are removed.

When the two objectives of each string (i.e., each rule set) in the current population are calculated, the secondary population is updated so that it includes all the non-dominated strings among examined ones during the execution of our algorithm. That is, each string in the current population is examined whether it is dominated by other strings in the current and secondary populations. If it is not dominated by any other strings, its copy is added to the secondary population. Then all strings dominated by the newly added one are removed from the secondary population. In this manner, the secondary population is updated at every generation.

The fitness value of each string S in the current population is defined from the two objectives as

$$fitness(S) = w_1 \cdot NCP(S) - w_2 \cdot |S|, \qquad (19)$$

where w_1 and w_2 are non-negative weights satisfying $w_1 + w_2 = 1$. These weights are randomly specified whenever a pair of parent strings is selected from the current population. In our former studies [8,16], we used a roulette wheel for selecting parent strings. Thus we had to calculate (19) for all strings in the current population to select a pair of parent strings. For selecting another pair of parent strings, we had to calculate (19) for all strings in the current population again using different weights. This is time-consuming especially when the population size is large. In this paper, we use binary tournament selection with replacement instead of roulette wheel selection to avoid such a time-consuming calculation. We have to calculate (19) for only four strings when a pair of parent strings is selected using the binary tournament selection (i.e., two strings for each parent string). A pre-specified number of pairs are selected from the current population using the binary tournament selection.

Uniform crossover is applied to each pair of parent strings to generate a new string. Biased mutation is applied to the generated string for efficiently decreasing the number of fuzzy rules included in each string. That is, different mutation probabilities are used for the mutation from 1 to 0 and that from 0 to 1. A much higher probability is assigned to the mutation from 1 to 0 than that from 0 to 1 for decreasing the number of fuzzy rules. In our computer simulations, the mutation probabilities from 1 to 0 and from 0 to 1 were specified as 0.1 and $1/N$, respectively (N is the number of candidate rules, i.e., string length). The next population consists of the newly generated strings by the selection, crossover, and mutation. Some non-dominated strings in the secondary population are randomly selected as elite solutions and their copies are added to the new population. Our two-objective genetic algorithm is summarized as follows:

Step 0: Parameter Specification.

Specify the population size N_{pop}, the number of elite solutions N_{elite} that are randomly selected from the secondary population and added to the current population, the crossover probability p_c, two mutation probabilities $p_m\ (1 \to 0)$ and $p_m\ (0 \to 1)$, and the stopping condition. In our computer simulations, the total number of generation updates was used as the stopping condition.

Step 1: Initialization.
Randomly generate N_{pop} binary strings of the length N as an initial population. Calculate the two objectives of each string in the initial population. In this calculation, unnecessary rules are removed from each string. Find non-dominated strings in the initial population. A secondary population consists of copies of those non-dominated strings.

Step 2: Genetic Operations.
Generate $(N_{pop} - N_{elite})$ strings by genetic operations (i.e., binary tournament selection, uniform crossover, biased mutation) from the current population.

Step 3: Evaluation.
Calculate the two objectives of each of the newly generated $(N_{pop} - N_{elite})$ strings. In this calculation, unnecessary rules are removed from each string. The current population consists of the modified strings.

Step 4: Secondary Population Update.
Update the secondary population by comparing each string in the current and secondary populations with one another.

Step 5: Elitist Strategy.
Randomly select N_{elite} strings from the secondary population and add their copies to the current population.

Step 6: Termination Test.
If the stopping condition is not satisfied, return to Step 2. Otherwise terminate the algorithm. All the non-dominated strings among examined ones during the execution of the algorithm are stored in the secondary population.

4.3 Computer Simulations

In computer simulations, we used the wine data because the performance of fuzzy systems on the wine data has already been reported by some authors in the literature as mentioned in the previous section (i.e., 98.3% classification rate on training data by three fuzzy rules in [25] and 96.76% classification rate on test data by 5.2 fuzzy rules in [30]). While those results suggest that a small number of fuzzy rules have high classification rates on the wine data, they did not clearly demonstrate a trade-off between classification rates and the number of fuzzy rules. This is because those studies were based on the framework of single-objective optimization. We examine the relation between classification rates and the number of fuzzy rules by computer simulations using our two-objective genetic algorithm.

We applied our rule selection method to the wine data. As candidate rules, we generated fuzzy rules of the length two or less. When all the 178 samples were used as training data, 14964 fuzzy rules were generated as candidate rules. Our two-objective genetic algorithm with the following parameter specifications was used for rule selection

Population size: 50,
Number of elite solutions: 5,
Crossover probability: 0.9,
Mutation probability: 0.1 from 1 to 0, and $1/N$ from 0 to 1,
Stopping condition: 10000 generations.

First we used all the 178 samples in the wine data set as training data. We performed 20 trials of our two-objective genetic algorithm from different initial populations. All the obtained rule sets from different trials were compared with one another. When a rule set was dominated by another rule set, the dominated one was removed. In this manner, we found non-dominated rule sets from the 20 trials of our two-objective genetic algorithm. Simulation results are summarized in Table 3. From this table, we can observe a clear trade-off between the number of fuzzy rules and classification rates on training data. We can also see that our results are comparable to the above-mentioned result on training data in the literature (i.e., 98.3% classification rate by three fuzzy rules in [15]).

Table 3. Non-dominated rule sets obtained from 20 trials. In computer simulations, all the 178 samples in the wine data were used as training data.

Number of rules	Classification rate
0	0.0 %
1	39.9 %
2	72.5 %
3	98.9 %
4	100.0 %

We also examined the relation between the number of fuzzy rules and classification rates on test data. We used the 10-fold cross-validation (10-CV) technique for evaluating classification rates on test data. In each trial of our two-objective genetic algorithm in the 10-CV technique, non-dominated rule sets were obtained from training data. Then the classification rate on test data was calculated for each non-dominated rule set. Average results were calculated over 20 iterations of the 10-CV technique. This means that our two-objective genetic algorithm was executed 20×10 times. Simulation results were summarized in Table 4. The last column of this table shows the number of trials from which rule sets with the corresponding number of fuzzy rules were obtained as non-dominated rule sets. For example, rule sets with four fuzzy rules were obtained by our two-objective genetic algorithm in 195 out of the 200 trails. The average classification rate on test data in this table was calculated over non-dominated rule sets with the same number of fuzzy rules.

Table 4. Simulation results of the 10-fold cross-validation.

Number of rules	Classification rate	Number of trials
0	0.0 %	200
1	37.8 %	200
2	69.3 %	200
3	93.3 %	200
4	94.1 %	195
5	94.8 %	139
6	95.9 %	26
7	100.0 %	2

Simulation results in Table 4 are slightly inferior to the reported result in [30] where the average classification rate on test data was 96.76% by 5.2 fuzzy rules. This may be because the classification rate on training data was maximized using the minimum number of fuzzy rules in our two-objective genetic algorithm. In the previous section, we obtained the 97.3% classification rate on test data using ten 4×4 fuzzy partitions with no rule selection. This high classification rate suggests that some additional mechanism in rule selection is necessary when our main aim is to maximize the generalization ability of fuzzy systems (i.e., classification rate on test data).

In our computer simulations, we used only short fuzzy rules generated from homogeneous fuzzy partitions with no modification of membership functions. Thus each fuzzy rule can be easily interpreted by human users. Moreover the number of fuzzy rules was minimized. Thus each fuzzy system consists of only a small number of simple fuzzy rules. In Fig. 14, we show three fuzzy rules that can classify 176 training patterns (i.e., 98.9% of the given 178 patterns: see Table 3). As shown in this figure, each fuzzy rule has only two antecedent fuzzy sets. All the other antecedent conditions are *don't care*, which is denoted by DC in Fig. 14. We can see from this figure that fuzzy rules can be easily understood by human users because they have only a few antecedent conditions and the shape of each antecedent fuzzy set is very simple. From Fig. 14, we can also see that some antecedent fuzzy sets are coarse and others are fine.

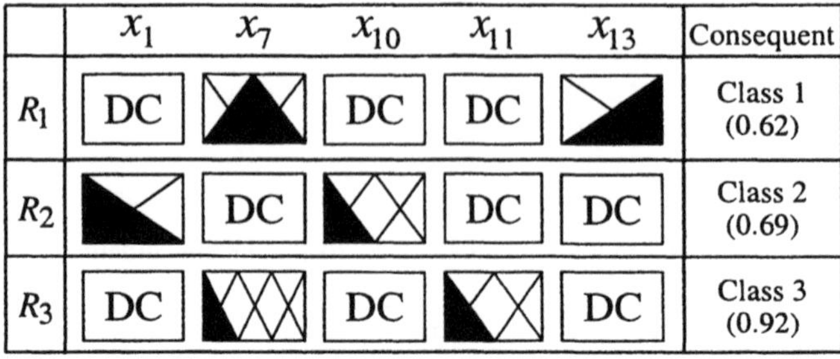

Fig. 14. An example of a non-dominated rule set obtained by our rule selection method.

5 Concluding Remarks

In this chapter, we examined the relation between the number of fuzzy rules and their classification performance. Many issues are related to the number of fuzzy rules. For example, the number of fuzzy rules depends on the specification of the fuzzy partition for each input variable. It also depends on input selection and rule selection. We first examined the effect of fuzzy partition and input selection on the classification performance of fuzzy systems. Simulation results showed that higher classification rates on training data were obtained from finer discretization of the pattern space into fuzzy subspaces. Finer discretization of the pattern space can be realized by the selection of more input variables and the use of a finer fuzzy partition for each input variable. Classification rates on test data, however, did not show such a simple monotonic relation. An appropriate specification of the number of input variables is problem-dependent. Five input variables of the wine data had the highest generalization ability while the best result was obtained by three input variables of the cancer data. An appropriate specification of the fuzzy partition for each input variable is also problem-dependent. A fine fuzzy partition with $K = 4$ had the highest generalization ability for the appendicitis data in Table 1 while the best result was obtained for the cancer data by a coarse fuzzy partition with $K = 2$ in Table 2.

Next the relation between the number of fuzzy rules and their classification performance was examined more directly by our two-objective rule selection method. Our method extended in this work can handle various fuzzy partitions for each input variable. This means that our method can simultaneously choose fuzzy partitions and fuzzy rules. Simulation results showed that high classification rates on training data were obtained by only a small number of fuzzy rules for the wine data set with 13 attributes. For example, our rule selection method could find four fuzzy rules with a 100% classification rate on all the 178 samples. Since only a small number of short fuzzy rules with simple antecedent fuzzy sets are selected by our method, generated fuzzy systems have high interpretability as shown in

Fig. 14. This is the main advantage of our rule selection method. Its drawback is that classification rates on test data are not always high. This is because the evolution of fuzzy systems is driven by the number of fuzzy rules and their classification performance on training data. If the maximization of generalization ability is the main objective in the design of fuzzy systems, some additional mechanism may be necessary for driving the evolution to fuzzy systems with high generalization ability. This is one of future research topics on fuzzy rule selection.

References

[1] M. Sugeno, "An introductory survey of fuzzy control," *Information Sciences*, vol. 36, pp. 59-83, 1985.

[2] C. C. Lee, "Fuzzy logic in control systems: Fuzzy logic controller - Part I and Part II," *IEEE Trans. on Systems, Man, and Cybernetics*, vol. 20, no. 2, pp. 404-435, 1990.

[3] C. T. Leondes (ed.), *Fuzzy Theory Systems: Techniques and Applications*, Academic Press, San Diego, 1999.

[4] J. C. Bezdek and S. K. Pal (eds.), *Fuzzy Models for Pattern Recognition: Methods That Search for Structures in Data*, IEEE Press, New York, 1992.

[5] L. I. Kuncheva, *Fuzzy Classifier Design*, Physica-Verlag, Heidelberg, 2000.

[6] D. E. Rumelhart, J. L. McClelland and the PDP Research Group: *Parallel Distributed Processing*, MIT Press, Cambridge, 1986.

[7] W. Pedrycz and V. de Oliveira, "Optimization of fuzzy models," *IEEE Trans. on Systems, Man, and Cybernetics - Part B: Cybernetics*, vol. 26, no. 4, pp. 627-637, August, 1996.

[8] H. Ishibuchi, T. Murata, and I. B. Turksen, "Single-objective and two-objective genetic algorithms for selecting linguistic rules for pattern classification problems," *Fuzzy Sets and Systems*, vol. 89, no. 2, pp. 135-150, 1997.

[9] M. Setnes, R. Babuska, and B. Verbruggen, "Rule-based modeling: Precision and transparency," *IEEE Trans. on Systems, Man, and Cybernetics - Part C: Applications and Reviews*, vol. 28, no. 1, pp. 165-169, 1998.

[10] M. Setnes, R. Babuska, U. Kaymak, and H. R. van Nauta Lemke, "Similarity measures in fuzzy rule base simplification," *IEEE Trans. on Systems, Man, and Cybernetics - Part B: Cybernetics*, vol. 28, no. 3, pp. 376-386, 1998.

[11] Y. Jin, W. von Seelen, and B. Sendhoff, "On generating FC^3 fuzzy rule systems from data using evolution strategies," *IEEE Trans. on Systems, Man and Cybernetics - Part B: Cybernetics*, vol. 29, no. 4, pp. 829-845, 1999.

[12] V. de Oliveira, "Semantic constraints for membership function optimization," *IEEE Trans. on Systems, Man, and Cybernetics - Part A: Systems and Humans*, vol. 29, no. 1, pp. 128-138, 1999.

[13] J. Yen and L. Wang, "Simplifying fuzzy rule-based models using orthogonal transformation methods," *IEEE Trans. on Systems, Man, and Cybernetics - Part B: Cybernetics*, vol. 29, no. 1, pp. 13-24, 1999.

[14] Y. Jin, "Fuzzy modeling of high-dimensional systems: Complexity reduction and interpretability improvement," *IEEE Trans. on Fuzzy Systems*, vol. 8, no. 2, pp. 212-221, 2000.

[15] M. Setnes and H. Roubos, "GA-based modeling and classification: Complexity and performance," *IEEE Trans. on Fuzzy Systems*, vol. 8, no. 5, pp. 509-522, 2000.

[16] H. Ishibuchi, T. Nakashima, and T. Murata, "Three-objective genetics-based machine learning for linguistic rule extraction," *Information Sciences*, vol. 136, no. 1-4, pp. 109-133, 2001.

[17] H. Ishibuchi, K. Nozaki, and H. Tanaka, "Distributed representation of fuzzy rules and its application to pattern classification," *Fuzzy Sets and Systems*, vol. 52, pp. 21-32, 1992.

[18] O. Cordon, M. J. del Jesus, and F. Herrera, "A proposal on reasoning methods in fuzzy rule-based classification systems," *International Journal of Approximate Reasoning*, vol. 20, pp. 21-45, 1999.

[19] H. Ishibuchi and T. Nakashima, "Effect of rule weights in fuzzy rule-based classification systems," *IEEE Trans. on Fuzzy Systems*, vol. 9, no. 4, pp. 506-515, 2001.

[20] H. Ishibuchi, T. Yamamoto, and T. Nakashima, "Fuzzy data mining: Effect of fuzzy discretization," *Proc. of 2001 IEEE International Conference on Data Mining*, pp. 241-248, 2001.

[21] H. Ishibuchi, T. Nakashima, and T. Morisawa, "Voting in fuzzy rule-based systems for pattern classification problems," *Fuzzy Sets and Systems*, vol. 103, no. 2, pp. 223-238, 1999.

[22] H. Ishibuchi, T. Nakashima, and T. Morisawa, "Simple fuzzy rule-based classification systems perform well on commonly used real world data sets," *Proc. of 1997 North American Fuzzy Information Processing Society Meeting*, pp. 251-256, 1997.

[23] R. Kohavi and G. H. John, "Wrappers for feature subset selection," *Artificial Intelligence*, vol. 97, no. 1-2, pp. 273-324, 1997.

[24] A. L. Blum and P. Langley, "Selection of relevant features and examples in machine learning," *Artificial Intelligence*, vol. 97, no. 1-2, pp. 245-271, 1997.

[25] H. Liu and H. Motoda, *Feature Selection for Knowledge Discovery and Data Mining*, Kluwer Academic Publishers, Boston, 1998.

[26] S. M. Weiss and C. A. Kulikowski, *Computer Systems That Learn*, Morgan Kaufmann Publishers, San Mateo, 1991.

[27] M. Grabisch and J. M. Nicolas, "Classification by fuzzy integral: Performance and tests," *Fuzzy Sets and Systems*, vol. 65, pp. 255-271, 1994.

[28] M. Grabisch, "The representation of importance and interaction of features by fuzzy measures," *Pattern Recognition Letters*, vol. 17, pp. 567-575, 1996.

[29] H. Ishibuchi, T. Nakashima, and T. Murata, "Performance evaluation of fuzzy classifier systems for multi-dimensional pattern classification problems," *IEEE Trans. on Systems, Man, and Cybernetics - Part B: Cybernetics*, vol. 29, no. 5, pp. 601-618, 1999.
[30] L. Castillo, A. Gonzalez, and R. Perez, "Including a simplicity criterion in the selection of the best rule in a genetic fuzzy learning algorithm," *Fuzzy Sets and Systems*, vol. 120, no. 2, pp. 309-321, 2001.
[31] R. C. Holte, "Very simple classification rules perform well on most commonly used dataset," *Machine Learning*, vol. 11, pp. 63-91, 1993.

Generating distinguishable, complete, consistent and compact fuzzy systems using evolutionary algorithms

Yaochu Jin

Future Technology Research, Honda R&D Europe
63073 Offenbach/Main, GERMANY
email: Yaochu_Jin@de.hrdeu.com

Abstract. Interpretability is one of the most important features for fuzzy rule systems. Unfortunately, fuzzy rules generated from data are usually difficult to understand for human beings if no special attention is paid to the interpretability during rule generation. This chapter proposes a method for generating interpretable fuzzy rules from data using evolutionary algorithms.

Two main aspects of interpretability are addressed. First, an index for distinguishability and completeness is suggested based on a similarity measure. Second, a degree of consistency between the fuzzy rules is proposed. Both indices are incorporated into the fitness function of the evolutionary algorithm to ensure that the generalized fuzzy rules are distinguishable, complete, consistent and compact without much loss of the performance.

1 Introduction

Fuzzy systems are believed to be suitable for knowledge representation and knowledge processing. One of the most important reasons to use fuzzy rules for system modeling and control is that they are comprehensible for human beings. This enables the fuzzy model to take advantage of *a priori* knowledge as well as of data. On the other hand, the resulting model can be interpreted and thus evaluated by users.

With the emergence of the techniques called soft computing or computational intelligence, fuzzy systems have obtained a new impetus. Backpropagation networks [18], RBF neural networks [6], hybrid pi-sigma networks [13,9], B-spline networks [5] and neuron-like structures [3] have been applied to the adaptation of the fuzzy membership functions and the consequent parameters. These methods are successful in that it is no longer necessary to determine exactly the parameters of fuzzy rules and the fuzzy partitions of the input variables beforehand.

Evolutionary algorithms have also been employed to design fuzzy systems. Since the first attempt to vary some parameters of a fuzzy rule base using genetic algorithms [14], large amount of work has been done to exploit the advantages of evolutionary algorithms for the design of fuzzy systems. Among different evolutionary methods, genetic algorithms (GAs) have been most popular mainly because they are capable of optimizing the parameters, the rule numbers as well as the rule structure simultaneously [16,17,7,4].

Fewer efforts have been made to date to design fuzzy systems using evolution strategies(ES). In [22], evolution strategies have been used to adjust the parameters of the fuzzy rules. In [10], an evolution strategy has also been used to optimize the fuzzy operations and scaling factors of a flexible structured fuzzy controller.

A common problem concerning adjustment of the membership parameters is that the shape of the membership functions can be modified so drastically that either some of the fuzzy subsets lose their corresponding physical meanings (refer to Fig. 1), or the fuzzy subsets do not cover the whole space of the input variable (Fig. 3 (b)), or the fuzzy sets are so similar that it no longer makes sense to associate them with proper semantic meanings (Fig. 2(b)). If the fuzzy partition is incomplete, the fuzzy system takes no action when the value of the variable falls in the uncovered region. Besides, no sufficient research work has been carried out to keep the consistency of the fuzzy rules in generating fuzzy rules from data. In most cases, only the rules that have the same antecedent but different consequents are considered to be inconsistent. In [21], a degree of belief is assigned to each generated rule and only the one with a maximal degree will be accepted if two rules have the same IF part but different THEN parts. In [7], the priority is given to the rule that first appears. In [12], flexibility of fuzzy operators, completeness of fuzzy partitions and consistency of fuzzy rules are taken into account using a multiobjective evolution strategy. Since all the above issues may prevent human beings from understanding the fuzzy rules, they have been considered to be the most important aspects of interpretability of fuzzy rules [11,8].

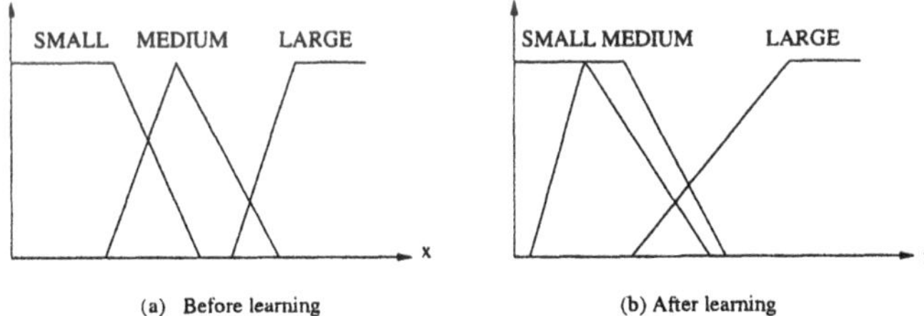

Fig. 1. Fuzzy partitions: (a) Before learning; (b) After learning. After learning, the location of the membership functions has been shifted so drastically that it is difficult to understand the meanings of the subsets "SMALL" and "MEDIUM".

In the following sections, distinguishability and completeness of fuzzy partitions, consistency of fuzzy rules, completeness of rule structure and compactness of the rule system are discussed. Based on these discussions, measures for distinguishability, completeness and consistency are suggested. With these measures being combined with the performance index of the fuzzy system, it is able to generate fuzzy rules from data using evolution strategies that are distinguishable, complete, compact and more consistent. The method is illustrated with an example on a driver-assistant system.

2 DC^3 Fuzzy Systems

2.1 Distinguishability of Fuzzy Partitions

Distinguishability is one of the main requirements for interpretability of fuzzy systems. When designing an interpretable fuzzy system, one usually associate a linguistic term to a fuzzy subset. However, there is an inherent condition for this, that is, the fuzzy subsets in a partition should be distinguishable. In Fig.2 (a), there are four well distinguishable fuzzy subsets, which can be labeled {ZERO, SMALL, MEDIUM, LARGE}. However, the eight fuzzy subsets in Fig.2(b) cannot easily be distinguished. In this case, it is hard to associate an understandable linguistic term to each fuzzy subset and thus fuzzy rules described by such fuzzy sets are hard to understand.

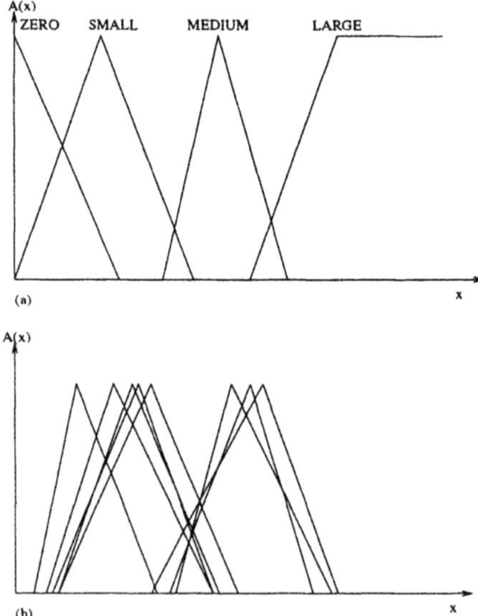

Fig. 2. Fuzzy partitions: (a) Distinguishable; (b) Indistinguishable.

To improve the distinguishability of fuzzy partitions, one can merge some of the similar fuzzy subsets, especially in adaptive fuzzy systems [19]. However, hard merging of fuzzy subsets may seriously degrade the performance of the fuzzy system. Therefore, an adaptive merging of subsets is suggested in [8].

The distinguishability of a fuzzy partition depends also on the number of fuzzy subsets within a fuzzy partition, refer to Fig.2(b). The smaller the number of fuzzy subsets is, the more distinguishable the fuzzy partition is, provided that it is complete. In fact, when one designs a fuzzy rule system based on expert knowledge or common sense, the number of fuzzy subset one uses is usually smaller than ten [20].

2.2 Completeness of Fuzzy Partitions

The completeness of a fuzzy rule system is composed of the completeness of the fuzzy partition of all input variables and the completeness of the rule structure. We first discuss the completeness of the fuzzy partitions. Suppose input variable x is partitioned into M fuzzy subsets $A_i(x), i = 1, 2, ..., M$, then the partition is said to be complete if the following condition holds:

$$\forall_{x \in U} \exists_{1 \leq i \leq M} A_i(x) > 0. \tag{1}$$

Furthermore, a measure of completeness for a fuzzy partition can be defined:

$$CM(x) = \sum_{i=1}^{M} A_i(x). \tag{2}$$

If $CM(x) = 1$ for all x, the fuzzy subsets construct a fuzzy partition in a strict sense and therefore it is said to be strictly complete. If $CM(x) = 0$, then the fuzzy partition is incomplete. Fig. 3 shows a complete and an incomplete partition.

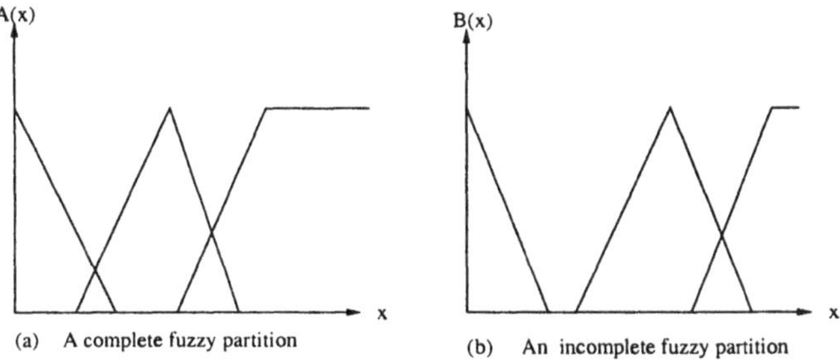

Fig. 3. Fuzzy partitions. (a) A complete fuzzy partition, and (b) an incomplete fuzzy partition.

The completeness of fuzzy partitions does not guarantee the completeness of the fuzzy system if the rule structure is not complete. Simply speaking, the completeness of the rule structure means that all subsets within the fuzzy partition of each input should be used by the fuzzy rule system at least once.

Similarly, a measure of completeness for the rule base can also be defined. Assume the rule base has N fuzzy rules and each rule has n antecedents, which are connected by t-norm T, then the completeness of the fuzzy rule system is given by:

$$CM(R) = \sum_{j=1}^{N} CM(R_j) \tag{3}$$

$$CM(R_j) = T_{i=1}^{n} A_{i,j}(x_i) \tag{4}$$

It is obvious that the completeness measure may vary for different inputs. If there exists an input where the completeness measure equals zero, the fuzzy system is said to be incomplete.

However, a complete rule structure is not necessarily a full grid rule structure. As shown in Fig. 4(b), the rule structure is complete, although there is no rule in some of the elements. On the contrary, the rule structure in Fig. 4(c) is incomplete. In fact, a compact but complete structure is not only feasible, but also very essential, especially for large systems.

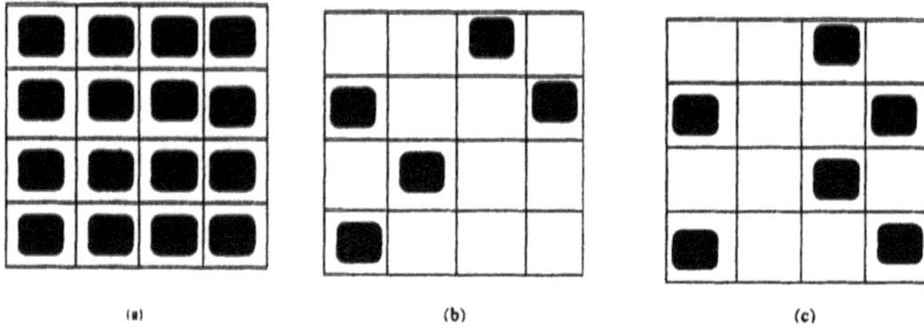

Fig. 4. Rule structures. A shaded area denotes a fuzzy rule. (a) A standard grid rule structure, (b) a non-standard but complete rule structure, and (c) an incomplete rule structure.

Incompleteness of fuzzy control systems may result in unpredictable actions produced by the fuzzy system, which is very undesirable in system control. If a fuzzy control system is found to be incomplete, certain measures have to be taken. One feasible idea is to make an interpolation between the neighboring fuzzy rules [15,24]. Suppose a fuzzy rule base has the following two rules:

R_1: If x is A_1, then y is B_1
R_2: If x is A_2, then y is B_2.

The fuzzy membership functions are illustrated in Fig.5. To make the rule base complete, a third fuzzy rule can be generated as follows:

R_3: If x is A', then y is B'.

There are different methods to determine the fuzzy membership functions A' and B'. However, the general idea is based on the assumption that the fuzzy rules are consistent.

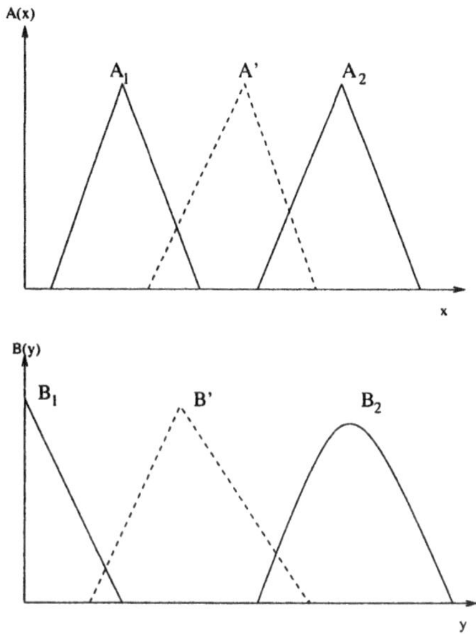

Fig. 5. An example to remedy incompleteness using interpolation.

2.3 Consistency of Fuzzy Rules

Consistency of the rules in a rule base concerns the consistency of the knowledge represented by the rule base. Let us have a look at the following two rules:

R_1: If x is positive Small, Then y is positive Large.
R_2: If x is positive Small, Then y is negative Large.

It is obvious that the two fuzzy rules are inconsistent because they have the same premise part, but their consequent parts are completely different. In our view, two fuzzy rules can be inconsistent even if their premise parts are not the same. A simple example showing this are the following two rules:

R_1: If x is positive Small, Then y is positive Large.
R_2: If x is positive Medium, Then y is negative Large.

In these two rules, neither the premise part nor the consequent part is the same. Nevertheless, these two rules look quite inconsistent because their premise parts are very similar, but their consequent parts are very different. However, if we rewrite the rules as:

R_1: If x is positive Small, Then y is positive Large.
R_2: If x is positive Medium, Then y is positive Medium.

In this case, the two fuzzy rules become again consistent. On the other hand, If the premise parts of two fuzzy rules are very different, then it is hard to say that they are inconsistent, no matter what consequent they have. For example:

R_1: If x is positive Large, Then y is positive Small.
R_2: If x is negative Large, Then y is positive Small.

In these two rules, their premise parts are completely different, although their consequent is the same. However, one cannot say that they are inconsistent.

From the previous discussions, we make the following observations concerning the consistency of two fuzzy rules.

- If two fuzzy rules have the same premise, but different consequent parts, they are inconsistent.
- If two fuzzy rules have very similar premises, but fully different consequent parts, they are inconsistent.
- If the premise parts of two rules are different, they are always considered to be consistent, no matter what consequent they have.
- If two rules have similar premises and similar consequents, they are consistent.

According to these observations, we can develop a consistency measure for fuzzy rules using a fuzzy similarity measure. Consider the following two fuzzy rules:

R_i: If x_1 is A_{i1}, and ... and x_n is A_{in},
Then y is B_i
R_k: If x_1 is A_{k1} and ... and x_n is A_{kn},
Then y is B_k

We first define *the Similarity of Rule Premise* (SRP) by:

$$SRP(R_i, R_k) = min_{j=1}^{n} S(A_{ij}, A_{kj}) \tag{5}$$

where $S(A_{ij}, A_{kj})$ is a fuzzy similarity measure between fuzzy sets A_{ij} and A_{kj}. We can define the fuzzy similarity measure as [18]:

$$S(A_{ij}, A_{kj}) = \frac{M(A_{ij} \cap A_{kj})}{M(A_{ij}) + M(A_{kj}) - M(A_{ij} \cap A_{kj})} \tag{6}$$

where, $M(A)$ is the size of fuzzy set A.

Following the Similarity of Rule Premise, we also define a *Similarity of Rule Consequent* (SRC) as follows:

$$SRC(R_i, R_k) = S(B_i, B_k) \tag{7}$$

where $S(B_i, B_k)$ is the similarity between fuzzy sets B_i and B_k. Thus, the consistency of the two fuzzy rules can be defined by:

$$CN(R_i, R_k) = \exp\left\{ -\frac{\left(\frac{SRP(R_i,R_k)}{SRC(R_i,R_k)} - 1.0\right)^2}{\left(\frac{1}{SRP(R_i,R_k)}\right)^2} \right\} \tag{8}$$

This definition for consistency has two fundamental properties. First, the consistency of fuzzy rules is always a degree, which depends mainly on the shape and location of the related fuzzy membership functions. The degree of consistency tends to be high when the SRP and SRC of the two rules are in proportion, provided that the SRP of the two rules is high. In the special case, if the rules have the same premise and the same consequent, the consistency has its highest value of 1. When the premises are the same but the consequent parts are different, the consistency degree varies from 0 to 1.0. Furthermore, if the premise parts of the two rules have low similarity, in other words, SRP is very small, then the degree of consistency is always high, no matter how SRC changes. This agrees with our previous discussion. One additional remark on the consistency definition is that it is mainly suitable for Mamdani-type fuzzy rules. For Takagi-Sugeno-Kang fuzzy rules, the consistency is harder to evaluate, when the rule consequent is a function of the input variables. However, interpretability of TSK rules may be investigated in terms of the physical meaning of each local model [23].

To give an illustrative example on the consistency measure, we assume that there are three linguistic variables x, y and z defined as in Fig. 6 and we have the following two fuzzy rules:

R_1: If x is Zero, y is Small, then z is Zero;
R_2: If x is Small, y is Small, then z is Small.

It is seen that both the premises and the consequents of the two rules are similar. According to our previous discussions, the two rules should be consistent. To confirm this, we calculate the consistency measure using the definition in Equation (8) and the result is $CN(R_1, R2) = 1.0$. If we change the consequent of rule R_2 so that the premises of the two rules are very similar, but the consequents are very different:

R_1: If x is Zero, y is Small, then z is Zero;
R_2: If x is Small, y is Small, then z is Very large.

From the definition, we have $CN(R_1, R2) = 0.0$. In other words, the two fuzzy rules are inconsistent even if they do not have the same rule premise (but very similar).

Now we change one of the rule premise in rule R_2, so that the premises of the two rules are not similar.

R_1: If x is Zero, y is Small, then z is Zero;
R_2: If x is Small, y is Large, then z is Small.

In this case, the rules should be always consistent, no matter which consequents they have. For the above two rules, $CN(R_1, R_2) = 1.0$. If we again change the consequent of rule R_2 to "Very large", we get $CN(R_1, R_2) = 0.96$, which denotes that the two rules are still consistent.

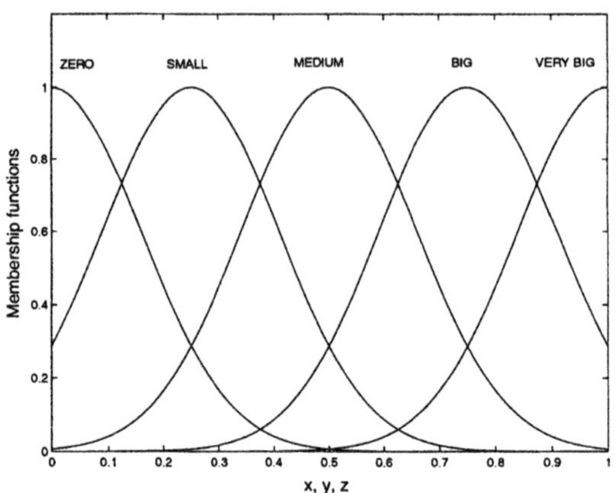

Fig. 6. The membership functions for x, y and z.

Consistency evaluation of fuzzy rules is sometimes difficult. This happens when a fuzzy system has more than one output variable and the relationship between these variables is unclear. Consider the following two fuzzy rules:

R_1: If x is Small, then y is Large
R_2: If x is Small, then z is Small

If we do not consider the relationship between y and z, then these two fuzzy rules are not contradictory. However, if y and z has some inherent relationship say if y is *Large*, z cannot be *Small*, then, these two rules are seriously inconsistent. In this case, more careful analysis is required.

2.4 Compactness of Fuzzy Rules

The compactness of fuzzy rules concerns the number of fuzzy rules in the rule base and the number of conditions in the premise part of the rules. The standard full grid structure (Fig. 4(a)) is not compact, because with this structure the number of fuzzy rules increases exponentially with the input number. Therefore, it is essential to optimize the structure of fuzzy rules, which determines both the number of conditions in the rule premise and the number of fuzzy rules.

3 DC[3] Fuzzy Rule Generation with Evolution Strategies

3.1 Evolution Strategy for Mixed Parameter Optimization

Evolution strategies (ES) are used to optimize the parameters as well as the structure of the fuzzy rules. Evolution strategies, instead of genetic algorithms are used here due to the following two considerations. One is that the coding of ES is direct for real valued parameter optimization and consequently the length of the object vector increases just linearly with the number of variables. Since the rule structure will be evolved together with the parameters, the number of variables to be optimized increases significantly wit the increase of the input number. If a binary genetic algorithm is used, the length of object vector will grow drastically. To alleviate resulting difficulties, either the range of the parameters needs to be limited, or special coding methods have to be developed. Generally speaking, evolution strategies have better performance for optimization tasks with real-valued parameters and relatively smooth fitness landscape [2]. However, this does not imply that ES is superior to GA, or vice versa.

Since both real and integer numbers are involved in the optimization, a slightly modified version of (μ,λ)-ES [1] is used here. An ES algorithm that is capable of dealing with mixed optimization can be described with the following notation:

$$(\mu, \lambda)\text{-ES} = (I, \mu, \lambda;\ m, s, \sigma;\ f, g) \tag{9}$$

where, I is a string of real or integer numbers representing an individual in the population, μ and λ are the numbers of the parents and offsprings respectively; σ is the parameter to control the step size, m represents the mutation operator, which is the main operator for ES. In ES algorithms, not only the variables, but also the step-size control parameter σ are mutated. In equation (9), the parameter s stands for the selection method and in this case, the parents will be selected only from the λ descendants; f is the objective function to be minimized, and g is the constraint function to which the object parameters are subject. The object parameters and the step-size control parameters are mutated in the following way:

$$\sigma_i' = \sigma_i \cdot exp(\tau_1 \cdot N(0,1) + \tau_2 \cdot N_i(0,1)),\ i = 1, 2, ..., Q \tag{10}$$

$$I_i' = I_i + \sigma_i' \cdot N_i(0,1),\ i = 1, 2, ..., Q_1 \tag{11}$$

$$I_i' = I_i + \lfloor \sigma_i' \cdot N_i(0,1) \rfloor, \; i = Q_1 + 1, Q_1 + 2, ..., Q \tag{12}$$

where $N(0,1)$ and $N_i(0,1)$ are normal distributions with zero mean and variance of 1, Q is the total number of object parameters, Q_1 is the number of real-valued object parameters and naturally $Q - Q_1$ denotes the number of the integer object parameters. In this work, the parameters representing the fuzzy membership functions are encoded with real variables, while the rule structure parameters are encoded with integer numbers. τ_1 and τ_2 are two global step size control parameters.

Several strategy parameters of ES, which have great influence on the performance of the algorithm, have to be fixed manually. These include the population size μ and λ, the global step control parameters τ_1 and τ_2, and the initial values of the step sizes σ_i. The optimization problems in the real world have normally a lot of local optima, in which a standard evolution strategy can get trapped. To acquire a best possible solution, it is desirable to improve the performance of the standard ES. In this study, a minor modification is made and, nevertheless, is proved to be effective. In practice, we find that the process of evolution stagnates when the step-size control σ converges to zero prematurely. To prevent the step-size from converging to zero, we re-initialize it with a value of, say, 1.0, when σ becomes very small. This enables the algorithm to escape from local minima on the one hand, on the other hand, it gives rise to some oscillations of the performance. Therefore, it is important to record the best individual that has been found so far. However, this best individual does not take part in the competition of selection if it does not belong to the current generation.

3.2 Fuzzy System Encoding and Fitness Setup

The coding of the parameters of the fuzzy system is straightforward. Without loss of generality, the following Gaussian membership functions are used:

$$A(x) = exp\{-\frac{(x-c)^2}{w^2}\} \tag{13}$$

Therefore, each membership function has two parameters, namely, center c and width w. In order to make sure that all the subsets of a fuzzy partition can distribute as freely as possible provided that the completeness condition is satisfied, the center of each fuzzy membership function can move on the universe of discourse of the corresponding variable, which is limited by the physical system. Of course, the fuzzy subsets should be ordered according to their center so that the checking of completeness can be done and that the mechanism of the rule structure optimization can work properly. The width of the membership functions is loosely limited so that it is larger than zero and naturally, not wider than the whole space. In fact, the width of the membership functions is also subject to the completeness conditions as well as the distinguishability requirements.

The coding of the rule structure is important because the size of a fuzzy system is fully specified by the rule structure. Suppose each input variable x_i has a maximal number of fuzzy subsets M_i, then the rule base has maximally $N = M_1 \times M_2 \times \cdots \times M_n$ fuzzy rules if there are n input variables. Thus, the premise structure of the rule system can be encoded by the following matrix:

$$Struc_{premise} = \begin{bmatrix} a_{11} & a_{12} & \cdots & a_{1n} \\ a_{21} & a_{22} & \cdots & a_{2n} \\ \cdots & \cdots & \cdots & \cdots \\ a_{N1} & a_{N2} & \cdots & a_{Nn} \end{bmatrix}, \tag{14}$$

where $a_{ji} \in \{0, 1, 2, ..., M_i\}, j = 1, 2, ..., N; i = 1, 2, ..., n$. The integer numbers, $1, 2, ..., M_i$ represent the corresponding fuzzy subsets in the fuzzy partition of x_i, while $a_{ji} = 0$ indicates that variable x_i does not appear in the jth rule. It is argued that the assumption of the maximal number of fuzzy partition will not harm the compactness of the rule system, because the redundant subsets will be discarded automatically by the algorithm during the checking of distinguishability. Similarly, the structure of the rule consequent can be encoded with a vector of positive integers:

$$Struc_{consequent} = [c_1, c_2, \cdots, c_N]^T, \tag{15}$$

where $c_j \in \{1, 2, ..., K\}, j = 1, 2, ..., N$, supposing that the consequent variable has maximal K fuzzy subsets. This works both for Mamdani type rules and Takagi-Sugeno fuzzy rules whose consequent is a constant. If the Takagi-Sugeno rules have a real function of input variables in the consequent, the coding of the consequent structure is not necessary.

As we mentioned before, a fuzzy system should exhibit not only good approximation, but also good interpretability. In the fitness function, extra terms will be added to guarantee the completeness and distinguishability of the fuzzy partitions and the consistency of the fuzzy rules. Completeness of the rule structure is ensured by discarding a fuzzy membership function that is not used by any of the fuzzy rules. The completeness of the fuzzy partition of each input variable is examined using a *fuzzy similarity measure*. One advantage of using the fuzzy similarity measure (FSM) is that by regulating the grade of FSM, the degree of overlap of two subsets can be properly controlled. If FSM of two neighboring fuzzy subsets is zero or too small, it means that either they do not overlap or do not overlap sufficiently. On the other hand, if FSM is too large, the two fuzzy subsets overlap too much and the distinguishability between them is lost. To maintain both distinguishability and completeness of the fuzzy partitions, the fuzzy similarity measure of any two neighboring membership functions is required to satisfy the following condition:

$$FSM^- \leq S(A_i, A_{i+1}) \leq FSM^+ \tag{16}$$

where A_i and A_{i+1} are two neighboring fuzzy sets, FSM^- and FSM^+ are the desired lower and upper bound of the fuzzy similarity measure. If this condition is not satisfied, the fitness of the generated fuzzy system will be assigned to such a large value that the corresponding individual is unlikely to survive. Notice that in this work, the optimization task is to minimize the fitness function, therefore the lower the fitness value is, the better.

The consistency index that has been suggested in Section 2 can not be applied directly to the evolutionary generation of fuzzy rules. To solve this problem, a degree of inconsistency of a rule base is suggested based on the consistency index. At first, a degree of *inconsistency* for the i-th rule is calculated as follows:

$$\begin{aligned} Incons(i) = & \sum_{\substack{1 \le k \le N \\ k \ne i}} [1.0 - Cons(R^1(i), R^1(k))] \\ & + \sum_{1 \le l \le L} [1.0 - Cons(R^1(i), R^2(l))] \\ & i = 1, 2, ..., N \end{aligned} \tag{17}$$

where, R^1 and R^2 denote the rule base generated from data and the rule base extracted from prior knowledge, N and L are the rule numbers of R^1 and R^2 respectively. The degree of inconsistency of each rule is then summed up to indicate the degree of inconsistency of a rule base:

$$f_{Incons} = \sum_{i=1}^{N} Incons(i) \tag{18}$$

which can be incorporated in the objective function of the evolutionary algorithm.

Note that no special measures have been taken to reduce the number of the fuzzy rules. In practice, it is found that the evolutionary algorithm tends to select a compact system. This implies that a standard fuzzy system with a full grid rule structure normally has worse performance than a compact system.

3.3 An Example: Fuzzy Driver Assistant Systems

In this section, we generate a fuzzy rule system for a drive assistant controller based on collected data using the proposed method.

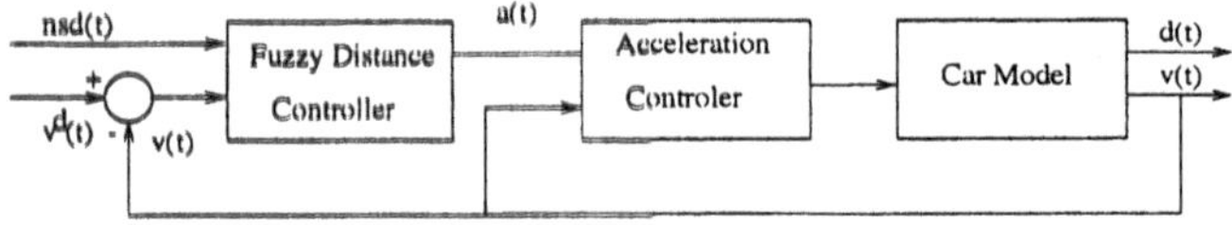

Fig. 7. Diagram of the control architecture.

A diagram of the control architecture of the driver assistant system is illustrated in Fig. 7, where $v(t)$ and $v_1(t)$ are the velocities of the controlled car and of the car in front of it, $d(t)$ is the distance between the two cars, $v_r(t) = v_1(t) - v(t)$ denotes the relative speed, and $nsd(t)$ is the normalized safety distance, which is calculated as follows:

$$nsd(t) = \frac{d(t) - s(t)}{s(t)}, \tag{19}$$

where, $s(t)$ is called the safety distance. It is found that $s(t)$ is basically in proportion to the speed $v(t)$, however, it seriously depends on the behavior of the driver. Not only have different drivers different views on safety distance, but the same driver also makes different decisions at different times. Our task is to design a fuzzy controller that is able to produce a correct acceleration based on the data collected from different drivers in different driving situations. In this study, data from one driving scene are used for training and the data from a second scene are used for test.

The following cost function is used to indicate the performance of the fuzzy controller:

$$f_E = \sum_{t=1}^{J} \sqrt{(v(t) - v^d(t))^2}, \tag{20}$$

where J is the total number of sampled data and v_d is the target velocity. To generate distinguishable, complete, consistent and compact fuzzy rules, the completeness and inconsistency indices are integrated into the fitness function:

$$f = f_E + \xi \cdot f_{Incons} + f_{Incompl}, \tag{21}$$

where $f_{Inconsis}$ denotes the inconsistency measure and $f_{Incompl}$ is a penalty term for the rule system if the completeness condition is not satisfied; ξ is a weighting constant to control the consistency level. In general, once the rule system is found to be incomplete, the penalty term $f_{Incompl}$ is so large that the individual is not able to survive. That is to say, the evolutionary algorithm tolerates some degree of inconsistency, but allows no incomplete fuzzy systems.

Based on the collected data, the meaningful range of normalized safety distance $nsd(t)$, relative speed $v_r(t)$ and the acceleration $a(t)$ are selected as [-1,5], [-10,10] and [-3,3] respectively. As mentioned before, the center of all the fuzzy membership functions is allowed to vary over the whole space of the corresponding variable. We suppose both nsd and v_r have a maximum of five fuzzy subsets and a has maximally eight fuzzy subsets. Therefore, the fuzzy system has maximally 25 fuzzy rules.

The following two rules have been used as the prior knowledge in our research to help check the consistency of the generated fuzzy rules.

If nsd is *Positive Big* and v_r is *Positive Big*, then a is *Positive Big*
If nsd is *Negative Big* and v_r is *Negative Big*, then a is *Negative Big*

No doubt, such prior knowledge is quite straightforward and is easy to obtain. Nevertheless, they play an important role in checking the rules produced from data.

Prior knowledge can not only be used in checking the consistency of the fuzzy system, but also be incorporated in the initialization of the evolutionary algorithm. For example, some individuals can be initialized with the parameters of a standard fuzzy rule system, while the others are initialized with randomly generated numbers.

In the beginning, we generate the fuzzy rule system using 316 data collected in the first driving situation. For comparison, fuzzy rule systems are generated with and without completeness and consistency checking. A fuzzy rule base is first generated without checking its completeness and consistency. Nevertheless, rules with the same IF-part but different THEN-part are avoided to assure the fairness of the comparison.

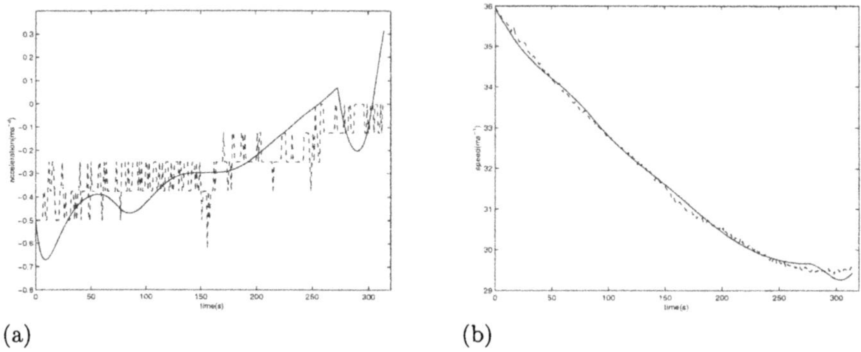

(a) (b)

Fig. 8. Results on the training data without interpretability constraints: (a) acceleration; (b) speed. Solid line: experiment data; dashed line: output of the fuzzy system.

The inconsistency degree of the fuzzy system is 13.4 and the speed and acceleration tracking results are illustrated in Fig. 8. In the figure, the dotted lines denote the acceleration and speed measured in the experiment, while the solid lines describe the results produced by the fuzzy controller. It is found that the fuzzy controller has good performance on the training data, where the root-mean-square (RMS) error on acceleration is $0.105ms^{-2}$ and the RMS speed error is $0.085ms^{-1}$.

However, a fuzzy system that exhibits smart performance for training data does not necessarily perform equally well on the test data. Before checking the fuzzy system with test data, we first have a look at the membership functions, see Fig. 9. We notice that some fuzzy subsets of $nsd(t)$ lack distinguishability, while the fuzzy partition of $v_r(t)$ is incomplete. This implies that over-fitting of the membership functions has occurred.

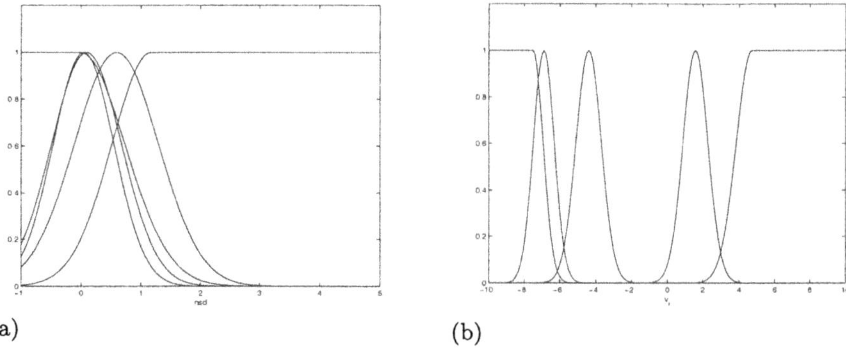

(a) (b)

Fig. 9. Fuzzy partitions without interpretability constraints: (a) *nsd*, some of the fuzzy membership functions are indistinguishable; (b) $v_r(t)$, the fuzzy partition is incomplete.

Now we evaluate the fuzzy system on the 276 test data obtained in the second driving situation. The results are presented in Fig.10, where the RMS error on acceleration is $0.233ms^{-2}$ and on speed is $1.044ms^{-1}$. It is noticed that the performance of the fuzzy system on the test data degraded seriously.

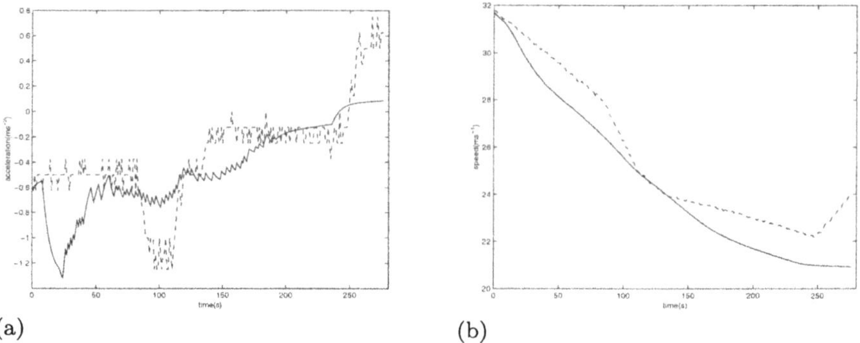

(a) (b)

Fig. 10. Results on test data without interpretability constraints: (a) Acceleration; (b) speed.

A fuzzy system is then generated with completeness and consistency checking. The inconsistency degree of the fuzzy system is now reduced to 0.63. Note first that the fuzzy partition of the two input variables (see Fig. 11) is now complete and the distribution of the membership functions seems to be more reasonable. The acceleration and speed tracking results for the training data are demonstrated in Fig. 12. Compared to the fuzzy rule system generated without consistency and completeness checking, the performance

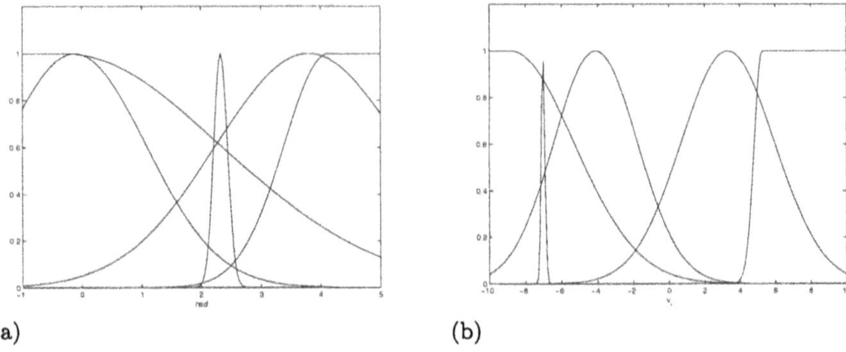

(a) (b)

Fig. 11. Fuzzy partitions with interpretability constraints: (a) *nsd*; (b) $v_r(t)$. Both fuzzy partitions are distinguishable and complete.

for the training data is quite similar. However, the performance on the test data is improved Fig. 13, where the RMS error on acceleration is $0.216ms^{-2}$ and on speed is $0.552ms^{-1}$.

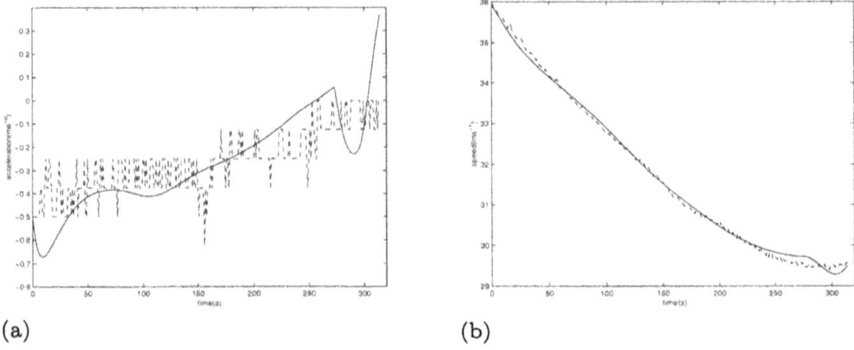

(a) (b)

Fig. 12. Results on the training data with interpretability constraints: (a) acceleration; (b) speed. Solid line: experimental data; dashed line: output of the fuzzy system.

From the above simulations, we see that a fuzzy system generated with completeness and consistency checking outperforms the one generated without checking on the test data. This means that improvement of the interpretability may also improve the generalization performance of the fuzzy systems.

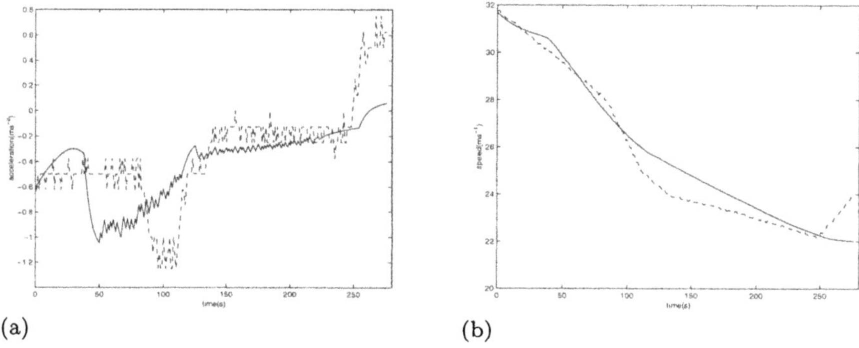

Fig. 13. Results on test data with interpretability constraints: (a) Acceleration (b) Speed.

4 Conclusion

A method for generating distinguishable, complete, consistent and compact fuzzy rules from data using evolution strategies has been proposed. Definitions for the completeness and distinguishability of the fuzzy partitions, the consistency of the fuzzy rules and the compactness of the fuzzy system, which are the main aspects of interpretability of fuzzy systems, have been suggested. The distinguishability, completeness and consistency measures are added to the fitness function of the evolutionary algorithms. In this way, not only will the performance of the fuzzy system be optimized, but also the interpretability of the extracted fuzzy rules is improved. Besides, it has been shown that improvement of the interpretability can also improve generalization performance when less training data are available.

References

1. T. Bäck. Parallel optimization of evolutionary algorithms. In *Parallel Problem Solving from Nature*, pages 418–427. Springer, 1994.
2. T. Bäck and H. Schwefel. An overview of evolutionary algorithms for parameter optimization. *Evolutionary Computation*, 1(1):1–23, 1993.
3. H.R. Berenji and P. Khedkar. Learning and tuning fuzzy controllers through reinforcement. *IEEE Transactions on Neural Networks*, 3(5):724–739, 1992.
4. O. Cordon, F. Herrera, F. Hoffmann, and L. Magdalena. *Genetic fuzzy systems: evolutionary tuning and learning of fuzzy knowledge bases.* World Scientific, 2001.
5. K.J. Hunt, R. Haas, and M. Brown. On the functional equivalence of fuzzy inference systems and spline-based networks. *Int. Journal of Neural Systems*, 6(2):171–184, 1995.
6. J.-S. R. Jang and C.-T. Sun. Self-learning fuzzy controller based on temporal back-propagation. *IEEE Transactions on Neural Networks*, 3:714–723, 1992.

7. Y. Jin. Decentralized adaptive fuzzy control for robot manipulators. *IEEE Transactions on Systems, Man, and Cybernetics*, 28(1):47–57, 1998.
8. Y. Jin. Fuzzy modeling of high-dimensional systems: Complexity reduction and interpretability improvement. *IEEE Transactions on Fuzzy Systems*, 8(2):212–221, 2000.
9. Y. Jin and J. Jiang. Techniques in neural network based fuzzy system identification and their applications to control of complex systems. In C.T. Leondes, editor, *Fuzzy Theory Systems: Techniques and Applications*, volume 1, chapter 5, pages 112–128. Academic Press, San Diego, CA, 1999.
10. Y. Jin and W. von Seelen. Evaluating flexible fuzzy controllers via evolution strategies. *Fuzzy Sets and Systems*, 108:243–252, 1999.
11. Y. Jin, W. von Seelen, and B. Sendhoff. An approach to rule-based knowledge extraction. In *IEEE Proceedings of International Conference on Fuzzy Systems*, pages 1188–1193, Anchorage, Alaska, 1998. IEEE Press.
12. Y. Jin, W. von Seelen, and B. Sendhoff. On generating fc^3 fuzzy rule systems from data using evolution strategies. *IEEE Transactions on Systems, Man and Cybernetics, Part B: Cybernetics*, 29(4):829–845, 1999.
13. Y. Jin, J. Zhu, and J.P. Jiang. Implemeting self-organizing fuzzy controllers with hybrid pi-sigma neural network. In *3rd International Workshop on Advanced Motion Control*, pages 893–902, Berkeley, CA, 1993.
14. C. Karr. Applying genetics to fuzzy logic. *IEEE AI Expert*, 6(2):26–33, 1992.
15. L.T. Koczy and K. Hirota. Size reduction by interpolation in fuzzy rules. *IEEE Transactions on Systems, Man and Cybernetics*, 27(1):14–25, 1997.
16. M.A. Lee and H. Takagi. Integrating design stages of fuzy systems using genetic algorithms. In *Proceedings of 2nd IEEE International Conference on Fuzzy Systems*, volume 1, pages 612–617, 1993.
17. D.D. Leitch. *A new genetic algorithm for evolution of fuzzy systems*. PhD thesis, Department of Engineering Science, University Oxford, Oxford, UK, 1995.
18. C.T. Lin and C.S.G. Lee. Real-time supervised structure/parameter learning for fuzzy neural networks. In *Proceedings of IEEE International Conference on Fuzzy Systems*, pages 1283–1290. IEEE Press, 1992.
19. M. Setnes, R. Babuska, U. Kaymak, and H.R. van Nauta Remke. Similarity measures in fuzzy rule base simplification. *IEEE Transactions on Systems, Man and Cybernetics*, 28(3):376–386, 1998.
20. J. Valente de Oliveira. On the optimization of fuzzy systems using bio-inspired strategies. In *IEEE Proceedings of International Conference on Fuzzy Systems*, pages 1129–1134, Anchorage, Alaska, 1998. IEEE Press.
21. L.-X. Wang and J. Mendel. Generating fuzzy rules by learning from examples. *IEEE Transactions on Systems, Man, and Cybernetics*, 22(6):1414–1427, 1992.
22. W. Wienholt. Improving a fuzzy inference system by means of evolution strategy. In B. Reusch, editor, *Fuzzy Logik*, pages 186–195. Springer, 1994.
23. J. Yen, L. Wang, and W. Gillespie. Improving the interpretability of tsk fuzzy models by combining global learning and local learning. *IEEE Transactions on Fuzzy Systems*, 6(4):530–537, 1998.
24. Q. Zhang and J. Gayko amd M. Kreutz. Optimization of a fuzzy controller for a driver assistant system. In *Proceedings of the Fuzzy-Neuro Systems*, pages 376–382, 1998.

Fuzzy CoCo: Balancing Accuracy and Interpretability of Fuzzy Models by Means of Coevolution.

Carlos-Andrés Peña-Reyes[1] and Moshe Sipper[2]

[1] Logic Systems Laboratory, Swiss Federal Institute of Technology in Lausanne, CH-1015 Lausanne, Switzerland, Email: Carlos.Pena@epfl.ch
[2] Department of Computer Science, Ben-Gurion University, Beer-Sheva 84105, Israel, Email: sipper@cs.bgu.ac.il

Abstract. In this chapter we present Fuzzy CoCo, a fuzzy modeling technique based on cooperative coevolution, conceived to provide high numeric precision (accuracy) while incurring as little a loss of linguistic descriptive power (interpretability) as possible. The search for interpretability is represented by several constraints taken into account when designing the evolutionary algorithm, which induce the drive for accuracy. Interpretability-oriented fuzzy modeling must conduct two separate but intertwined search processes: (1) the search for membership functions, and (2) the search for rules. Towards this end, Fuzzy CoCo employs two coevolving species: database (membership functions) and rule base. Coevolution allows to overcome limitations presented by single-population evolutionary algorithms when confronted with fuzzy modeling, including stagnation, convergence to local optima, and computational costliness. We demonstrate the efficacy of Fuzzy CoCo by applying it to a hard, real-world problem—prediction of breast-cancer malignancy—obtaining excellent results.

1 Introduction

The earliest fuzzy systems were constructed using knowledge provided by human experts, and were thus linguistically correct. However, the difficulty of applying such an approach for high-dimension, ill-known, or data-intensive models, led to the coming of new data-driven fuzzy modeling techniques.

These techniques initially concentrated on solving a parameter-optimization problem based on the numeric performance of the systems. Unfortunately, they paid little attention to linguistic aspects, thus neglecting one of the most important advantages offered by fuzzy systems. Recently, as fuzzy modeling techniques have concentrated more on linguistic issues, the difficulty of improving system interpretability without losing (numeric) performance has become evident. This accuracy-interpretability trade-off is currently one of the most active research lines in fuzzy modeling.

In this chapter we present Fuzzy CoCo, a fuzzy modeling technique based on cooperative coevolution, conceived to provide high accuracy while incurring as little a loss of interpretability as possible. With Fuzzy CoCo the user is in control of the balance between accuracy and interpretability thanks to the method's configurability.

This chapter is organized as follows: In the next section we provide an overview of evolutionary computation and evolutionary fuzzy modeling. Then, Section 3 discusses some aspects related to interpretability requirements. Section 4 presents Fuzzy CoCo, our cooperative coevolutionary approach to fuzzy modeling. Section 5 then describes a sample application of Fuzzy CoCo to a hard problem: breast-cancer assessment by mammography interpretation. The results obtained are presented in Section 6. Finally, we present concluding remarks in Section 7.

2 Background

2.1 Evolutionary computation

The domain of evolutionary computation involves the study of the foundations and the applications of computational techniques based on the principles of natural evolution. Evolution in nature is responsible for the "design" of all organisms on earth, and for the strategies they use to interact with each other. Evolutionary algorithms employ this powerful design philosophy to find solutions to hard problems.

Generally speaking, evolutionary techniques can be viewed either as search methods, or as optimization techniques. As written by Michalewicz [10]:

> Any abstract task to be accomplished can be thought of as solving a problem, which, in turn, can be perceived as a search through a space of potential solutions. Since usually we are after 'the best' solution, we can view this task as an optimization process.

Three basic mechanisms drive natural evolution: *reproduction*, *mutation*, and *selection*. The first two act on the *chromosomes* containing the genetic information of the *individual* (the *genotype*), rather than on the individual itself (the *phenotype*) while selection acts on the phenotype. Reproduction is the process whereby new individuals are introduced into a *population*. During sexual reproduction, *recombination* (or *crossover*) occurs, transmitting to the offspring chromosomes that are a melange of both parents' genetic information. Mutation introduces small changes into the inherited chromosomes; it often results from copying errors during reproduction. Selection, acting on the phenotype, is a process guided by the Darwinian principle of survival of the fittest. The fittest individuals are those best adapted to their environment, which thus survive and reproduce.

Evolutionary computation makes use of a metaphor of natural evolution, according to which a problem plays the role of an environment wherein lives a population of individuals, each representing a possible solution to the problem. The degree of adaptation of each individual (i.e., candidate solution) to its environment is expressed by an adequacy measure known as the *fitness function*. The phenotype of each individual, i.e., the candidate solution itself, is generally encoded in some manner into its *genome* (genotype). Evolutionary algorithms potentially produce progressively better solutions to the problem. This is possible, thanks to the constant introduction of new "genetic" material into the population, by applying so-called genetic operators which are the computational equivalents of natural evolutionary mechanisms.

There are several types of evolutionary algorithms, among which the best known are *genetic algorithms*, *genetic programming*, *evolution strategies*, and *evolutionary programming*; though different in the specifics they are all based on the same general principles. The archetypal evolutionary algorithm proceeds as follows: an initial population of individuals, $P(0)$, is generated at random or heuristically. Every evolutionary step t, known as a *generation*, the individuals in the current population, $P(t)$, are *decoded* and *evaluated* according to some predefined quality criterion, referred to as the fitness, or fitness function. Then, a subset of individuals, $P'(t)$—known as the *mating pool*—is selected to reproduce, with selection of individuals done according to their fitness. Thus, high-fitness ("good") individuals stand a better chance of "reproducing," while low-fitness ones are more likely to disappear.

Selection alone cannot introduce any new individuals into the population, i.e., it cannot find new points in the search space. These points are generated by altering the selected population $P'(t)$ via the application of crossover and mutation, so as to produce a new population, $P''(t)$. Crossover tends to enable the evolutionary process to move toward "promising" regions of the search space. Mutation is introduced to prevent premature convergence to local optima, by randomly sampling new points in the search space. Finally, the new individuals $P''(t)$ are introduced into the next-generation population, $P(t+1)$; usually $P''(t)$ simply becomes $P(t+1)$. The termination condition may be specified as some fixed, maximal number of generations or as the attainment of an acceptable fitness level. Figure 1 presents the structure of a generic evolutionary algorithm in pseudo-code format.

```
begin EA
  t:=0
  Initialize population P(t)
  while not done do
    Evaluate P(t)
    P'(t) := Select[P(t)]
    P''(t) := ApplyGeneticOperators[P'(t)]
    P(t+1) := Introduce[P''(t),P(t)]
    t:=t+1
  end while
end EA
```

Fig. 1. Pseudo-code of a standard evolutionary algorithm.

Because they combine elements of directed and stochastic search, evolutionary techniques exhibit a number of advantages over other search methods. First, they usually require less knowledge and fewer assumptions about the characteristics of the search space. Second, they can more easily avoid getting stuck in local optima. Finally, they strike a good balance between *exploitation* of the best solutions, and *exploration* of the search space. The strength of evolutionary algorithms derives from their population-based search, and from the use of the genetic mechanisms described above. The existence of a population of candidate solutions entails a parallel search, with the selection mechanism directing the search to the most promising regions, the crossover operator encouraging the exchange of information between these search-space regions, and the mutation operator enabling the exploration of new directions.

The application of an evolutionary algorithm involves a number of important considerations. The first decision to take when applying such an algorithm is how to encode candidate solutions within the genome. The representation must allow for the encoding of all possible solutions while being sufficiently simple to be searched in a reasonable amount of time. Next, an appropriate fitness function must be defined for evaluating the individuals. The (usually scalar) fitness value must reflect the criteria to be optimized and their relative importance. Representation and fitness are thus clearly problem-dependent, in contrast to selection, crossover, and mutation, which seem *prima facie* more problem-independent. Practice has shown, however, that while standard genetic operators can be used, one often needs to tailor these to the problem as well.

2.2 Evolutionary fuzzy modeling

Fuzzy modeling is the task of identifying the parameters of a fuzzy inference system so that a desired behavior is attained [25]. With the *direct* approach a fuzzy model is constructed using knowledge from a human expert. This task becomes difficult when the available knowledge is incomplete or when the problem space is very large, thus motivating the use of *automatic* approaches to fuzzy modeling. Selection of relevant variables and adequate rules is critical for obtaining a good system. One of the major problems in fuzzy modeling is the *curse of dimensionality*, meaning that the computation requirements grow exponentially with the number of variables.

A *fuzzy inference system* is a rule-based system that uses fuzzy logic to reason about data [26]. Its basic structure consists of four main components, as depicted in Figure 2: (1) a fuzzifier, which translates crisp (real-valued) inputs into fuzzy values; (2) an inference engine that applies a fuzzy reasoning mechanism to obtain a fuzzy output; (3) a defuzzifier, which translates this latter output into a crisp value; and (4) a knowledge base, which contains both an ensemble of fuzzy rules, known as the rule base, and an ensemble of membership functions known as the database.

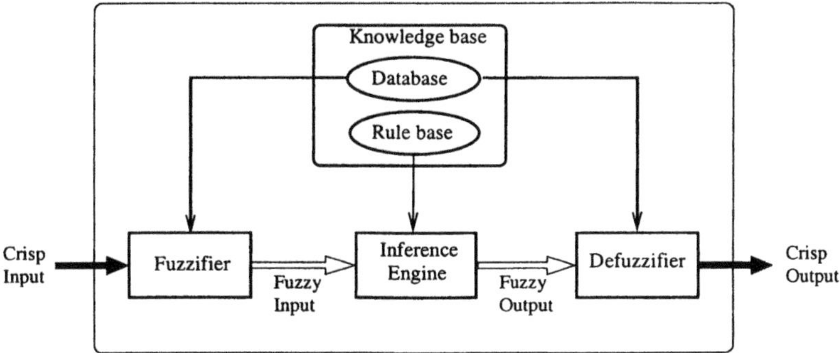

Fig. 2. Basic structure of a fuzzy inference system.

The parameters of fuzzy inference systems can be classified into four categories (Table 1) [13, 15]: logical, structural, connective, and operational.

Logical parameters are usually predefined by the designer based on experience and on problem characteristics. Structural, connective, and operational parameters may be either predefined, or obtained by synthesis or search methodologies. Generally, the search space, and thus the computational effort, grows exponentially with the number of parameters. Therefore, one can either invest more resources in the chosen search methodology, or infuse more *a priori*, expert knowledge into the system (thereby effectively reducing the search space). The aforementioned trade-off between accuracy and interpretability is usually expressed as a set of constraints on the parameter values, thus complexifying the search process.

Evolutionary algorithms are used to search large, and often complex, search spaces. They have proven worthwhile on numerous diverse problems, able to find near-optimal solutions given an adequate performance (fitness) measure. Fuzzy modeling can be considered as an optimization process where part or all of the parameters of a fuzzy system constitute the search space. Works investigating the application of evolutionary techniques in the do-

Table 1. Parameter classification of fuzzy inference systems.

Class	Parameters
Logical	Reasoning mechanism Fuzzy operators Membership function types Defuzzification method
Structural	Relevant variables Number of membership functions Number of rules
Connective	Antecedents of rules Consequents of rules Rule weights
Operational	Membership-function values

main of fuzzy modeling had first appeared about a decade ago [8, 9]. These focused mainly on the tuning of fuzzy inference systems involved in control tasks (e.g., cart-pole balancing, liquid-level system, and spacecraft rendezvous operation). Evolutionary fuzzy modeling has since been applied to an ever-growing number of domains, branching into areas as diverse as chemistry, medicine, telecommunications, biology, and geophysics. For a detailed bibliography on evolutionary fuzzy modeling up to 1996, the reader is referred to [1, 3].

Depending on several criteria—including the available *a priori* knowledge about the system, the size of the parameter set, and the availability and completeness of input-output data—artificial evolution can be applied in different stages of the fuzzy-parameter search. Three of the four categories of fuzzy parameters in Table 1 can be used to define targets for evolutionary fuzzy modeling: structural parameters, connective parameters, and operational parameters [13, 15]. As noted before, logical parameters are usually predefined by the designer based on experience.

Knowledge tuning (operational parameters). The evolutionary algorithm is used to tune the knowledge contained in the fuzzy system by finding membership-function values. An initial fuzzy system is defined by an expert. Then, the membership-function values are encoded in a genome, and an evolutionary algorithm is used to find systems with high performance. Evolution often overcomes the local-minima problem present in gradient descent-based methods. One of the major shortcomings of knowledge tuning is its dependency on the initial setting of the knowledge base.

Behavior learning (connective parameters). In this approach, one assumes that expert knowledge is sufficient in order to define the membership functions; this determines, in fact, the maximum number of rules [25]. The genetic algorithm is used to find either the rule consequents, or an adequate subset of rules to be included in the rule base.

As the membership functions are fixed and predefined, this approach lacks the flexibility to modify substantially the system behavior. Furthermore, as the number of variables and membership functions increases, the curse of dimensionality becomes more pronounced and the interpretability of the system decreases rapidly.

Structure learning (structural parameters). In many cases, the available information about the system is composed almost exclusively of input-output data, and specific knowledge about the system structure is scant. In such a case, evolution has to deal with the simultaneous design of rules, membership functions, and structural parameters. Some methods use a fixed-length genome encoding a fixed number of fuzzy rules along with the membership-function values. In this case the designer defines structural constraints according to the available knowledge of the problem characteristics. Other methods use variable-length genomes to allow evolution to discover the optimal size of the rule base.

Both behavior and structure learning can be viewed as rule-base learning processes with different levels of complexity. They can thus be assimilated within other methods from machine learning, taking advantage of experience gained in this latter domain. In the evolutionary-algorithm community there are two major approaches for evolving such rule systems: the Michigan approach and the Pittsburgh approach [10]. A more recent method has been proposed specifically for fuzzy modeling: the iterative rule learning approach [6]. These three approaches are briefly described below.

The Michigan approach. Each individual represents a single rule. The fuzzy inference system is represented by the entire population. Since several rules participate in the inference process, the rules are in constant competition for the best action to be proposed, and cooperate to form an efficient fuzzy system. The cooperative-competitive nature of this approach renders difficult the decision of which rules are ultimately responsible for good system behavior. It necessitates an effective credit-assignment policy to ascribe fitness values to individual rules.

The Pittsburgh approach. Here, the evolutionary algorithm maintains a population of candidate fuzzy systems, each individual representing an entire fuzzy system. Selection and genetic operators are used to produce new generations of fuzzy systems. Since evaluation is applied to the entire system, the credit-assignment problem is eschewed. This approach allows to include additional optimization criteria in the fitness function, thus affording the implementation of multi-objective optimization. The main shortcoming of this approach is its computational cost, since a population of full-fledged fuzzy systems has to be evaluated each generation.

The iterative rule learning approach. As in the Michigan approach, each individual encodes a single rule. An evolutionary algorithm is used to find a single rule, thus providing a partial solution. The evolutionary algorithm is used iteratively for the discovery of new rules, until an appropriate rule base is built. To prevent the process from finding redundant rules (i.e., rules covering the same input subspace), a penalization scheme is applied each time a new rule is added. This approach combines the speed of the Michigan approach with the simplicity of fitness evaluation of the Pittsburgh approach. However, as with other incremental rule-base construction methods, it can lead to a non-optimal partitioning of the antecedent space.

3 Interpretability Considerations

As mentioned before, the fuzzy-modeling process has to deal with an important trade-off between the *accuracy* and the *interpretability* of the model. The model is expected to provide high numeric precision while incurring as little a loss of linguistic descriptive power as possible. Currently, there exist no well-established definitions for interpretability of fuzzy systems, mainly due to the subjective nature of such a concept. However, some works have

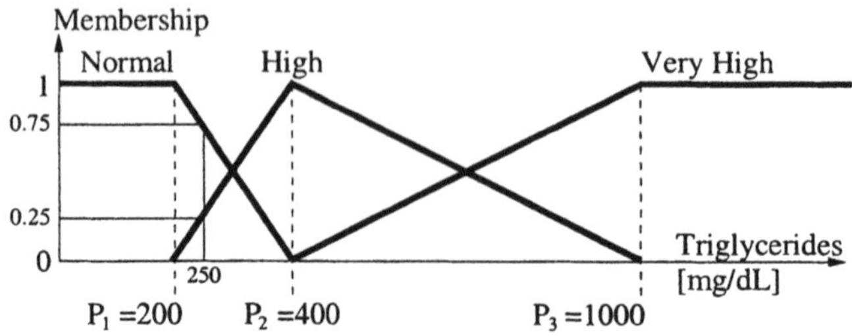

Fig. 3. Example of a fuzzy variable: *Triglycerides* has three possible fuzzy values, labeled **Normal**, **High**, and **Very High**, plotted above as degree of membership versus input value. The values P_i, setting the trapezoid and triangle apices, define the membership functions. In the figure, an example input value 250 mg/dL is assigned the membership values $\mu_{Normal}(250) = 0.75$, $\mu_{High}(250) = 0.25$, and $\mu_{VeryHigh}(250) = 0$. Note that $\mu_{Normal}(250) + \mu_{High}(250) + \mu_{VeryHigh}(250) = 1$.

attempted to define objective criteria that facilitate the automatic modeling of interpretable fuzzy systems [5,21].

The fuzzy system of Figure 2 processes information in three stages: the input interface (fuzzifier), the processing stage (inference engine), and the output interface (defuzzifier). The interface deals with linguistic variables and their corresponding labels. These linguistic variables define the *semantics* of the system. The inference process is performed using fuzzy rules that define the connection between input and output fuzzy variables. These fuzzy rules define the *syntax* of the fuzzy system. Fuzzy modelers must thus take into account both semantic and syntactic criteria to obtain interpretable systems. Below, we present some criteria that represent conditions driving fuzzy modeling toward human-interpretable systems together with strategies to satisfy them.

3.1 Semantic criteria

The notion of "linguistic variable" formally requires associating a meaning to each fuzzy label [27]. Hence, each membership function should represent a linguistic term with a clear semantic meaning. For example, in Figure 3, the fuzzy variable *Triglycerides* has three meaningful labels: **Normal**, **High**, and **Very High**. The following semantic criteria describe a set of properties that the membership functions of a fuzzy variable should possess in order to facilitate the task of assigning linguistic terms [13,15,17]. The focus is on the meaning of the ensemble of labels instead of the absolute meaning of each term in isolation.

- *Distinguishability.* Each linguistic label should have semantic meaning and the fuzzy set should clearly define a range in the universe of discourse of the variable. In the example of Figure 3, to describe variable *Triglycerides* we used three meaningful labels: **Normal**, **High**, and **Very**

High. Their membership functions are defined using parameters P_1, P_2, and P_3.

- *Justifiable number of elements.* The number of membership functions of a variable should be compatible with the number of conceptual entities a human being can handle. This number should not exceed the limit of 7 ± 2 distinct terms.
- *Coverage.* Any element from the universe of discourse should belong to at least one of the fuzzy sets. That is, its membership value must be different than zero for at least one of the linguistic labels. More generally, a minimum level of coverage ϵ may be defined, giving rise to the concept of *strong coverage.* Referring to Figure 3, we see that any value along the x-axis belongs to at least one fuzzy set; no value lies outside the range of all sets.
- *Normalization.* Since all labels have semantic meaning, then, for each label, at least one element of the universe of discourse should have a membership value equal to one. In Figure 3, we observe that all three sets **Normal**, **High**, and **Very High** have elements with membership value equal to 1.
- *Complementarity.* For each element of the universe of discourse, the sum of all its membership values should be equal to one (as in the example in Figure 3). This guarantees uniform distribution of meaning among the elements.

3.2 Syntactic criteria

A fuzzy rule relates one or more input-variable conditions, called antecedents, to their corresponding output fuzzy conclusions, called consequents. The example rule presented in Figure 4, associates the conditions **High** and **Middle** of the input variables *Triglycerides* and *Age*, respectively, with the conclusion **Moderate** of the output variable *Cardiac risk.* The linguistic adequacy of a fuzzy rule base lies on the interpretability of each rule as well as on that of the whole set of rules. The following syntactic criteria define some conditions which—if satisfied by the rule base—reinforce the interpretability of a fuzzy system [5].

- *Completeness.* For any possible input vector, at least one rule should be fired to prevent the fuzzy system from getting blocked.
- *Rule-base simplicity.* The set of rules must be as small as possible. If, however, the rule base is still large, rendering hard a global understanding of the system, the number of rules that fire simultaneously for any input vector must remain low in order to furnish a simple local view of the behavior.
- *Single-rule readability.* The number of conditions implied in the antecedent of a rule should be compatible with the aforementioned number of conceptual entities a human being can handle (i.e., $\leq 7 \pm 2$).

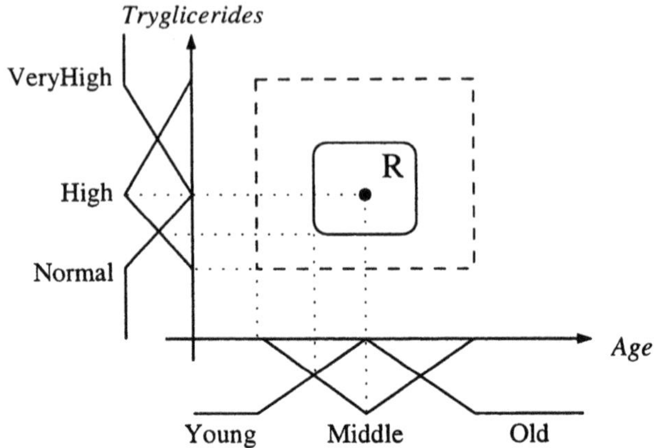

Fig. 4. Example of a fuzzy rule and its firing range. The rule **if** *Triglycerides* **is** *High* **and** *Age* **is** *Middle* **then** *Cardiac risk* **is** *Moderate*, marked as **R**, is (partially) fired by input values into the dashed-line rectangle (i.e. $\mu(R) > 0$). The solid-line rectangle denotes the region where $\mu(R) > 0.5$.

- *Consistency.* If two or more rules are simultaneously fired, their consequents should not be contradictory, i.e., they should be semantically close.

3.3 Strategies to satisfy semantic and syntactic criteria

The criteria presented above, intended to assess interpretability of a fuzzy system, define a number of restrictions on the definition of fuzzy parameters. Semantic criteria limit the choice of membership functions, while syntactic criteria bind the fuzzy rule base. We present below some strategies to apply these restrictions when defining a fuzzy model.

- *Linguistic labels shared by all rules.* A number of fuzzy sets is defined for each variable, which are interpreted as linguistic labels and shared by all the rules [5]. In other words, each variable has a unique semantic definition. This results in a grid partition of the input space as illustrated in Figure 5. To satisfy the completeness criterion, it is normally used a fully defined rule base, meaning that it contains all the possible rules, as in the example shown in Figure 6a. Label sharing by itself facilitates but does not guarantee the semantic integrity of each variable. More conditions are necessary.
- *Normal, orthogonal membership functions.* The membership functions of two successive labels must be complementary (i.e., their sum must be equal to one) in their overlapping region, whatever form they have [4,21]. Moreover, in such regions each label must ascend from zero to unity membership values [13, 15]. The variables presented in Figures 3 and 5, satisfy these requirements.

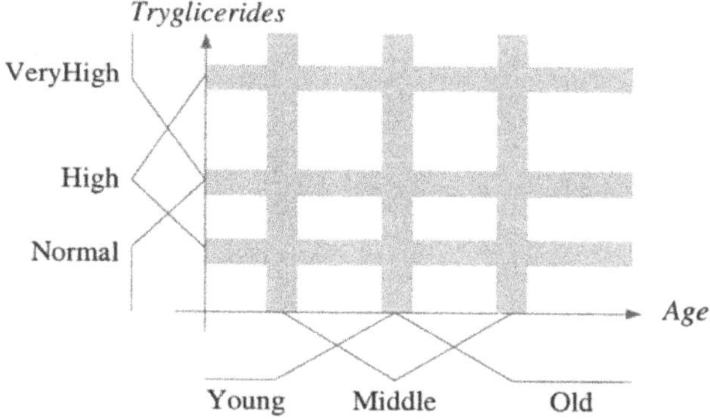

Fig. 5. Grid partition of the input space. In this example, two semantically correct input variables, each with three labels, divide the input space into a grid of nine regions.

- *Don't-care conditions.* A fully defined rule base, as that shown in Figure 6a, becomes impractical for high-dimension systems. The number of rules in a fully defined rule base exponentially increases as the number of input variables increases (e.g., a system with five variables, each with three labels, would contain $3^5 = 243$ rules). Moreover, given that each rule antecedent contains a condition for each variable, the rules might be too lengthy to be understandable, and too specific to describe general circumstances.
  To tackle these two problems some authors use "*don't-care*" as a valid input label [7, 13, 15]. Variables in a given rule that are marked with a *don't-care* label are considered as irrelevant. For example, in the rule base shown in Figure 6b the rule R_A:

  if *Triglycerides* **is** *don't-care* **and** *Age* **is** *Old* **then** *Cardiac risk* **is** *Moderate*,

  replaces three rules (i.e., R_3, R_6, and R_9 in Figure 6a) and is interpreted as:

  if *Age* **is** *Old* **then** *Cardiac risk* **is** *Moderate.*

  Although *don't-care* labels allow a reduction of rule-base size, their main advantage is the improvement of rule readability.
- *Default rule.* In many cases, the behavior of a system exhibits only a few regions of interest, which can be described by a small number of rules (e.g., R_5, R_A, and R_B in Figure 6b). To describe the rest of the input space, a simple default action, provided by the default rule, would be enough [23]. The example in Figure 6c shows that the default rule, named R_0, covers the space of rules R_1, R_2, and R_4. By definition, a default condition is true when all other rule conditions are false. In a fuzzy context, the default rule is as true as all the others are false. Consequently,

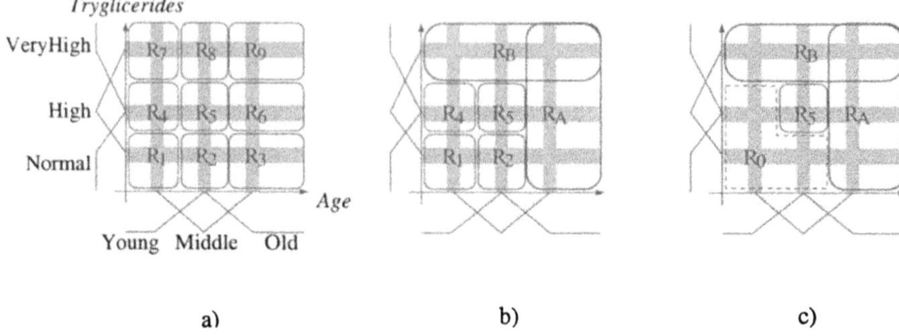

a) b) c)

Fig. 6. Strategies to define the rule base. a) Fully defined rule base: the system contains all nine possible rules of the form **if** *Triglycerides* **is** *label* **and** *Age* **is** *label* **then** .... b) *don't-care* labels: two rules, R_A and R_B, containing *don't-care* labels cover almost half of the input space. Variables marked with a *don't-care* label are considered as irrelevant for the rule in question. R_A is thus interpreted as **if** *Age* **is** *Old* **then** ..., and R_B as **if** *Triglycerides* **is** *VeryHigh* **then** .... c) Default rule, called here R_0, defines a default action to be performed when none of the so-called active rules—R_5, R_A, and R_B—apply. By definition, the activation of the (fuzzy) default rule is $\mu(R_0) = 1 - max(\mu(R_i))$, with $i = \{1, 2, 3, \ldots\}$. The rectangles denote the region where $\mu(R_i) > 0.5$.

the activation degree of the default rule, $\mu(R_0)$, is thus given by $\mu(R_0) = 1 - max(\mu(R_i))$, where $\mu(R_i)$ is the activation degree of the i-th rule.

4 Fuzzy CoCo: A Cooperative Coevolutionary Approach to Fuzzy Modeling

As mentioned earlier, the accuracy-interpretability trade-off fuzzy modelers face implies the assumption of constraints acting on the parameter values, mainly on the membership-function shapes.

In this section we present Fuzzy CoCo, a cooperative coevolutionary approach, capable of obtaining high-performance, interpretable systems. We have conceived Fuzzy CoCo to allow a high degree of freedom in the type of fuzzy systems it can design in order to allow the user to manage the trade-off between performance and interpretability. The next subsection briefly explains cooperative coevolution, after which Section 4.2 presents Fuzzy CoCo. Finally, Section 4.3 shows how the interpretability criteria are considered in Fuzzy CoCo.

4.1 Cooperative coevolution

Coevolution refers to the simultaneous evolution of two or more species with coupled fitness. Such coupled evolution favors the discovery of complex solutions whenever complex solutions are required [12]. Simplistically speak-

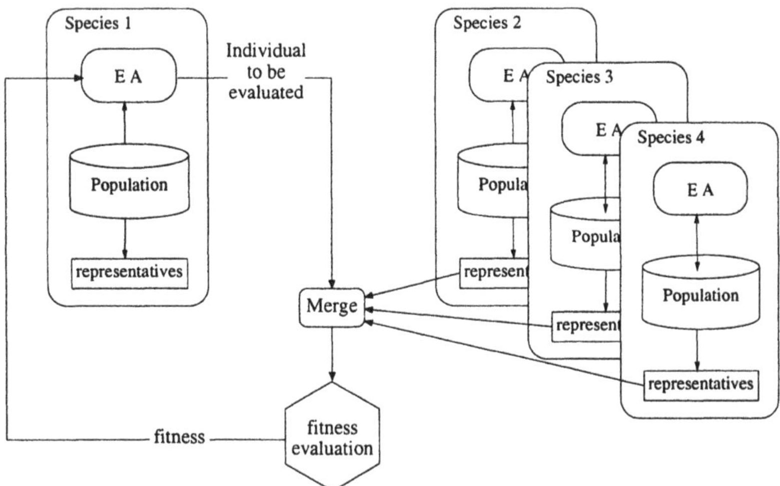

Fig. 7. Potter's cooperative coevolutionary system. The figure shows the evolutionary process from the perspective of Species 1. The individual being evaluated is combined with one or more *representatives* of the other species so as to construct several solutions which are tested on the problem. The individual's fitness depends on the quality of these solutions.

ing, one can say that coevolving species can either compete (e.g., to obtain exclusivity on a limited resource) or cooperate (e.g., to gain access to some hard-to-attain resource). Cooperative (also called symbiotic) coevolutionary algorithms involve a number of independently evolving species which together form complex structures, well-suited to solve a problem. The fitness of an individual depends on its ability to collaborate with individuals from other species. In this way, the evolutionary pressure stemming from the difficulty of the problem favors the development of cooperative strategies and individuals. Single-population evolutionary algorithms often perform poorly—manifesting stagnation, convergence to local optima, and computational costliness—when confronted with problems presenting one or more of the following features: (1) the sought-after solution is complex, (2) the problem or its solution is clearly decomposable, (3) the genome encodes different types of values, (4) strong interdependencies among the components of the solution, (5) component-ordering drastically affects fitness. Cooperative coevolution effectively addresses these issues, consequently widening the range of applications of evolutionary computation. Potter [19, 20] developed a model in which a number of populations explore different decompositions of the problem. Below we detail this framework as it forms the basis of our own approach.

In Potter's system, each species represents a subcomponent of a potential solution. Complete solutions are obtained by assembling *representative* members of each of the species (populations). The fitness of each individual depends on the quality of (some of) the complete solutions it participated in, thus measuring how well it cooperates to solve the problem. The evolution of each species is controlled by a separate, independent evolutionary algorithm. Figure 7 shows the general architecture of Potter's cooperative coevolutionary framework, and the way each evolutionary algorithm computes the fitness of its individuals by combining them with selected representatives from the other species. A greedy strategy for the choice of representatives of a species is to use one or more of the fittest individuals from the last generation.

4.2 The coevolutionary algorithm

Fuzzy CoCo is a cooperative coevolutionary approach to fuzzy modeling wherein two coevolving species are defined: database (membership functions) and rule base [13,15]. This approach is based primarily on the framework defined by Potter [19,20].

A fuzzy modeling process usually deals with the simultaneous search for operational and connective parameters (Table 1). These parameters provide an almost complete definition of the linguistic knowledge describing the behavior of a system, and the values mapping this symbolic description into a real-valued world (a complete definition also requires logical and structural parameters whose definition is best suited for human skills). Thus, fuzzy modeling can be thought of as two separate but intertwined search processes: (1) the search for the membership functions (i.e., operational parameters) that define the fuzzy variables, and (2) the search for the rules (i.e., connective parameters) used to perform the inference.

Fuzzy modeling presents several features discussed earlier which justify the application of a cooperative-coevolutionary approach: (1) The required solutions can be very complex, since fuzzy systems with a few dozen variables may call for hundreds of parameters to be defined. (2) The proposed solution—a fuzzy inference system—can be decomposed into two distinct components: rules and membership functions. (3) Membership functions are continuous and real-valued, while rules are discrete and symbolic. (4) These two components are interdependent because the membership functions defined by the first group of values are indexed by the second group (rules).

Consequently, in Fuzzy CoCo, the fuzzy modeling problem is solved by two coevolving, cooperating species. Individuals of the first species encode values which define completely all the membership functions for all the variables of the system. For example, with respect to the variable *Triglycerides* shown in Figure 3, this problem is equivalent to finding the values of P_1, P_2, and P_3.

Individuals of the second species define a set of rules of the form:

$$\textbf{if } (v_1 \textbf{ is } A_1) \textbf{ and } \ldots \textbf{ and } (v_n \textbf{ is } A_n) \textbf{ then } (\textit{output} \textbf{ is } C),$$

where the term A_v indicates which of the linguistic labels of fuzzy variable v is used by the rule. For example, a valid rule could contain the expression:

if ... and (*Triglycerides* **is** *High*) **and ... then ...**

which includes the membership function *High* whose defining parameters are contained in the first species (population).

The two evolutionary algorithms used to control the evolution of the two populations are instances of a simple genetic algorithm [22]. Figure 8 presents the Fuzzy CoCo algorithm in pseudo-code format. The genetic algorithms apply fitness-proportionate selection to choose the mating pool, and apply an elitist strategy with an elitism rate E_r to allow some of the best individuals to survive into the next generation. The elitism strategy extracts E_S individuals—the so-called elite—to be reinserted into the population after evolutionary operators have been applied (i.e., selection, crossover, and mutation). Note that the elite is not removed from the population, participating thus in the reproduction process. Standard crossover and mutation operators are applied with probabilities P_c and P_m, respectively.

```
begin Fuzzy CoCo
   g:=0
   for each species S
      Initialize populations P_S(0)
      Evaluate population P_S(0)
   end for
   while not done do
      for each species S
         g:=g+1
         E_S(g) = elite-select P_S(g − 1)
         P'_S(g) = select P_S(g − 1)
         P''_S(g) = crossover P'_S(g)
         P'''_S(g) = mutate P''_S(g)
         P_S(g) = P'''_s(g) + E_S(g)
         Evaluate population P_S(g)
      end for
   end while
end Fuzzy CoCo
```

Fig. 8. Pseudo-code of Fuzzy CoCo. Two species coevolve in Fuzzy CoCo: membership functions and rules. The elitism strategy extracts E_S individuals to be reinserted into the population after evolutionary operators have been applied (i.e., selection, crossover, and mutation). Selection results in a reduced population $P'_S(g)$ (usually, the size of $P'_S(g)$ is $\|P'_S\| = \|P_S\| - \|E_S\|$). The line "Evaluate population $P_S(g)$" is elaborated in Figure 9.

We introduced elitism to avoid the divergent behavior of Fuzzy CoCo, observed in preliminary trial runs. Non-elitist versions of Fuzzy CoCo tended to lose the genetic information of good individuals found during evolution, consequently producing populations with mediocre individuals scattered through-

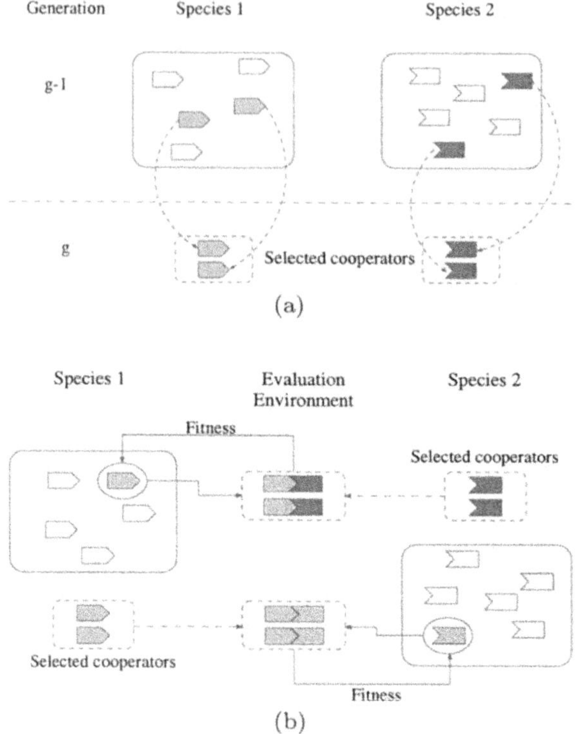

Fig. 9. Fitness evaluation in Fuzzy CoCo. (a) Several individuals from generation $g-1$ of each species are selected according to their fitness to be the representatives of their species during generation g; these representatives are called "cooperators." (b) During the evaluation stage of generation g (after selection, crossover, and mutation—see Figure 8), individuals are combined with the selected cooperators of the other species to construct fuzzy systems. These systems are then evaluated on the problem domain and serve as a basis for assigning the final fitness to the individual being evaluated.

out the search space. This is probably due to the relatively small size of the population which renders difficult the preservation of good solutions while exploring the search space. The introduction of simple elitism produces an undesirable effect on Fuzzy CoCo's performance: populations converge prematurely even with reduced values of the elitism rate E_r. To offset this effect without losing the advantages of elitism, it was necessary to increase the mutation probability P_m by an order of magnitude so as to improve the exploration capabilities of the algorithm. (Increased mutation rates were also reported by Potter [19, 20] in his coevolutionary experiments.)

A more detailed view of the fitness evaluation process is depicted in Figure 9. An individual undergoing fitness evaluation establishes cooperations with one or more representatives of the other species, i.e., it is combined with individuals from the other species to construct fuzzy systems. The fitness value assigned to the individual depends on the performance of the fuzzy systems it participated in (specifically, either the average or the maximal value).

Representatives, called here *cooperators*, are selected both fitness-proportionally and randomly from the previous generation since they have already been assigned a fitness value (see Figure 8). In Fuzzy CoCo, N_{cf} cooperators are selected according to their fitness, usually the fittest individuals, thus favoring the exploitation of known good solutions. The other N_{cr} cooperators are selected randomly from the population to represent the diversity of the species, maintaining in this way exploration of the search space.

Fuzzy CoCo compares favorably with non-coevolutionary approaches to fuzzy modeling. In our previous work, Fuzzy CoCo attained higher performance and required less computation to design the fuzzy systems [14,15]

4.3 Introducing interpretability in Fuzzy CoCo

As mentioned before, Fuzzy CoCo allows a high degree of freedom in the type of fuzzy systems it can design, letting the user determine the accuracy-interpretability trade-off. When the interest is to preserve as much as possible the interpretability of the evolved systems, the fuzzy model should satisfy the semantic and syntactic criteria presented in Section 3.

The aforementioned strategies—label sharing, orthogonal membership functions, don't-care conditions, and default rule—must guide both the design of the fuzzy inference system and the definition of both species' genomes. Besides, one or more of the linguistic criteria may participate in the fitness function as a way to reinforce the selection pressure towards interpretable systems. The example presented in the next section illustrates the way interpretability criteria are introduced in Fuzzy CoCo.

5 Application Example: The Catalonia Online Breast-Cancer Risk Assessor

Mammography remains the principal technique for detecting breast cancer. Its undoubtable value in reducing mortality notwithstanding, mammography's positive predictive value (PPV) is low: only between 15 and 35% of mammographic-detected lesions are cancerous [11,18]. The remaining 65 to 85% of biopsies, besides being costly and time-consuming, cause understandable stress on women facing the doubt of cancer. A computer-based tool that assists radiologists during mammographic interpretation would contribute to increasing the PPV of biopsy recommendations.

5.1 The database

The *Catalonia Mammography Database*, which is the object of our study, was collected at the Duran y Reynals hospital in Barcelona. It consists of 15 input attributes and a diagnostic result indicating whether or not a carcinoma was detected after a biopsy. The 15 input attributes include three clinical characteristics (Table 2) and two groups of six radiologic features, according to the type of lesion found in the mammography: mass or microcalcifications (Table 3).

Table 2. Variables corresponding to a patient's clinical data.

v_1	Age	[28-82] years	
v_2	Menstrual history	1	Premenopausal
		2	Postmenopausal
v_3	Family history	1	None
		2	Second familiar
		3	First familiar
		4	Contralateral
		5	Homolateral

A radiologist fills out a reading form for each mammography, assigning values for the clinical characteristics and for one of the groups of radiologic features. Then, the radiologist interprets the case using a five-point scale: (1) benign; (2) probably benign; (3) indeterminate; (4) probably malignant; (5) malignant. According to this interpretation a decision is made on whether to practice a biopsy on the patient or not. The Catalonia database contains data corresponding to 227 cases, all of them suspect enough to justify a biopsy recommendation. For the purpose of this study, each case was further examined by three different readers—for a total of 681 readings—but only diverging readings were kept. The actual number of readings in the database is 516, among which 187 are positive (malignant) cases and 329 are negative (benign) cases.

5.2 Proposed system

The solution scheme we propose is depicted in Figure 10. It consists of a reading form, a fuzzy subsystem, and a threshold unit. Based on the 15 input attributes collected with the reading form, the fuzzy system computes a continuous appraisal value of the malignancy of a case. The threshold unit then outputs a biopsy recommendation according to the fuzzy system's output. The threshold value used in this system is 3, which corresponds to the "indeterminate" diagnostic. Fuzzy CoCo is applied to design the fuzzy system in charge of appraising malignancy.

Table 3. Variables corresponding to radiologic features. There are two groups of variables used to describe the mammographic existence of microcalcifications and masses (left and right columns respectively).

Microcalcifications			Mass		
v_4		Disposition	v_{10}		Morphology
	1	Round		1	Oval
	2	Indefinite		2	Round
	3	Triangular or Trapezoidal		3	Lobulated
	4	Linear or Ramified		4	Polilobulated
v_5		Other signs of group form		5	Irregular
	1	None	v_{11}		Margins
	2	Major axis in direction of nipple		1	Well delimited
	3	Undulating contour		2	Partially well delimited
	4	Both previous		3	Poorly delimited
v_6		Maximum diameter of group		4	Spiculated
		[3-120] mm	v_{12}		Density greater than parenchyma
v_7		Number		1	Not
	1	<10		2	Yes
	2	10 to 30	v_{13}		Focal distortion
	3	>30		1	Not
v_8		Morphology		2	Yes
	1	Ring shaped	v_{14}		Focal asymmetry
	2	Regular sharp-pointed		1	Not
	3	Too small to determine		2	Yes
	4	Irregular sharp-pointed	v_{15}		Maximum diameter
	5	Vermicular, ramified			[5-80] mm
v_9		Size irregularity			
	1	Very regular			
	2	Sparingly regular			
	3	Very irregular			

5.3 Fuzzy-parameter setup

We used prior knowledge about the Catalonia database to guide our choice of fuzzy parameters. In addition, we took into account the interpretability criteria presented in Section 3 to define constraints on the fuzzy parameters. Referring to Table 1, we delineate below the fuzzy system's set-up:

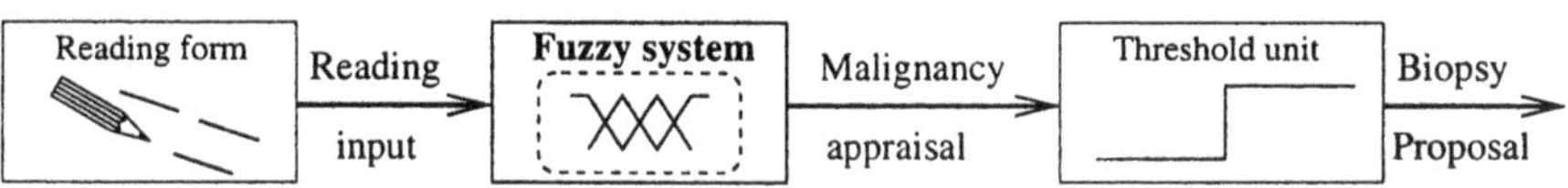

Fig. 10. Proposed system. Note that the fuzzy system displayed in the middle is in fact the entire fuzzy inference system of Figure 2.

- Logical parameters: singleton-type fuzzy systems; min-max fuzzy operators; orthogonal, trapezoidal input membership functions (see Figure 11); weighted-average defuzzification.
- Structural parameters: two input membership functions (*Low* and *High*; see Figure 11); two output singletons (*benign* and *malignant*); a user-configurable number of rules. The relevant variables are one of Fuzzy CoCo's evolutionary objectives. *Low* and *High* linguistic labels may be further replaced by labels having medical meaning according to the specific context of each variable.

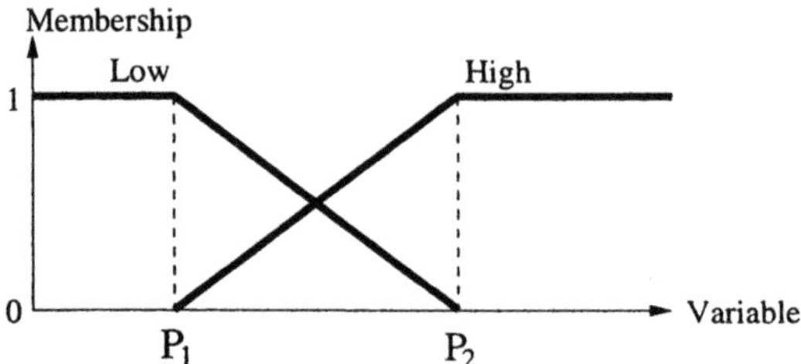

Fig. 11. Input fuzzy variables. Each fuzzy variable has two possible fuzzy values labeled **Low** and **High**, and orthogonal membership functions, plotted above as degree of membership versus input value. P_1 and P_2 define the membership-function apices.

- Connective parameters: the antecedents and the consequent of the rules are searched by Fuzzy CoCo. The algorithm also searches for the consequent of the default rule. All rules have unitary weight.
- Operational parameters: the input membership-function values are to be found by Fuzzy CoCo. The values of P_1 and P_2 (Figure 11) are restricted to the universe of each variable to satisfy the interpretability criteria presented in Section 3. For the output singletons we used the values 1 and 5, for *benign* and *malignant*, respectively.

5.4 Genome encodings

Fuzzy CoCo thus searches for four parameters: input membership-function values, relevant input variables, and antecedents and consequents of rules. To encode these parameters into both species' genomes, which together describe an entire fuzzy system, it is necessary to take into account the heterogeneity of the input variables as explained below.

- Species 1: Membership functions. The fifteen input variables (v_1 - v_{15}) present three different types of values: continuous (v_1,v_6, and v_{15}), discrete (v_3 - v_5 and v_7 - v_{11}), and binary (v_2 and v_{12} - v_{14}). It is not necessary to encode membership functions for binary variables as they can only take on two values. The membership-function genome encodes

Table 4. Genome encoding of parameters for membership-function species. Genome length is 106 bits.

Variable type	Qty	Parameters	Bits	Total bits
Continuous	3	2	7	42
Discrete	8	2	4	64
		Total Genome Length		106

the remaining 11 variables—three continuous and eight discrete—each with two parameters P_1 and P_2, defining the membership-function apices (Figure 11). Table 4 delineates the parameters encoding the membership-function genome.

- Species 2: Rules. The i-th rule has the form:

  if (v_1 **is** A_1^i) **and** ... **and** (v_{15} **is** A_{15}^i) **then** (*output* **is** C^i),

  where A_j^i can take on the values: 1 (*Low*), 2 (*High*), or 0 or 3 (*don't-care*). C^i can take on the values: 1 (*benign*) or 2 (*malignant*). However, as mentioned before, each database case presents three clinical characteristics and six radiologic features according to the type of lesion found: mass or microcalcifications (note that only a few special cases contain data for both groups). To take advantage of this fact, the rule-base genome encodes, for each rule, 11 parameters: the three antecedents of the clinical-data variables, the six antecedents of one radiological-feature group, an extra bit to indicate whether the rule applies for mass or microcalcifications, and the rule consequent. Furthermore, the genome contains an additional parameter corresponding to the consequent of the default rule. Relevant variables are searched for implicitly by allowing the algorithm to choose non-existent membership functions as valid antecedents ($A_j^i = 0$ or $A_j^i = 3$); in such case the respective variable is considered irrelevant, and removed from the rule. Table 5 delineates the parameters encoding the rules genome.

Table 5. Genome encoding of parameters for rules species. Genome length is $(20 \times N_r) + 1$ bits, where N_r denotes the number of rules.

Parameters	Qty	Bits	Total bits
Clinical antecedents	$3 \times N_r$	2	$6 \times N_r$
Radiologic antecedents	$6 \times N_r$	2	$12 \times N_r$
Rule-type selector	N_r	1	N_r
Consequents	$N_r + 1$	1	$N_r + 1$
	Total Genome Length		$(20 \times N_r) + 1$

5.5 Evolutionary parameters

Table 6 delineates values and ranges of values of the evolutionary parameters. The algorithm terminates when the maximum number of generations, G_{max},

Table 6. Fuzzy CoCo set-up.

Parameter	Values
Population size N_p	90
Maximum generations G_{max}	$700 + 200N_r$
Crossover probability P_c	1
Mutation probability P_m	{0.005,0.01}
Elitism rate E_r	{0.1,0.2}
"Fit" cooperators N_{cf}	1
Random cooperators N_{cr}	1

is reached (we set $G_{max} = 700 + 200 \times N_r$, i.e., dependent on the number of rules used in the run), or when the increase in fitness of the best individual over five successive generations falls below a certain threshold (10^{-4} in our experiments). Note that mutation rates are relatively higher than those used with a simple genetic algorithm.

5.6 Fitness function

As the main function of the proposed system is the assessment of a medical diagnosis, our fitness definition takes into account medical diagnostic criteria. The most commonly employed measures of the validity of diagnostic procedures are the sensitivity and specificity, the likelihood ratios, the predictive values, and the overall classification (accuracy) [2]. Table 7 provides expressions for four of these measures which are important for evaluating the performance of our systems. Three of them are used in the fitness function, the last one is used in Section 6 to support the analysis of the results. Besides these criteria, the fitness function provides extra selective pressure based on two syntactic criteria: simplicity and readability (see Section 3).

Table 7. Diagnostic performance measures. The values used to compute the expressions are: True positive (TP): the number of positive cases correctly detected, true negative (TN): the number of negative cases correctly detected, false positive (FP): the number of negative cases diagnosed as positive, and false negative (FN): the number of positive cases diagnosed as negative.

Measure	Expression
Sensitivity	$\frac{TP}{TP+FN}$
Specificity	$\frac{TN}{TN+FP}$
Accuracy	$\frac{TP+TN}{TP+TN+FP+FN}$
Positive predictive value (PPV)	$\frac{TP}{TP+FP}$

Our fitness function combines the following five criteria: 1) F_{sens}: sensitivity, or true-positive ratio, computed as the percentage of positive cases correctly classified; 2) F_{spec}: specificity, or true-negative ratio, computed as the percentage of negative cases correctly classified (note that there is usually an important trade-off between sensitivity and specificity which renders difficult the satisfaction of both criteria); 3) F_{acc}: classification performance, computed as the percentage of cases correctly classified; 4) F_r: rule-base size fitness, computed as the percentage of unused rules (i.e., the number of rules that are never fired and can thus be removed altogether from the system); and 5) F_v: rule-length fitness, computed as the average number of *don't-care* antecedents (i.e., unused variables) per rule. This order also represents their relative importance in the final fitness function, from most important (F_{sens}) to least important (F_r and F_v).

The fitness function is computed in three steps—basic fitness, accuracy reinforcement, and size reduction—as explained below:

1. Basic fitness. Based on sensitivity and specificity, it is given by
$$F_1 = \frac{F_{sens} + \alpha F_{spec}}{1+\alpha},$$
where the weight factor $\alpha = 0.3$ reflects the greater importance of sensitivity.
2. Accuracy reinforcement. Given by
$$F_2 = \frac{F_1 + \beta F'_{acc}}{1+\beta},$$
where $\beta = 0.01$. $F'_{acc} = F_{acc}$ when $F_{acc} > 0.7$; $F'_{acc} = 0$ elsewhere. This step slightly reinforces the fitness of high-accuracy systems.
3. Size reduction. Based on the size of the fuzzy system, it is given by
$$F = \frac{F_2 + \gamma F_{size}}{1+2\gamma},$$
where $\gamma = 0.01$. $F_{size} = (F_r + F_v)$ if $F_{acc} > 0.7$ and $F_{sens} > 0.98$; $F_{size} = 0$ elsewhere. This step rewards top systems exhibiting a concise rule set, thus directing evolution toward more interpretable systems.

6 Results

This section describes the results obtained when applying the methodology described in Section 5. We first delineate the success statistics relating to the evolutionary algorithm. Then, we present the diagnostic performance of two selected evolved fuzzy systems that exemplify our approach.

A total of 65 evolutionary runs were performed, all of which found systems whose fitness exceeds 0.83. In particular, considering the best individual per

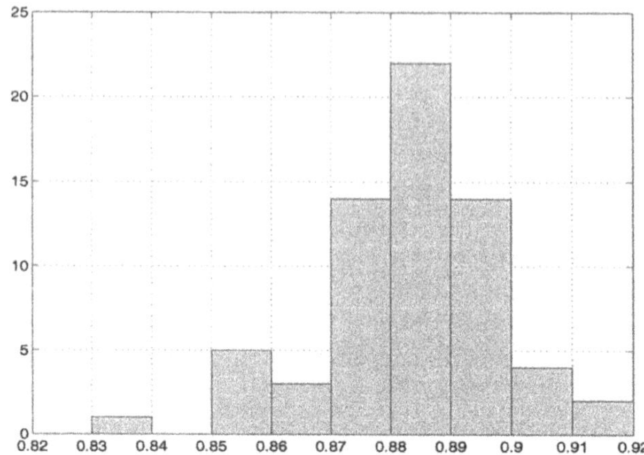

Fig. 12. Summary of results of 65 evolutionary runs. The histogram depicts the number of systems exhibiting a given fitness value at the end of the evolutionary run. The fitness considered is that of the best individual of the run.

run (i.e., the evolved system with the highest fitness value), 42 runs led to a fuzzy system whose fitness exceeds 0.88 , and of these, 6 runs found systems whose fitness exceeds 0.9; these results are summarized in Figure 12.

Table 8 shows the results of the best systems obtained. The maximum number of rules per system was fixed at the outset to be between ten and twenty-five.

Table 8. Results of the best systems evolved. Results are divided into four classes, in accordance with the maximum number of rules-per-system, going from 10-rule systems to 25-rule ones. Shown below are the fitness values of the top systems as well as the average fitness per class, along with the number of rules which effectively used by the system (R_{eff}) and the average number of variables per rule (V_r).

Maximum number	Best individual			Average per class		
of rules	Fitness	R_{eff}	V_r	Fitness	R_{eff}	V_r
10	0.8910	9	2.22	0.8754	9.17	2.52
15	0.8978	12	2.50	0.8786	12.03	2.62
20	0.9109	17	2.41	0.8934	14.15	2.59
25	0.9154	17	2.70	0.8947	15.78	2.76

As mentioned before, our fitness function includes two syntactic criteria to favor the evolution of good diagnostic systems exhibiting interpretable rule bases (see Section 3). Concerning the simplicity of the rule base, rules that are encoded in a genotype but that never fire are removed from the phenotype (the final system), rendering it more interpretable. Moreover, to improve readability, the rules are allowed (and encouraged) to contain *don't-care* conditions. The relatively low values of R_{eff} and V_r in Table 8 confirm the reinforced interpretability of the evolved fuzzy systems.

Table 9. Results of ten-fold cross-validation. Results are divided into three classes, in accordance with the maximum number of rules-per-system, going from 3-rule systems to 30-rule ones. Shown below is the average fitness per class obtained in both training and test sets (the fitness represents the average of the average fitnesses for the six cross-validation runs), along with the difference between training and test performances, expressed both in absolute value and percentage.

Maximum number of rules	Training set	Test set	Difference	Percentage
3	0.8269	0.7865	0.0404	4.89
7	0.8511	0.8036	0.0474	5.58
10	0.8712	0.8030	0.0682	7.83
15	0.8791	0.8125	0.0666	7.58
20	0.8814	0.8244	0.0569	6.47
30	0.8867	0.8104	0.0763	8.60

To assess the generalization capabilities of the fuzzy systems obtained, we apply ten-fold cross-validation [24]. For each ten-fold test the data set is first partitioned into ten equal-sized sets, then each set is in turn used as the test set while Fuzzy CoCo trains on the other nine sets. Table 9 presents the average results obtained performing six cross-validation runs for several rule-base sizes. The percent difference between training- and test-sets is relatively low, even for large rule bases, indicating low overfitting. Further generalization analysis, which are not in the scope of this chapter, have confirmed this statement.

Table 10 shows the diagnostic performance of two selected evolved systems. The first system, which is the top one over all 65 Fuzzy CoCo runs, is a 17-rule system exhibiting a sensitivity of 99.47% (i.e., it detects all but one of the positive cases), and a specificity of 68.69% (i.e., 226 of the 329 negative cases are correctly detected as benign). The second system is the best found when searching for ten-rule systems. The sensitivity and the specificity of this 9-rule system are, respectively, 98.40% and 64.13%. As mentioned in Section 5, the usual positive predictive value (PPV) of mammography ranges between 15 and 35%. As shown in Table 10, Fuzzy CoCo increases this value beyond 60%—64.36% for the 17-rule system—while still exhibiting a very high sensitivity.

Table 10. Diagnostic performance of two selected evolved systems. Shown below are the sensitivity, the specificity, the accuracy, and the positive predictive value (PPV) of two selected systems. In parentheses are the values, expressed in number of cases, leading to such performance measures. The 17-rule system is the top system. The 9-rule system is the best found when searching for ten-rule systems.

	17-rule	9-rule
Sensitivity	99.47% (186/187)	98.40% (184/187)
Specificity	68.69% (226/329)	64.13% (211/329)
Accuracy	79.84% (412/516)	76.55% (395/516)
PPV	64.36% (186/289)	60.93% (184/302)

7 Concluding Remarks

We presented Fuzzy CoCo, a fuzzy modeling technique based on cooperative coevolution, along with an application to breast-cancer diagnosis. In fuzzy modeling, the interpretability-accuracy trade-off is of crucial import, imposing several conditions on the input and output membership functions as well as on the rule definition. In evolutionary fuzzy modeling these conditions are translated both into restrictions on the choice of fuzzy parameters and into criteria included in the fitness function. We designed Fuzzy CoCo to be highly configurable, thus facilitating the management of the interpretability-accuracy trade-off.

Applying Fuzzy CoCo to breast-cancer diagnosis we concentrated on increasing the interpretability of solutions by the mechanisms proposed earlier, obtaining excellent results. We note, however, that the consistency of the entire rule base and its compatibility with the specific domain knowledge can only be assessed by further interaction with medical experts (radiologists, oncologists).

In the future we wish to test some novel ideas that could improve Fuzzy CoCo: 1) Coevolution of $N_r + 1$ species, one species for each of the N_r rules in addition to the membership-function species. 2) Coexistence of several Fuzzy CoCo instances (each one set to evolve systems with a different number of rules), permitting migration of individuals among them so as to increase the exploration and the diversity of the search process. 3) Apply the strategy of rising and death of species proposed by Potter and DeJong [20] in order to evolve systems with variable numbers of rules and membership functions.

References

1. J. T. Alander. An indexed bibliography of genetic algorithms with fuzzy logic. In Pedrycz [16], pages 299–318.
2. H. Brenner. Measures of differential diagnostic value of diagnostic procedures. *Journal of Clinical Epidemiology*, 49(12):1435–1439, December 1996.
3. O. Cordón, F. Herrera, and M. Lozano. On the combination of fuzzy logic and evolutionary computation: A short review and bibliography. In Pedrycz [16], pages 33–56.
4. J. Espinosa and J. Vandewalle. Constructing fuzzy models with linguistic integrity from numerical data-AFRELI algorithm. *IEEE Transactions on Fuzzy Systems*, 8(5):591, October 2000.
5. S. Guillaume. Designing fuzzy inference systems from data: An interpretability-oriented review. *IEEE Transactions on Fuzzy Systems*, 9(3):426–443, June 2001.
6. F. Herrera, M. Lozano, and J. L. Verdegay. Generating fuzzy rules from examples using genetic algorithms. In B. Bouchon-Meunier, R. R. Yager, and L. A. Zadeh, editors, *Fuzzy Logic and Soft Computing*, pages 11–20. World Scientific, 1995.

7. H. Ishibushi, T. Nakashima, and T. Murata. Performance evaluation of fuzzy classifier systems for multidimensional pattern classification problems. *IEEE Transactions on Systems, Man, and Cybernetics. Part B: Cybernetics*, 29(5):601–618, October 1999.
8. C. L. Karr. Genetic algorithms for fuzzy controllers. *AI Expert*, 6(2):26–33, February 1991.
9. C. L. Karr, L. M. Freeman, and D. L. Meredith. Improved fuzzy process control of spacecraft terminal rendezvous using a genetic algorithm. In G. Rodriguez, editor, *Proceedings of Intelligent Control and Adaptive Systems Conference*, volume 1196, pages 274–288. SPIE, February 1990.
10. Z. Michalewicz. *Genetic Algorithms + Data Structures = Evolution Programs.* Springer-Verlag, Heidelberg, third edition, 1996.
11. S. G. Orel, N. Kay, C. Reynolds, and D. C. Sullivan. Bi-rads categorization as a predictor of malignancy. *Radiology*, 211(3):845–880, June 1999.
12. J. Paredis. Coevolutionary computation. *Artificial Life*, 2:355–375, 1995.
13. C. A. Peña-Reyes and M. Sipper. A fuzzy-genetic approach to breast cancer diagnosis. *Artificial Intelligence in Medicine*, 17(2):131–155, October 1999.
14. C. A. Peña-Reyes and M. Sipper. The flowering of Fuzzy CoCo: Evolving fuzzy iris classifiers. In V. Kurková, N. C. Steele, R. Neruda, and M. Kárný, editors, *Proceedings of the 5th International Conference on Artificial Neural Networks and Genetic Algorithms (ICANNGA 2001)*, pages 304-307. Springer-Verlag, 2001.
15. C. A. Peña-Reyes and M. Sipper. Fuzzy CoCo: A cooperative-coevolutionary approach to fuzzy modeling. *IEEE Transactions on Fuzzy Systems*, 9(5):727–737, October 2001.
16. W. Pedrycz, editor. *Fuzzy Evolutionary Computation.* Kluwer Academic Publishers, 1997.
17. W. Pedrycz and J. Valente de Oliveira. Optimization of fuzzy models. *IEEE Transactions on Systems, Man and Cybernetics, Part B: Cybernetics*, 26(4):627–636, August 1996.
18. L. Porta, R. Villa, E. Andia, and E. Valderrama. Infraclinic breast carcinoma: Application of neural networks techniques for the indication of radioguided biopsias. In J. Mira, R. Moreno-Díaz, and J. Cabestany, editors, *Biological and Artificial Computation: From Neuroscience to Technology*, volume 1240 of *Lecture Notes in Computer Science*, pages 978–985. Springer, 1997.
19. M. A. Potter. *The Design and Analysis of a Computational Model of Cooperative Coevolution.* PhD thesis, George Mason University, 1997.
20. M. A. Potter and K. A. DeJong. Cooperative coevolution: An architecture for evolving coadapted subcomponents. *Evolutionary Computation*, 8(1):1–29, spring 2000.
21. J. Valente de Oliveira. Semantic constraint for membership function optimization. *IEEE Transactions on Systems, Man, and Cybernetics. Part A: Systems and Humans*, 29(1):128–138, January 1999.
22. M. D. Vose. *The Simple Genetic Algorithm.* MIT Press, Cambridge, MA, August 1999.
23. P. Vuorimaa. Fuzzy self-organizing map. *Fuzzy Sets and Systems*, 66:223–231, 1994.
24. I. H. Witten and E. Frank. *Data Mining: Practical Machine Learning Tools and Techniques with Java Implementations.* Data Management Systems. Morgan Kaufmann, San Francisco, USA, 2000.

25. R. R. Yager and D. P. Filev. *Essentials of Fuzzy Modeling and Control.* John Wiley & Sons., New York, 1994.
26. R. R. Yager and L. A. Zadeh. *Fuzzy Sets, Neural Networks, and Soft Computing.* Van Nostrand Reinhold, New York, 1994.
27. L. A. Zadeh. The concept of a linguistic variable and its applications to approximate reasoning. *Information Science*, Part I, 8:199-249; Part II, 8:301-357; Part III, 9:43-80., 1975.

On the Achievement of Both Accurate and Interpretable Fuzzy Systems Using Data-Driven Design Processes*

José Valente de Oliveira[1] and Paulo Fazendeiro[2]

[1] Universidade do Algarve, Faculdade de Ciências e Tecnologia, Centro de Sistemas Inteligentes, Campus de Gambelas, 8000-062 Faro, PORTUGAL. jvolivei@ualg.pt

[2] Universidade da Beira Interior, Departamento de Informática, Rua Marquês d'Ávila e Bolama, 6200-001 Covilhã, PORTUGAL. pandre@noe.ubi.pt

Abstract. In this chapter it is argued that, one of the most interesting features of fuzzy system is the insight provided on the linguistic relationship between their variables, or in other words, is the possibility of interpret their parameters as a set of linguistic rules. Both the notions of accuracy and interpretability are reviewed. A general design policy where both accuracy and interpretation can be taken into account is reviewed. It is argued that the lost of accuracy does not necessary occurs when this type of data-driven design policy is applied. For illustration purposes, simulation results are given from the realistic control problem of neuromuscular relaxation of patients under surgery using continuous infusion of atracurium.

1 Introduction

Membership function is a core concept in what concerns fuzzy system semantics. Membership functions represent linguistic terms, which are used by a human expert either to describe his or her knowledge and actions or to reason about the world.

When data-driven unconstrained design processes are used, such as neural networks learning rules, or evolutionary optimisation techniques like genetic algorithms, one cannot ensure that at the end humans can interpret membership functions as linguistic terms. In this case, we are neglecting one of the most interesting features of fuzzy systems, i.e., the insight provided on the linguistic relationship between system's variables.

The chapter is organised as follows. First, it is stated what we mean by accuracy and interpretation. Then the proposed concepts are translated into functionals to be optimised. It is argued that the functionals relative to interpretation can be viewed as a form of including human knowledge into the design process. From this perspective it is claimed that, depending on the available data (e.g., partially unknown data) an optimisation process that takes into account the semantic dimension can perform better than other that does not take this dimension into account. In general, fuzzy systems have very many degrees of freedom, what can easily trap an unconstrained design process, due to local minima.

* The work presented here was partially supported by FCT – Fundação para a Ciência e Tecnologia under Project Hipocrates, PRAXIS/P/EEI/10149/1998.

For the sake of illustration, simulation results from the control of neuromuscular relaxation of patients under surgery using continuous infusion of atracurium are presented. In this kind of problems, the controller design should be guided by criteria such as robustness, reliability, and performance in a clinical-like environment. Moreover, the control system should behave in a way that is easily understood by the anaesthetist, allowing the eventual accommodation of its clinical experience in peculiar clinical situations. These requirements make the proposed design policy a particularly interesting approach to this problem.

2 On accuracy

A considerable amount of interesting physical systems can be represented by vector differential equations as the following:

$$\begin{aligned} \frac{dx(t)}{dt} &= \Phi[x(t), u(t)] \\ y(t) &= \Psi[x(t)] \end{aligned} \tag{1}$$

where $x(t) \in \mathcal{X} \subset \Re^n$ represents the state variables at time t, $u(t) \in \mathcal{U} \subset \Re^p$ represents the inputs at time t, and $y(t) \in \mathcal{Y} \subset \Re^m$ represents the outputs at time t of a p input, m output system of order n. This representation is called input-state-output representation of a continuous-time system.

In the following, the focus is on the discrete-time version of the differential equations (1). A discrete-time (or sampled data) system is a system whose variables vary only at given time instants. Discrete systems appear either because the problem in study corresponds to a discrete time phenomenon or because the study of the continuous-time system (1) is made using digital computers.

To attempt a clear presentation the system is assumed as single input single output (SISO). The discrete version of (1) is then given by:

$$\begin{aligned} x[k+1] &= \Phi(x[k], u[k]) \\ y[k] &= \Psi(x[k]) \end{aligned} \tag{2}$$

where $x[k] = [x_1[k] \; x_2[k] \; \ldots x_n[k]]$ represents the state variables of the system at the discrete time instant k, $u[k]$, and $y[k]$ are discrete time signals (or sequences) of the input and output, respectively, at discrete time k.

The identification problem arises whenever the functions Φ and Ψ in (2) are unknown. In this problem, the task is to construct a suitable model that when excited with the same inputs $u[k]$ as the system, produces an output $\hat{y}[k]$ that approximates $y[k]$ in the sense that

$$\lim_{k \to \infty} |y[k] - \hat{y}[k]| \leq \epsilon \tag{3}$$

for some specified $\epsilon > 0$.

For formulating a well-posed identification problem several assumptions are required, cf. [10]. Here it is assumed that the system is time-invariant, causal and stable with known parameterisation but with unknown parameters. Furthermore, the input $u[k]$ is assumed as an uniformly bounded function of time k.

Observability is a key issue in system identification. Informally, this concept refers to the ability of estimating the state of a system (or identifying it) based on a sequence of input-output observations of finite length. In general, there are no constructive methods for observer design in non-linear systems, thus one has to assume again some (strong) assumptions. For instance here we will assume that, if a SISO system is described by (2) then its state can be reconstructed by (at least) N measurements of its input and output. This yields (at least) N non-linear equations in N unknowns $x[k]$. Again, one has to assume that, this has a unique solution for any set of (at least) N inputs $u[k]$ within an compact (closed and bounded) domain.

Under these assumptions one can state that the input-output behaviour of (2) can be accurately approximated, in the sense of (3), by an input-output model of the form:

$$y[k+1] = \Upsilon(y[k], y[k-1], \ldots, y[k-p+1], u[k], u[k-1], \ldots, u[k-q+1]) \quad (4)$$

where p and q are positive constants such that $p, q \geq n$, being the order of (2).

During identification two set-up schemes are usually considered, giving rise to the so-called parallel model, and series-parallel model. In the parallel model, the output of model at a given time instant, say $\hat{y}[k]$, depends on its own past values $\hat{y}[k-i]; i = 1, \ldots, p$, and of those of the input. In the series-parallel model, the model output $\hat{y}[k]$ depends on the past values of the physical system output $y[k-i]; i = 1, \ldots, p$, (instead of those of the model), and of those of the input.

The series-parallel model exhibits some advantages over the parallel model scheme. Namely, since the physical system is assumed bounded-input bounded-output (BIBO) stable, all the signals fed back into the model are uniformly bounded. This cannot be guaranteed for the parallel case without showing first that the model is BIBO stable. In addition, since there is no feedback within a series-parallel model identification scheme, a static optimisation procedure, such as neural networks learning rules, or evolutionary optimisation techniques like genetic algorithms, can be used obviating the computational overhead of dynamic optimisation processes. Furthermore for avoiding unmodeled dynamics the structure of the model is chosen to be identical to that of the system.

There are a number of theoretical results on the universal approximation property of static fuzzy system, e.g., [1,2,4,11,14,18].

These results cover nearly all the types of fuzzy system used in fuzzy modelling, and fuzzy control practice and show that these are able to approximate any real-valued continuous function on a compact domain to any degree of accuracy.

Relatively to these theoretical results two issues are worth mention: i) nothing is said on how to build such systems, and ii) the number of rules (and thus membership functions) is assumed unbounded.

This last point is a difficulty since, at least for Sugeno type of fuzzy systems, the property of universal approximation does not hold any longer for a bounded (arbitrarily large) number of rules [9]. This result seems to be applicable for all kind of fuzzy systems with a fixed number of rules, including naturally those that can be interpretable (and perhaps to some particular type of neural networks).

This reinforces our claim that interpretability is one of the most interesting features of fuzzy systems.

3 On interpretability

The notion of linguistic variable formally requires a *semantic rule* for associating a meaning to each one of its values, the linguistic terms represented by the referential or primary fuzzy sets [17]. Hence, each fuzzy set should represent a linguistic term with a clear semantic meaning.

In this section, a set of semantic properties that the membership functions of the referential fuzzy sets should have to obviate the subjective task of assigning linguistic terms to them is reviewed [12,13].

1. *A moderate number of membership functions:* The number of membership functions should not be arbitrary. Although this number has been recognised as application dependent, there are some reasons for imposing an upper bound on the number of membership functions.
   One reason pertains to Zadeh's principle of incompatibility. As soon as the number of membership functions increases the precision of the system may increase too, but its relevance will decrease. In the limit, when the number of membership functions approaches the number of data, a fuzzy system becomes a numeric system. What would be then an appropriated limit?
   Miller [8] suggested that the typical number of different chunks efficiently handled at the short-term memory is 7 ± 2. This result was based on stimulus tests involving visual control panels, human separation of different tones, and on experiments with taste intensities. The maximum number of chunks for each stimulus correctly handled consistently falls around 7 ± 2. The results also suggest that the capacity of short-term memory seems to increase with familiarity, and decreases with anxiety.
   This observation on the human information process limitations has guided most of the strategies for segmenting, factoring, or decomposing problems in Computer Sciences and other fields. The problem is that above this number the errors generated by us while juggling, dealing with, or handling objects, entities, or concepts rise disproportionately, i.e., in a strong non-linear fashion.
   Based on the observation of the limited capacity and high volatile short-term memory, one can adopt the Miller "magic" number 7 ± 2 as an upper-bound for the number of membership functions. Curiously enough, it is interesting to verify that most of the successful industrial and commercial applications of fuzzy systems do not exceed this limit.
2. *Distinguishability:* By definition of linguistic variable, there should be a semantic rule associating a meaning to each value of the variable. That is, a membership function should represent a linguistic term with a clear meaning. To obviate the subjective establishment of this membership functions/linguistic term association, membership functions of a given linguistic variable should be distinct enough from each other.
   In the absence of this property, a variable could have equal membership functions what would be superfluous, tautological, and misleading. The existence of close or nearly identical membership functions (linguistic terms) may cause inconsistencies at the processing stage.
3. *Normality:* Semantically, to consider non-normal membership functions is to question the meaning of the represented linguistic terms. On the other hand,

normality allows one to use the entire range of membership grades, usually [0, 1]; therefore, membership functions should be normal, i.e.,

$$\exists_{v \in \mathbf{V}} \forall_{i=1,2,\ldots,n} \mu_i(v) = 1.0 \tag{5}$$

where v is any external variable defined in $\mathbf{V}$, n is the number of membership functions and μ_i is the i-th membership function.

4. *Coverage:* Membership functions μ_i of the referential sets $\mathcal{X}_i$ should cover the entire universe of discourse, so that every piece of data may have a linguistic representation. For any $\epsilon \geq 0$

$$\forall_{v \in \mathbf{V}} \exists_{i:1 \leq i \leq n} \mu_i(v) > \epsilon. \tag{6}$$

5. *Natural zero positioning:* If required by the nature of the problem, one of the membership functions should be unimodal, convex, and centred in zero, representing the linguistic term *nearly zero*. This applies to most fuzzy controllers essentially because a *nearly zero* tracking error (exactly zero, if possible) is desirable. It should be emphasised the difference between the location of the linguistic term *nearly zero*, and its necessity for solving a given problem.

Two remarks are worth noting: i)None of the previous properties violates the assumptions for accuracy presented in section 2; ii) the properties can be ensured using a variety of constraints.

This allows us to pursuit our claim that, depending on the design goal and the constraints used to represent the properties it is possible to find a fuzzy system both accurate and interpretable.

4 On the design goals

Some of the semantic properties of membership function can be satisfied by an *a priori* selection of certain membership function features. However the coverage and distinguishability properties should be monitored and enforced during the design process of the overall system.

It should be stressed that, as stated previously, these properties are general enough to accommodate a number of different monitorisation and correction techniques. These should be selected and used in such way that allows both accurate and interpretable fuzzy system to be achieved.

One of such techniques is to use a constraint based directly on condition (6). This constraint can be expressed in terms of the sigma-count measure of the internal representation of the data. The sigma-count gives a measure of the cardinality of a fuzzy set. Let A be a fuzzy set defined in a discrete UoD, such as

$$A = [\mu_A(x_1)\ \mu_A(x_2)\ \cdots\ \mu_A(x_n)]' = [a_1\ a_2\ \cdots\ a_n]'$$

where n is the number of elements x_i in the support of A, and a_i $(i = 1, \ldots, n)$ are the degrees of membership defining the fuzzy set A. The sigma-count of A, $M_p(A)$ is given by [4]:

$$M_p(A) = \sqrt[p]{a_1^p + \ldots + a_n^p} \tag{7}$$

Therefore (6) can be ensured by

$$\forall_{v \in \mathbf{V}} M_p(\mathcal{L}(v)) > \epsilon \tag{8}$$

where $\mathcal{L}(v)$ gives the internal representation of the extern variable v, a fuzzy (sub)set.

Another general constraint is based on the concept of *optimal interfaces.* This concept can be presented as follows, cf. [12]: let $\mathbf{V}= [a,b] \subset \Re$. The ordered pair $(\mathcal{L}_v, \mathcal{N}_v)$ is said a pair of optimal interfaces (or simply optimal interfaces) iff:

$$\forall_{v \in \mathbf{V}} \mathcal{N}_v(\mathcal{L}_v(v)) = v \tag{9}$$

where $\mathcal{L}_v : \mathbf{V} \rightarrow [0,1]^n$ is the mapping provided by the input interface of the variable v, $\mathcal{N}_v[0,1]^n \rightarrow \mathbf{V}$ with $\mathcal{N}_v([0,\ldots,0]) = \emptyset$ being the mapping provided by the output interface associated with the variable v. See [12] for an analysis of the properties of optimal interfaces.

For ensuring distinguishability we will adopt a constraint on the region of the hypercube where the internal (fuzzy) representation of the fuzzy system may exist. This can be done using again the sigma-count measure, as follows:

$$\forall_{v \in \mathbf{V}} M_p(\mathcal{L}_v(v)) \leq 1 \tag{10}$$

Notice that asking for a sigma-count always strictly less than 1 is not feasible since we are assuming normal membership functions. On the other hand, asking for a sigma-count greater than 1 is equivalent to remove the constraint.

The parameter p in (10) is used for controlling the strength of the constraint. For $p = 1$ we have a strong constraint whereas it can be eliminated as $p \rightarrow +\infty$.

Numerical solutions of optimisation problems with constraints are one of the main concerns of the Mathematical Programming area. There are a variety of techniques for dealing with equality and inequality constraints. Here we use one of these techniques, known as the penalty function method, which enables to solve constrained optimisation problems using unconstrained techniques such as the back-propagation algorithm or genetic algorithms.

5 Illustrative results: the case of neuromuscular blockade control

The control of neuromuscular relaxation (or blockade) is characterised by a high degree of uncertainty associated with the dynamic behaviour of the bio-system under control. The variability of patient responses implies the frequent adaptation of infusion rates, increasing the anaesthetic workload. Automatic feedback control of infusion has been shown to outperform manual control [6,7].

In this application, the merits of a controller should be primarily related to its robustness, reliability, and performance in a clinical environment. Moreover, the control system should allow the introduction of modifications in the controller in a way that is easily understood by the anaesthetist, for accommodating clinical experience and peculiar clinical situations. This requirement makes of fuzzy control a particularly interesting approach to this problem.

The dynamic response of neuromuscular blockade may be modelled by a linear pharmacokinetic model relating the drug infusion rate $u(t)$ with the plasma concentration $c_p(t)$ and a non-linear model relating $c_p(t)$ with the induced pharmacodynamic response $r(t)$ [15,16].

This overall model, with atracurium as muscle relaxant agent, may be described by the state equations:

$$\left.\begin{aligned} \dot{x}_i(t) &= -\lambda_i\, x_i(t) + a_i\, u(t) \qquad i=1,2 \\ c_p(t) &= \textstyle\sum_{i=1}^{2} x_i(t) \end{aligned}\right\} \text{pharmacokinetic}$$
$$\left.\begin{aligned} \dot{c}_{\mathrm{E}}(t) &= -k_{\mathrm{E0}}\, c_{\mathrm{E}}(t) + k_{\mathrm{E0}}\, c_p(t) \\ r(t) &= \frac{100 c_{p50}^{\gamma}}{c_{p50}^{\gamma} + c_{\mathrm{E}}^{\gamma}(t)} \end{aligned}\right\} \text{pharmacodynamic} \tag{11}$$

where λ_i, a_i, k_{E0}, c_{p50} and γ are patient-dependent parameters. The variable $r(t)$, normalised between 0 and 100, correspond to the level of muscular activity, with 0 corresponding to full paralysis and 100 to full muscular activity (no paralysis at all).

For clinical reasons, the patient must undergo an initial bolus in order to induce relaxation in a short period of time (smaller than five minutes).

In order to analyse the resulting data the following indices were used:

- The mean values of the relaxation level (My) and of the drug consumption (Mu).
- Their standard deviation (SDy and SDu).
- The root mean square deviation, which gives an indication of the spread of relaxation around the target value, defined by $RMSD = \sqrt{\frac{1}{N}\sum(y_i - ref_i)^2}$.
- Maximum value of the response (Overshoot) and settling time (t_s).
- A measure to gauge both the tracking error and control effort, defined by

$$Q = \sum_{i=1}^{N} \frac{1}{2}[(ref_i - y_i)^2 + \rho(u_i - u_{i-1})^2]$$

Two different control strategies are studied below. In section 5.1 we analyse the case of an off-line parameter estimation process for fuzzy PID controllers. In this case the parameters are optimised in advance and kept constant during the reported control experiments.

The second control strategy, reported in section 5.2, represents the class of adaptive control schemes. More specifically, it is an instance of a model–based predictive control scheme. In the presented results, the parameters of the controllers are optimised on–line.

5.1 Fuzzy PID controller

In this section, two Mamdani controllers configured as PIDs with equal parameterisation and initial conditions were optimised using different strategies. One of the controllers is tuned without constraints (the typical case found in the literature) while the other is tuned using the proposed constraints.

For each fuzzy controller, five triangular membership functions are assumed for each universe of discourse. The same initial conditions and parameters of the GA were employed: A population with 100 individuals. The maximum number of generations was 80. The two best individuals of a generation are kept in the next generation (elitism). Probability of mutation is 0,04 while the probability of crossover is 0,9.

The fitness function f, for the previous minimisation problem is (i) $f = 1/(1 + J_1)$ for the unconstrained case, and (ii) $f = 1/(1 + J)$ for the constrained case; J is given as follows accordingly to the penalty function method:

$$J = \sum_i J_i \tag{12}$$

where

$J_1 = K_1 \sum_{i=1}^{N} \frac{1}{2}[(ref_i - y_i)^2 + \rho(u_i - u_{i-1})^2]$, tracking error and control effort

$J_2 = K_2 \|c_{ZE}\|^2 step(\|c_{ZE}\|)$, natural zero

$J_3 = K_3 + \sum_x [(M_p(x) - 1)^2 step(M_p(x) - 1)]$, distinguishability

$J_4 = K_4 + \sum_x [(M_p(x) - 0.2)^2 step(0.2 - M_p(x))]$, coverage

with the positive penalty factors $K_1 = 1$, $K_2 = 10000$, and $K_3 = K_4 = 1000$; c_{ZE} stands for the center of the membership function of the linguistic term zero. The unit step function is defined as:

$$\text{step}(x) = \begin{cases} 1 \text{ if } x > 0 \\ 0 \text{ if } x \leq 0 \end{cases} \tag{13}$$

To stress that it is possible to find a set of membership function parameters providing good performance results and still be able to interpret them as linguistic terms, in the reported results, the fuzzy rules common to both controllers are not changed during the tuning process.

Instead of a regular fuzzy PID controller, a parallel combination of a fuzzy PI and a fuzzy PD was considered, see Fig.1 for an illustration.

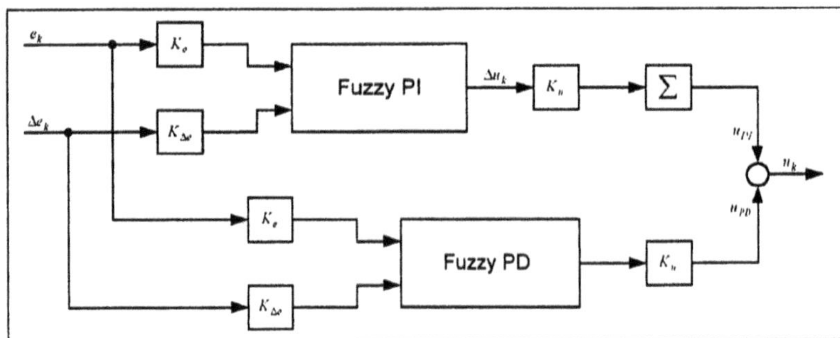

Fig. 1. Parallel fuzzy PID structure.

This has some advantages such as it is easier to define and understand the control rules, while for the same behaviour requires less number of rules. The output of such structure is given by $u_k^{PID} = u_k^{PD} + u_k^{PI}$.

The rule base for each component of the structure is given in Table 1. See also [3].

Table 1. RULE BASES FOR THE PI AND PD COMPONENTS OF THE FUZZY PID CONTROLLER.

Fuzzy PI		$\Delta e(k)$				
		MN	SN	Z	SP	MP
$e(k)$	MN	MP	MP	MP	SP	Z
	SN	MP	SP	SP	Z	SN
	Z	MP	SP	Z	SN	MN
	SP	SP	Z	SN	SN	MN
	MP	Z	SN	MN	MN	MN

Fuzzy PD		$\Delta e(k)$				
		MN	SN	ZE	SP	MP
$e(k)$	MN	VBP	BP	MP	MP	SP
	SN	BP	MP	MP	SP	ZE
	ZE	MP	MP	SP	ZE	ZE
	SP	MP	SP	ZE	ZE	ZE
	MP	ZE	ZE	ZE	ZE	ZE

Figures 2 and 3 show the control results for the same family of simulated patients and initial conditions. Figure 2 refers to a fuzzy PID tuned without constraints on the membership functions, while fig. 3 refers to the case where the proposed constraints were enforced.

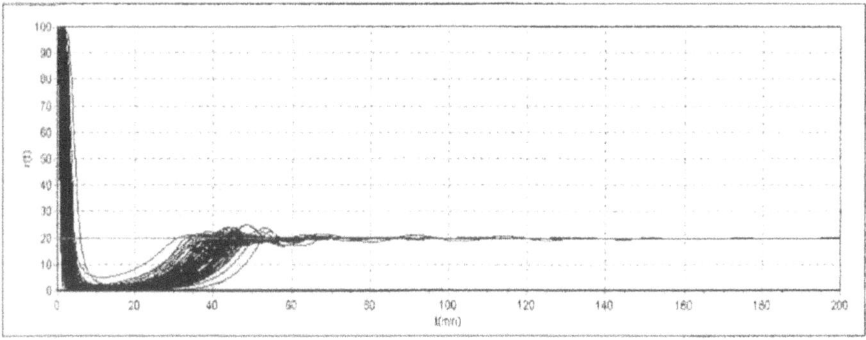

Fig. 2. Control results for a fuzzy PID tuned without constraints.

As show in table 2, no significant difference on the results are observed, not even in terms of quantitative performance measures such as Q.

Figure 4 shows an example of what can happen to the membership functions when tuned by an unconstrained optimisation process. The membership functions in the figure refers to the error's UoD. One can see an uncovered point in the UoD, and two identical membership functions at the right. Similar results are often found for the others UoDs. Without appropriated constraints, an usually large number of successive initialisations is required for the design process to find a satisfactory membership function configuration.

Figure 5 shows the set of membership functions found by the constrained optimisation process, for the same UoD, i.e., error. As expected the membership functions

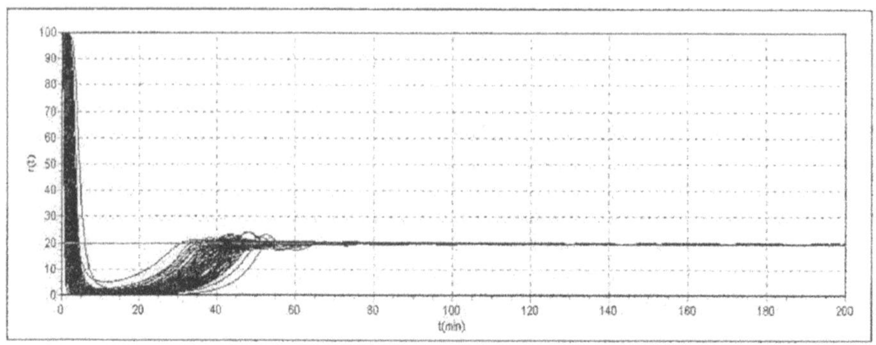

Fig. 3. Control results for a fuzzy PID tuned with the proposed constraints.

Table 2. SUMMARY OF PERFORMANCE RESULTS.

	With Constraints	Without Constraints
Q	2469,277	2469,466
My (%)	19,248	19,204
SDy (%)	2,806	2,804
RMSD	2,909	2,917
Mu (μg/kg/min)	4,706	4,710
SDu (μg/kg/min)	1,373	1,274
$t_s(min)$	128,482	129,368
Overshoot (%)	0,038	0,049

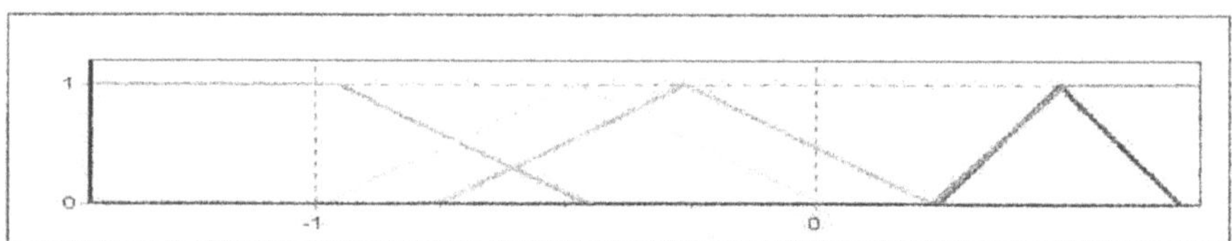

Fig. 4. Membership functions tuned by the unconstrained optimisation process.

found for the others UoDs could be also easily interpretable as medium negative (MN), small negative (SN), zero (ZE), small positive (SP) and medium positive (MP).

5.2 Fuzzy adaptive controller

In the following, two Mamdani controllers configured as PDs with equal parameterisation and initial conditions were submitted to an adaptive process of on-line tuning of the position and width of membership functions of the output linguistic variable.

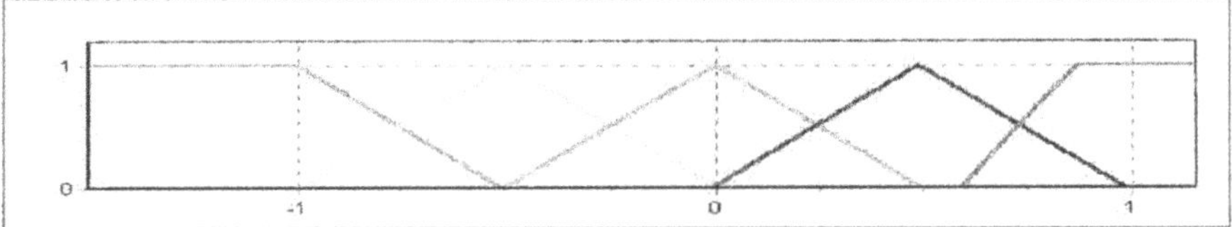

Fig. 5. Membership functions tuned by the constrained optimisation process.

This optimisation is guided by the principles of model-based predictive control cf. [11], i.e., the optimal sequence of position and width of consequent membership functions depends on the predicted level of muscular activity.

This predicted level is accomplished by a model exhibiting the following features: (i) the structure is equal to the model of the simulated patient (11); (ii) the parameters are given by the mean value of the parameters within the set of simulated patients and (iii) a incremental version is used, resulting from the linearisation around the current working point.

This incremental version is described in (14), where r_0 stands for the last observed sample of the system output. The resulting linearised model allows us to accommodate data from the patient under control into the model:

$$r(t) \approx r_0 - \frac{\gamma}{c_{p50}} \frac{r_0^2}{100} \left(\frac{100}{r_0} - 1 \right)^{\frac{\gamma-1}{\gamma}} (c_{\mathrm{E}} - c_{\mathrm{E}}^0). \tag{14}$$

No on–line identification is performed in the predictive model.

Two control experiments are reported. One of the controllers is tuned without constraints (only relies on the minimisation of $J1$ – see below) while the other is tuned using the proposed constraints.

For each fuzzy controller, five symmetrical triangular membership functions are assumed for each universe of discourse. The same initial conditions and parameters of the GA were employed: the center and width of each membership function were coded as a binary bit-string of length 16. A population with 300 individuals (each one formed by concatenation of the preceding five strings). The maximum number of generations was 50. The two best individuals of a generation are kept in the next generation (elitism). Probability of mutation is 0,04 while the probability of crossover is 0,9.

In this case, and as before, the fitness function f is (i) $f = 1/(1 + J_1)$ for the unconstrained case, and (ii) $f = 1/(1 + J)$ for the constrained case; the J components are:

$J_1 = K_1 \sum_{i=1}^{H_p} \frac{1}{2}[(ref_i - \hat{y}_i)^2 + \rho(u_i - u_{i-1})^2]$,
predicted tracking error and control effort

$J_2 = K_2 \sum_{i=1}^{4} step(c_i - c_{i+1})$, natural order of linguistic terms

$J_3 = K_3 + \sum_x [(M_1(x) - 1)^2 step(M_1(x) - 1)]$, distinguishability

$J_4 = K_4 + \sum_x [(M_1(x) - 0.4)^2 step(0.4 - M_1(x))]$, coverage

with c_i the center of the membership function representing the i-th member from the term set $\{ZE, SP, MP, BP, VBP\}$. Since we are employing a GA with fixed gene structure for each individual, $J2$ is used as a constraint of natural order between linguistic terms. In plain words, $J2$ penalise exchanges in the relative order of linguistic terms. The positive penalty factors are $K_1 = 1$, $K_2 = 10000$, and $K_3 = K_4 = 2000$.

Figures 6 and 7 show the control results for the two optimisation processes applied to the same patient randomly chosen.

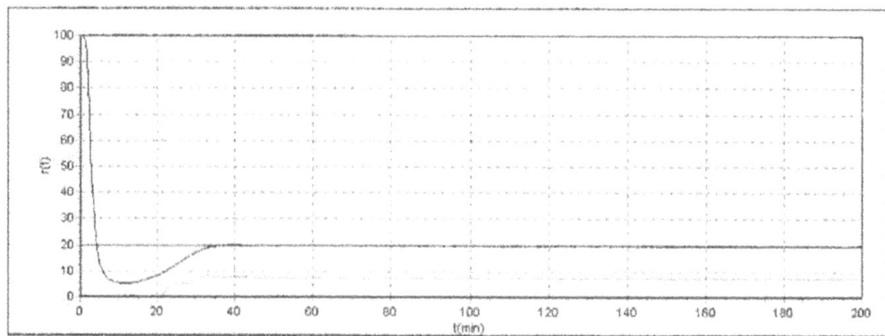

Fig. 6. Typical control results with the proposed constraints.

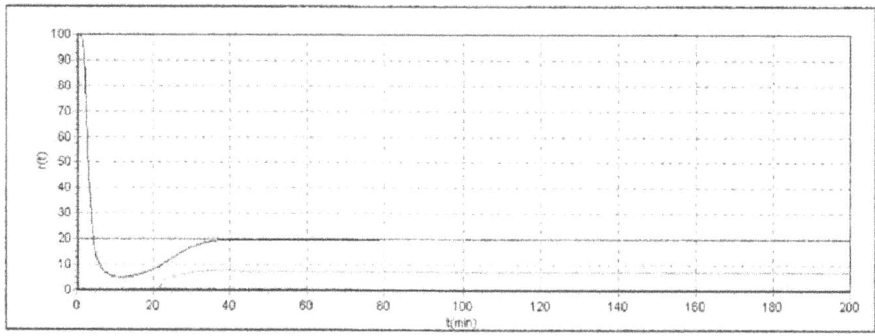

Fig. 7. Typical control results without constraints.

Although no significant difference on the results are observed, Table 3 shows that for some quantitative performance measures (such as Q, RMSD and t_s) the inclusion of the proposed constraints gives raise to better results.

We could say from all tests done that the two approaches are quite similar from the accuracy perspective.

In what concerns interpretability, Fig.8 shows the evolution over time of the position of membership functions when tuned by the unconstrained optimisation process. Figure 9 shows a top view of the linguistic consequent value of the predominant rule, i.e. in each time instant it is represented both the function center and its width.

Table 3. SUMMARY OF PERFORMANCE RESULTS.

	With Constraints	Without Constraints
Q	26,055	39,852
My (%)	19,890	19,853
SDy (%)	0,299	0,366
RMSD	0,318	0,395
Mu (μg/kg/min)	7,340	7,330
SDu (μg/kg/min)	0,128	0,069
$t_s(min)$	31,333	32,333
Overshoot (%)	0,004	0,001

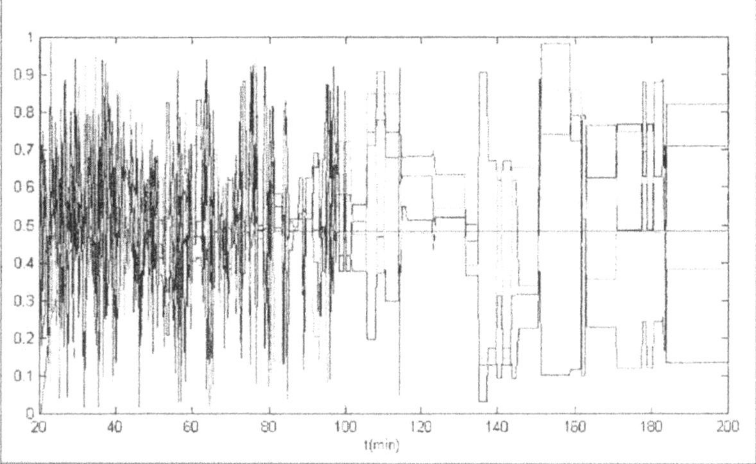

Fig. 8. Evolution of centres without constraints.

One can see that it's almost impossible to get some insight at the control strategy being followed. It's a very difficult task for the anaesthetist both to verify whether or not the controller is being consistent with his own medical experience and to introduce eventual modifications.

Comparing the former plots with the ones presented in Figs. 10 and 11 (referring to the constrained optimisation process) we see in Fig. 10 a smooth, coherent, and convergent evolution of membership function centres. The same pattern is found for the evolution of the predominant linguistic term in Fig. 11.

In our application, the constrained optimisation process has performed better, than the unconstrained one and has reduced the running time of the optimisation process.

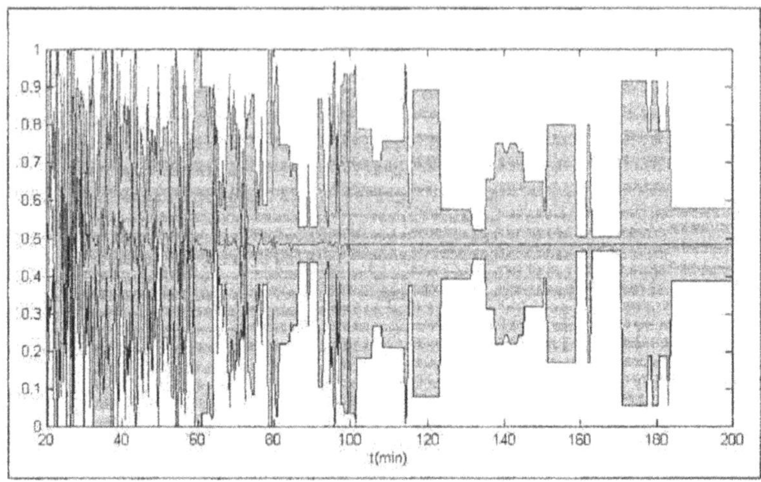

Fig. 9. Evolution of predominant linguistic term without constraints.

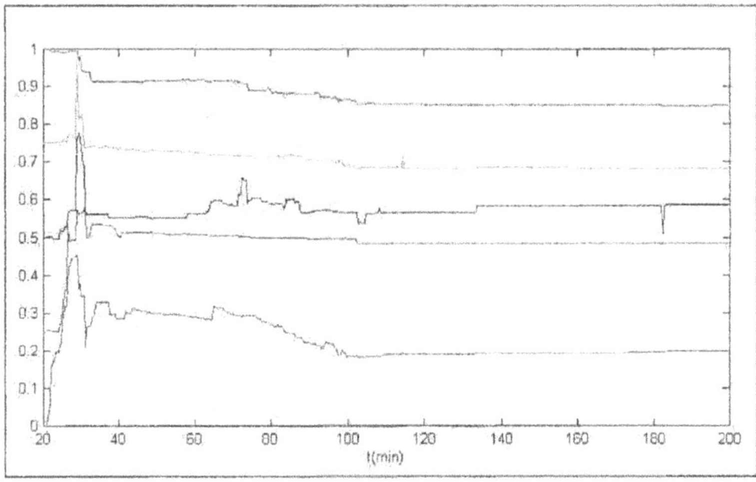

Fig. 10. Evolution of centres with the proposed constraints.

6 Conclusions

In this paper is argued that one of most interesting characteristics of fuzzy systems is the ability to read their internal parameter structure as a set of if-then (linguistic) rules.

It can be argued that the functionals relative to interpretation can be viewed as a form of including human knowledge into the design process. These are used to define the area in the parameter space where the solutions of the optimization problem can be found. It is argued also that taken interpretation into account in the design process does not neccessarily implies a degradation in accuracy.

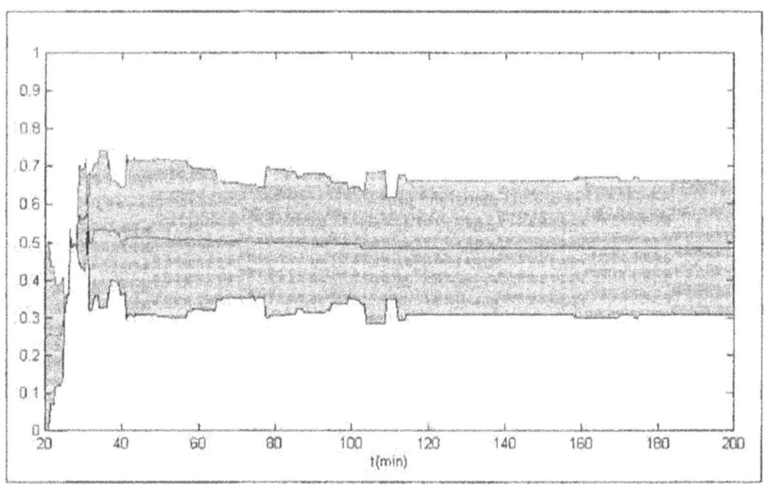

Fig. 11. Evolution of predominant linguistic term with the proposed constraints.

Simulation results showing the application of our design policy to the control problem of neuromuscular relaxation of patients under surgery using continuous infusion of atracurium are presented.

References

1. J. J. Buckley, "Universal fuzzy controller", *Automatica*, vol. 28, pp.1245-1248, 1992.
2. J. J. Buckley, "Sugeno type controllers are universal controllers", *Fuzzy Sets and Systems*, **53**, pp. 299-304, 1993.
3. P. Fazendeiro and J. Valente de Oliveira, "Semantic issues in the tuning of fuzzy systems using bio-inspired strategies: Application to the control of neuromuscular blockade", *Proc. of the Symposium on Intelligent Systems in Control and Measurement*, Veszprém, Hungary, September 9-14, 2000.
4. B. Kosko, "Fuzzy systems as universal approximators", *Proc. of IEEE Inter. Conf. on Fuzzy Systems.* San Diego, USA, 1153-1162, 1992.
5. B. Kosko, *Neural Networks and Fuzzy Systems* (Englewood Cliffs, NJ: Prentice Hall, 1992).
6. D.A. Linkens (ed.) *Intelligent Control in Biomedicine*, Taylor Francis, 1994.
7. T. Mendonça, *Methods and Algorithms for control of biological systems*, PhD thesis (in Portuguese), University of Oporto, 1992.
8. G. A. Miller, "The magic number seven, plus or minus two: Some limits on our capacity for processing information", *Psychological Review*, vol. 63 (1956), pp. 81-97.
9. B. Moser, "Sugeno controllers with a bounded number of rules are nowhere dense", *Fuzzy Sets and Systems* 104, pp. 269-277, 1999.
10. H. Nijmeijer and Van der Schaft, A.J., (1990) *Non-linear Dynamical Control of Systems*, New York: Springer-Verlag.

11. J. Valente de Oliveira and J. M. Lemos, "A simplified fuzzy framework for adaptive long-range predictive control", in *Advances in Fuzzy Control*, D. Driankov, R. Palm (eds), Physica-Verlag, Germany 1998, pp. 317-336.
12. J. Valente de Oliveira, "Towards Neuro-Linguistic Modelling: Constraints for optimisation of membership functions", *Fuzzy Sets and Systems*, **106**, pp. 357-380.
13. J. Valente de Oliveira, "Semantic constraints for membership function optimisation", *IEEE Trans. on Systems, Man, and Cybernetics, Part A: Systems and Man*, Vol. 29, No. 1 (1999), pp. 128-138.
14. L.X. Wang, "Fuzzy systems are universal approximators", *Proc. of IEEE Inter. Conf. on Fuzzy Systems.* San Diego, USA, pp. 1163-1170, 1992.
15. S. Ward, A. Neil, B. Weatherley, and M. Coral, "Pharmacokinetics, of atracurium besylate in healthy patients (after a single i.v. bolus dose)", *British Journal Anaesthesia* (1983), **55**, 113-116.
16. B. Weatherley, S. Williams and E. Neil, "Pharmacokinetics, Pharmcodynamics and Dose-Response Relationships of Atracurium Administration i.v.", *British Journal of Anaesthesia* (1983), **55**, 39-45.
17. L. A. Zadeh, "The concept of a linguistic variable and its applications to approximate reasoning I, II, and III", Information Science, **8**, 1975, pp. 199-249, 301-357, **9**, pp. 43-80.
18. X.-J. Zeng and M. Singh, "Approximation theory of fuzzy systems - SISO case", *IEEE Trans. on Fuzzy Systems*, vol. 2, no. 2, pp. 162-176.

SECTION 3

EXTENDING THE MODELING PROCESS TO IMPROVE THE ACCURACY

Linguistic Hedges and Fuzzy Rule Based Systems

Chuen-Yau Chen[1] and Bin-Da Liu[2]

[1] Department of Electronic Engineering, I-Shou University, Kaohsiung County, Taiwan 84008, R. O. C.
e-mail: cychen@ieee.org

[2] Department of Electrical Engineering, National Cheng Kung University, Tainan, Taiwan 70101, R. O. C.
e-mail: bdliu@cad.ee.ncku.edu.tw

Abstract. In a fuzzy rule based system, the finer the rule base is defined, the better the results are inferred. However, it induces a rule base of a large size that takes much time to establish. The linguistic hedges can adjust the shape of the system membership functions by either stressing or suppressing the membership degrees of the variables, which leads to either a coarser tuning manner or a finer tuning manner and speeds up the inference result to fit the system requirements. Combining the linguistic hedges with the fuzzy rule based systems emerges the following advantages: 1) it needs only the simple-shape membership functions rather than the carefully designed ones for characterizing the related variables; 2) it is sufficient to adopt less number of rules for inference; 3) the rules are developed intuitionally without heavily depending on the endeavor of experts.

1 Introduction

Fuzzy logic, proposed by Zadeh in 1965, is a logic with fuzzy truth, fuzzy connectives and fuzzy rules of inference rather than the conventional two-valued or even multi-valued logic [1]. In a fuzzy rule base system (FRBS), the inference engine plays the role of a kernel. It explores the fuzzy rules preconstructed by experts to accomplish inference. Since the rules specify the implication relationships between the input variables and output variables characterized by their corresponding membership functions, the choice of the rules along with the membership functions makes significant impacts on the final performance of the FRBS and therefore becomes the major strategy in FRBS design. The greater the number of membership functions, the greater the number of rules, and the finer the inference results.

With regard to the membership functions, instead of having the membership functions constructed manually by skilled operators or experts, several researchers have proposed methods for automatically selecting the high performance membership functions for FRBSs. Karr [2] properly specified the membership functions to ensure efficient FRBS performance by using the genetic algorithms (GA) [3], [4]. Krishnapuram [5] relied on the properties of possibilistic clustering to develop an approach for generating membership functions. Kim [6] applied the concept of inductive reasoning to generate the membership function. In his scheme, no extra information is needed except the experienced data describing the input and output relationships.

With regard to the rule construction in an FRBS design, there has been many studies recently. Wang [7] proposed general methods to generate the fuzzy rules automatically according to the input-output data pairs of the control system. With these methods, the rule extraction process can be done by one-pass to reduce the system construction complexity, and the experience of a human expert can be combined with the rules obtained from automatic learning. On the other hand, the tree structure is such a simple and easily understood method for modeling the problem that many applications are solved [8]–[10]. Turksen [11] proposed a two-level tree search method for a fuzzy expert system. With his method, taking advantage of the tree approach, the computational complexity and the search time of the fuzzy system can be reduced. In addition, Liu [12] proposed a tree-based fuzzy logic controller design methodology. By means of this method, not only can the rules be extracted automatically but the search time can also be much reduced.

On the other hand, Zadeh [13] proposed the fuzzy linguistic hedges such as *very, more or less, much, essentially,* and *slightly* to modify the membership functions of the fuzzy sets. Since the linguistic hedges proposed by Zadeh in 1973, only a small amount of literature dealing with these concepts has been published [14]–[20]. Banks [14] used hedge operations to better qualify and emphasis the crisp variables to mix crisp and fuzzy logic in applications. Bouchon-Meunier [15] investigated several interesting properties of linguistic hedge. Modifying the existing linguistic-hedge models, Novák [16] proposed a horizon shifting model of linguistic hedges, by which the membership function can be shifted as well as its steepness modified. To maintain the completeness of the set of the linguistic hedges, Liu [17]–[20] proposed several hedge operators.

Although the rule developments and the membership function constructions can be accomplished automatically by a variety of algorithms, this major strategy in an FRBS design still becomes a considerable challenge when the number of mentioned variables is increased, when the parameters of the controlled plants are varied, or when the conditions of the environment are changed. In this chapter, we introduce a novel FRBS called a *linguistic-hedge fuzzy logic controller* (LHFLC) [21], [22]. The LHFLC takes advantage of the superior characteristics inherent in the linguistic hedges and the search ability of GA. In this controller, each variable utilized is characterized by only three fuzzy sets with simple-shape membership functions; therefore, the maximum number of fuzzy rules required for a system with N input variables is 3^N. Moreover, a module called the *linguistic-hedge module* embedded in this controller plays the role of a linguistic modifier. It is used to dynamically modify the shape of simple-shape membership functions according to the feedback signal from the controlled plants. This modification action allows the LHFLC to utilize only fewer rules and simple-shape membership functions without the satisfactory performance degrading.

This chapter is composed of five sections, of which this introduction is the first. In section 2, we will review the fuzzy set theory and introduce the concept of fuzzy linguistic hedges. Besides, the fuzzy rule base control system will be mentioned. We will also give the architecture of the LHFLC. In section 3 we will introduce the modified simple-genetic-algorithm (MSGA) used for searching the optimal linguistic-hedge combination vectors. In section 4, we will show the evaluations of this LHFLC architecture. Section 5 will conclude this chapter.

2 Linguistic-Hedge Fuzzy Logic Controller

2.1 Fuzzy Sets and Fuzzy Linguistic Hedges

In contrast to the crisp set logic that uses a characteristic function with the values $\{0, 1\}$ to describe the membership between an element and a set, we interpret fuzzy sets as membership functions μ_F that associate with each element x of the universe of discourse X a number $\mu_F(x)$ in the interval $[0, 1]$. In the continuous case, a fuzzy set F may be represented as

$$F = \int_X \mu_F(x)/x. \tag{1}$$

Its discrete counterpart in which F has a finite support $x_1, x_2, \ldots, x_n$ is

$$F = \sum_{i=1}^{n} \frac{\mu_i}{x_i} = \frac{\mu_1}{x_1} + \cdots + \frac{\mu_n}{x_n} \tag{2}$$

where μ_i $(i = 1, \cdots, n)$ is the grade of membership of x_i in F.

In an FRBS, the information is described linguistically. The linguistic hedge is an operator used to modify the shape of membership functions. According to the statement in [13], linguistic-hedge operations can be classified into three categories: *concentration*, *dilation*, and *contrast intensification*. In this chapter, we only focus on the *concentration* type and the *dilation* type hedge operations.

Concentration Operating a concentration operator on a fuzzy set A results in the reduction in the magnitude of the grade of membership of x in A which is relatively small for those x with a high grade of membership in A and relatively large for those x with low membership. The hedge operation of "*concentration* $\mu_A(x)$" defined by Zadeh [13] is

$$\mathrm{CON}(\mu_A(x)) \triangleq \mu_A^{\alpha}(x); \qquad \alpha > 1. \tag{3}$$

Based on the above definition, a few related hedge operations such as *absolutely*, *very*, *much more*, *more*, and *plus* can be defined by specifying the values of α in (3) as 4, 2, 1.75, 1.5, and 1.25, respectively as [17], [20].

Dilation In contrast, the effect of dilation is opposite to that of concentration. The hedge operation of "*dilation* $\mu_A(x)$" defined by Zadeh [13] is

$$\mathrm{DIL}(\mu_A(x)) \triangleq \mu_A^{\alpha}(x); \qquad 0 < \alpha < 1. \tag{4}$$

Similarly, some related hedge operations such as *minus*, *more or less*, and *slightly* can be defined by specifying the values of α in (4) as 0.75, 0.5, and 0.25, respectively [19], [20].

In order to consider how a hedge operator affects the fuzzy set, we take two typical hedge operators to demonstrate it. The hedge operator *very* is used to stand for the *concentration* type operation; the hedge operator *more or less* is used to stand for the *dilation* type operation. Figure 1 shows the fuzzy sets *cold*, *very cold*, and *more or less cold* characterized by their membership functions $\mu_{cold}(t)$, $\mu_{very\ cold}(t)$, and $\mu_{more\ or\ less\ cold}(t)$. The membership function

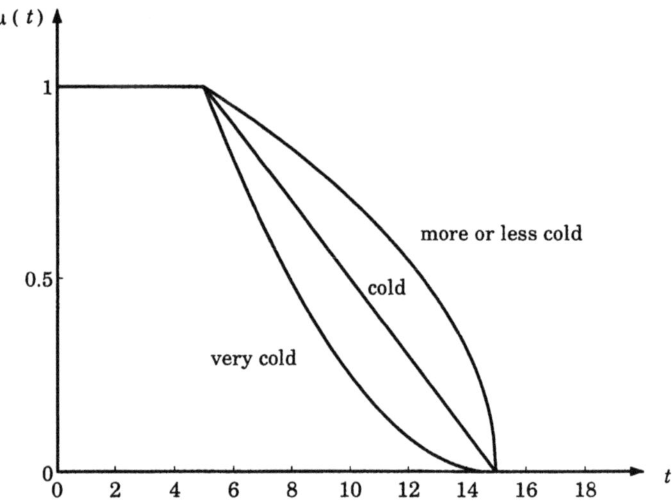

Fig. 1. Effects of the fuzzy linguistic hedge "*very*" and "*more or less*" on the fuzzy set "*cold*".

of the fuzzy set *very cold* is generated by applying the hedge operator *very* to that of the fuzzy set *cold* while the membership function of the fuzzy set *more or less cold* is generated by applying the hedge operator *more or less* to that of the fuzzy set *cold*. Obviously, the linguistic hedge *very* tends to narrow the shape of the membership function and decrease the membership degree; the linguistic hedge *more or less* tends to widen the shape of the membership function and increase the membership degree. That is, the members in the fuzzy set *very cold* are closer to the temperature of *cold* while the members in the fuzzy set *more or less cold* are farther far away from the temperature of *cold*.

2.2 Fuzzy Rule Based Control System

A fuzzy rule based system designed on the basis of the fuzzy logic is an approximate reasoning-based system, which does not require exactly analytical models and is much closer in spirit to human thinking and natural language than the traditional logic system. Figure 2 shows the block diagram of an FRBS in control applications, usually called fuzzy logic controller (FLC), consisting of four principal units: the *fuzzifier module*, the *fuzzy inference engine*, the *knowledge base*, and the *defuzzifier module*. Fuzzification related to

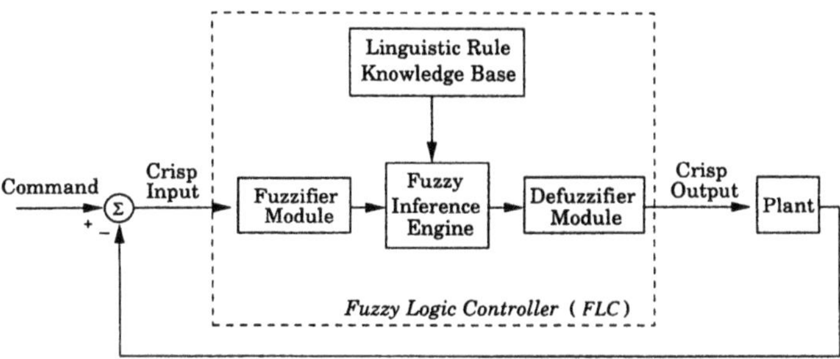

Fig. 2. Basic configuration of a fuzzy rule based system in control applications.

the vagueness and imprecision in a natural language is necessary during an earlier stage to translate the input crisp data into the fuzzy representation for further processing. A fuzzy system is characterized by a set of linguistic statements according to expert knowledge which is usually represented in the form of "IF-THEN" rules expressed as

$$\begin{aligned}&\text{IF (a set of conditions are satisfied) THEN}\\&\qquad\text{(a set of consequences can be inferred).}\end{aligned} \tag{5}$$

The antecedent is a condition in its application domain and the consequence is an action for the system under control. Above all, the rule base provides a convenient way for expressing control policy and domain knowledge. The knowledge base module is used to specify the rule base, which comprises a knowledge of the application domain and the attendant goals. Moreover, to deal with the fuzzy information described above, the fuzzy inference engine employs the fuzzy knowledge base to simulate human decision making and infer actions. Finally, the defuzzifier module is used to translate the processed fuzzy data into the crisp data suited to real world applications.

When designing an FRBS, the control strategies have to be based on the determination of the fuzzy membership function of control variables and the linguistic control rules. Therefore, after finishing the design of an FRBS, if the control result fails to meet the system requirements due to a change in the outside environment of the control system, the system control strategies have to be modified to fit the control objective. The possible solution to this problem is that we can adjust either the membership function of the fuzzy sets or the control rules to achieve the control objective. To explain the linguistic hedges from a more physical perspective, we consider their effects on the inference performance of an FRBS whose goal is to produce the suitable control actions to control the plants reaching the desired situations. In general, the schedule of the membership functions of the fuzzy sets in this case are in the form shown in Fig. 3, in which the fuzzy sets are labeled with linguistic variables Negative-Big (NB), Negative-Medium (NM), Negative-Small (NS), Zero (ZE), Positive-Small (PS), Positive-Medium (PM), and Positive-Big (PB). What an FRBS should do is lead the plant to the state such

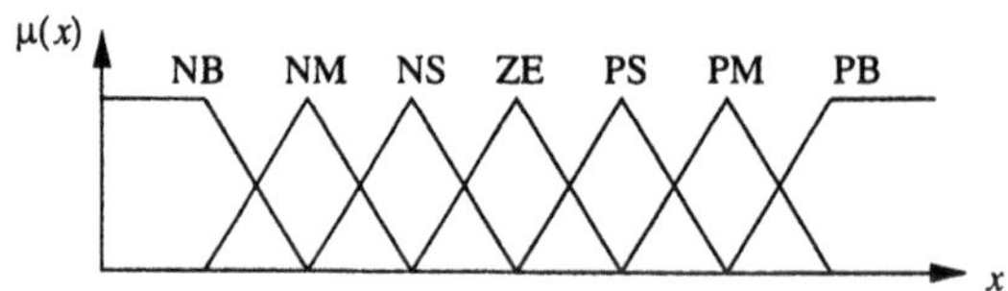

Fig. 3. Membership functions of the fuzzy sets NB, NM, NS, ZE, PS, PM, and PB.

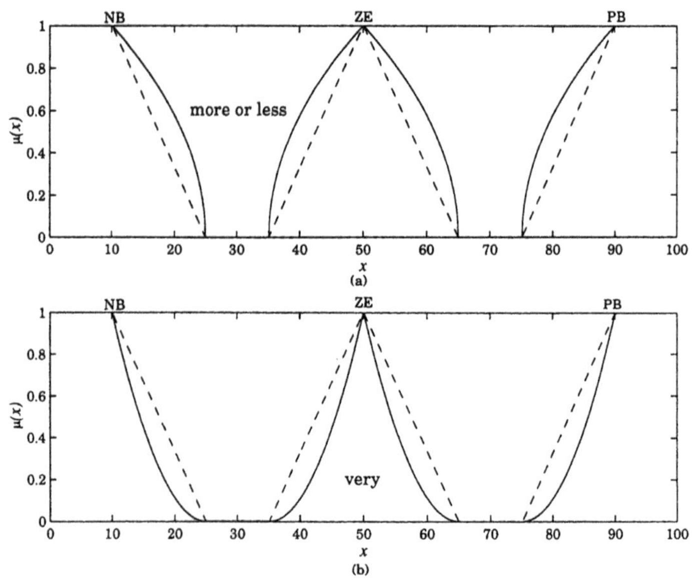

Fig. 4. Diagrammatic sketch of effects with *very* and *more or less*. (a) Effect of *more or less*. (b) Effect of *very*.

that the input variables and the output variables of this FRBS enter the range around the fuzzy set ZE or reach ZE, which indicates that a balanced condition is met. For simply explaining the effect of linguistic hedges on the membership functions, we concentrate only on the fuzzy sets NB, ZE, and PB. The dashed lines in Fig. 4 are the membership functions of the fuzzy sets NB, ZE, and PB. The solid lines in Fig. 4(a) represent the effect of the hedge operator *more or less* on the fuzzy sets NB, ZE, and PB, while those in Fig. 4(b) reveal the effect owning to the hedge operator *very*. Clearly, the effect of *more or less* can be viewed as the stress of fuzzy sets in physical meaning. Alternatively, if x is located in NB or PB, the *more or less* effect increases the membership degrees and stresses the meaning of NB or PB. This action forces FRBS to consider that the input state is still far from the target ZE. Furthermore, the FRBS stresses the output control action to reach the target earlier. In other words, this coarser tune on control action leads to a shorter transient time. The situation of x located in ZE is the same as stated above. The stress of ZE forces FRBS to consider that the input state is very close to the target and to tune the output control action in a finer manner to fit system demand that leads to a smaller overshoot. On the other hand, opposite to that of *more or less*, the effect of *very* suppresses the input state. This means that if x is located in NB or PB, the suppression forces FRBS to consider that the input state is not far from the target ZE and to control the output action to approach the target in a finer manner that prevent from occurring a large overshoot. As usual, if x is located in ZE, suppression forces FRBS to consider that the input state is not close to the target and to tune the output action in a coarser manner to fit the system demand that shortens the transient time. This phenomenon implies that we can use less number of rules accompanying with the linguistic hedges in an FLC to achieve a performance as good as that of an FLC with more rules. In a quantization system, the finer the step size, the lower the quantization error; on the other hand, the coarser the step size, the higher the quantization error. An FLC with more rules is like an quantization system with a finer step size, which leads the plant to the goal in a finer and more stable manner. On the other hand, an FLC with less number of rules is like an quantization system with a coarser step size, which leads the plant to the goal in a nearly abrupt manner. However, this unpleasant behavior can be kept away by the linguistic hedges.

2.3 Linguistic-Hedge Fuzzy Logic Controller Architecture

The major difference between LHFLC and the conventional FLC is that a module called *linguistic-hedge module* is inserted into the conventional one to adjust the shape of fuzzy membership functions dynamically according to the feedback signal from the plant. The emerged interesting result is that this LHFLC maintains better performance even though the number of the inference rules is reduced to a number as small as possible such as that only nine rules are used. Figure 5 is the block diagram of this LHFLC, which consists of several modules similar to those in a conventional FLC except for the linguistic-hedge module attached to the fuzzifier module. Relying on

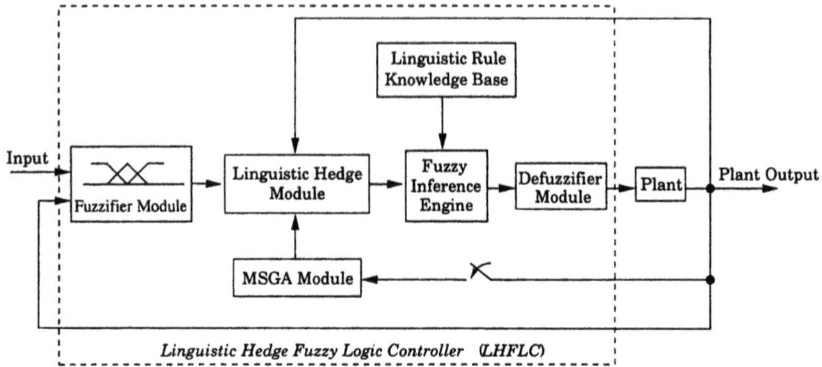

Fig. 5. LHFLC architecture.

the benefits described, the number of inference rules used in this LHFLC is nine. These rules are usually scheduled in a 3 × 3 rule table. As shown in Fig. 6, three fuzzy sets labeled NB, ZE, and PB are used in this architecture, which are the most general and universal representations of membership functions used in FLCs. Because the domain of each input variable and the co-domain of each output variable of any system can be scaled and shifted to a range about the origin. Therefore, the quantity in the left side of the origin can be categorized in NB, and the quantity in the right side of the origin can be categorized in PB. The mathematical expressions of the Z-shape

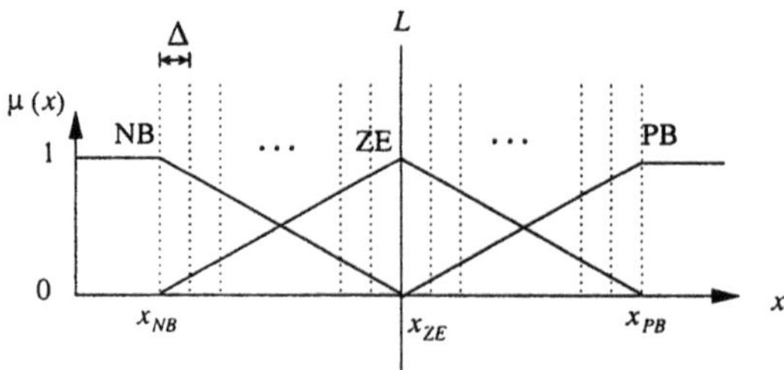

Fig. 6. Membership functions in LHFLC.

membership function $\mu_{NB}(x)$ of fuzzy subset NB, the $\wedge$-shape membership function $\mu_{ZE}(x)$ of fuzzy subset ZE, and the S-shape membership function $\mu_{PB}(x)$ of fuzzy set PB are listed in Table 1. In order to apply the linguistic-hedge operations to the proposed FLC, the domains of the input variables are

Table 1. Mathematical expressions of the membership functions $\mu_{NB}(x)$, $\mu_{ZE}(x)$, and $\mu_{PB}(x)$.

	$\mu_{NB}(x)$	$\mu_{ZE}(x)$	$\mu_{PB}(x)$
$-\infty < x \leq x_{NB}$	1	0	0
$x_{NB} \leq x \leq x_{ZE}$	$-\frac{x - x_{NB}}{x_{ZE} - x_{NB}} + 1$	$-\frac{\lvert x - x_{ZE}\rvert}{x_{PB} - x_{ZE}} + 1$	0
$x_{ZE} \leq x \leq x_{PB}$	0	$-\frac{\lvert x - x_{ZE}\rvert}{x_{PB} - x_{ZE}} + 1$	$\frac{x - x_{PB}}{x_{PB} - x_{ZE}} + 1$
$x_{PB} \leq x < +\infty$	0	0	1

partitioned into n intervals. From the mathematical point of view, the membership functions $\mu_{NB}(x)$, $\mu_{ZE}(x)$, and $\mu_{PB}(x)$ seem to be assembled by n piecewise linear functions. These partitioned membership functions denoted as $\mu_{NB'}(x)$, $\mu_{ZE'}(x)$, and $\mu_{PB'}(x)$ can be expressed as

$$[\mu_{NB'}(x)\ \mu_{ZE'}(x)\ \mu_{PB'}(x)] = tr(\mathbf{P}(x))[\mu_{NB}(x)\ \mu_{ZE}(x)\ \mu_{PB}(x)] \tag{6}$$

where $tr(\cdot)$ denotes the trace of a matrix, and $\mathbf{P}(x)$ is the partition matrix defined as

$$\begin{aligned}\mathbf{P}(x) = [&(u(x - x_{NB}) - u(x - x_{NB} - \triangle))\mathbf{v}_1 \cdots \\ &(u(x - x_{NB} - (n-1)\triangle) - u(x - x_{NB} - n\triangle))\mathbf{v}_n]_{n\times n}\end{aligned} \tag{7}$$

in which $\triangle$ denotes the step size of the input domain partition, and $u(x)$ is the unit step function of x defined as

$$u(x) = \begin{cases} 0 & -\infty < x < 0 \\ 1 & 0 \leq x < +\infty \end{cases} \tag{8}$$

and $\mathbf{v}_i$ is the basis of n dimensional vector space, which is defined as

$$\mathbf{v}_i = [v_1\ \cdots\ v_n]^T \tag{9}$$

where

$$v_j = \begin{cases} 1 & j = i \\ 0 & \text{otherwise} \end{cases} . \tag{10}$$

The membership function in each interval i ($i = 1, \cdots, n$) is now modified by its corresponding linguistic-hedge operator h_i which is the i-th element of the linguistic-hedge combination vector $\mathbf{h} = [h_1 \cdots h_n]$ defining the proper linguistic-hedge operators of the n intervals of the whole input domain. For the sake of the convenience of mathematical expression, we define the linguistic-hedge combination matrix $\mathbf{H}$ as

$$\mathbf{H} = [h_1\mathbf{v}_1 \cdots h_n\mathbf{v}_n]_{n\times n}. \tag{11}$$

That is, every entry of $\mathbf{H}$ is 0 except the diagonal entries h_is which give the linguistic-hedge operators of the corresponding interval of membership function. Since the matrix $\mathbf{H}$ is diagonal, the membership functions $\mu_{H \circ NB}(x)$, $\mu_{H \circ ZE}(x)$, and $\mu_{H \circ PB}(x)$ resulting from modification by corresponding linguistic-hedge operators can be expressed as

$$\begin{bmatrix} \mu_{H \circ NB}(x) \\ \mu_{H \circ ZE}(x) \\ \mu_{H \circ PB}(x) \end{bmatrix} = \begin{bmatrix} tr(\mathbf{P}(x)((\mu_{NB}(x))^{\mathbf{H}}) \\ tr(\mathbf{P}(x)((\mu_{ZE}(x))^{\mathbf{H}}) \\ tr(\mathbf{P}(x)((\mu_{PB}(x))^{\mathbf{H}}) \end{bmatrix} . \tag{12}$$

After processing in the fuzzifier module and the linguistic-hedge module, we send the resulting signals to the succeeding stage referred to as the inference engine. This stage infers the fuzzy control actions employing fuzzy implication and rules constructed by the expert experience. The fuzzy reasoning method adopted in the LHFLC is Mamdani's Minimum Operation Rule [25]. The final stage is the defuzzifier module whose function is to transfer the signal from the fuzzy set into the real world for obtaining the actual control actions. The widely used method, central of gravity (COG), is adopted in the proposed LHFLC. In this case, the crisp output z_0 can be derived by [23]

$$z_0 = (\sum_{i=1}^{n} \mu_z(w_i) \cdot w_i) / \sum_{i=1}^{n} \mu_z(w_i) \tag{13}$$

where w_i is the abscissa at which the membership function reaches the maximum value $\mu_z(w_i)$ and n is the number of quantization levels of the output variable z.

According to the above descriptions, we can find that the characteristic of this architecture simplifies the complexity of the LHFLC design. From the viewpoint of the LHFLC architecture, inserting a linguistic-hedge module allows us to use the simple triangle-like membership functions and a fewer number of rules instead of the more carefully designed membership functions and the large number of rules to reach the control goals. As a result, the membership function constructions and the rule developments become simpler works.

3 Modified Simple-Genetic-Algorithm in Search of the Linguistic-Hedge Combination Vectors

In an LHFLC, we must tune the linguistic-hedge combinations which are difficult to be contributed according to human experience and knowledge. There are nine possible linguistic-hedge operations including the *empty-hedge* operation defined as

$$empty\text{-}hedge\ \mu(x) \triangleq \mu^1(x), \tag{14}$$

to be adopted in the linguistic-hedge module for modifying the corresponding membership functions. Consider the input domain partitioned into n intervals with their own linguistic-hedge operations. In the case of a system with three fuzzy sets, the overall number of linguistic-hedge combinations is as many as 3×9^n. To simplify this search problem, we investigate these three mentioned membership functions as shown in Fig. 6 again. The demonstrated significant result is their symmetrical property. That is, the membership function $\mu_{ZE}(x)$ is self-symmetrical about the line $L : x = x_{ZE}$; the membership functions $\mu_{NB}(x)$ and $\mu_{PB}(x)$ are symmetrical to each other about the line L. Therefore, the linguistic-hedge combination of the fuzzy set ZE ranging from the first interval to the $\frac{n}{2}$-th interval and those ranging from the $(\frac{n}{2}+1)$-st interval to the n-th interval must be symmetrical about the line L. Accordingly, the linguistic-hedge combination vector $\mathbf{h}$ used to specify the membership function of fuzzy set ZE must be in the form of

$$\mathbf{h} = [\mathbf{A} | \mathbf{A}^r] \tag{15}$$

where $\mathbf{A} = [a_1 \cdots a_{\frac{n}{2}}]$ specifies the linguistic-hedge operators corresponding to the $\frac{n}{2}$ intervals to the left of the line L, and $\mathbf{A}^r$ is the vector whose elements are in the reverse order with respect to those of $\mathbf{A}$, i.e.,

$$\mathbf{A}^r = \mathbf{A}\mathbf{R} = [a_{\frac{n}{2}} \cdots a_1] \tag{16}$$

where the transfer matrix $\mathbf{R}$ is defined as

$$\mathbf{R} = [\mathbf{v}_{\frac{n}{2}} \cdots \mathbf{v}_1]_{\frac{n}{2} \times \frac{n}{2}}. \tag{17}$$

Similarly, consider the fuzzy sets NB and PB. The linguistic-hedge combination vector $\mathbf{B}$ specifying the linguistic-hedge operators ranging from the first interval to the $\frac{n}{2}$-th interval of the fuzzy set NB and the linguistic-hedge combination vector $\mathbf{C}$ specifying the linguistic-hedge operators ranging from the $(\frac{n}{2}+1)$-st interval to the n-th interval of the fuzzy set PB must be also symmetrical about the line L. That is,

$$\mathbf{C} = \mathbf{B}^r = \mathbf{B}\mathbf{R} = [b_{\frac{n}{2}} \cdots b_1]. \tag{18}$$

Obviously, once the vector $\mathbf{A}$ and the vector $\mathbf{B}$ are determined, the vectors $\mathbf{A}^r$ and $\mathbf{B}^r (= \mathbf{C})$ can be also obtained in turn. The linguistic-hedge combination vector $\mathbf{h}$ used to specify the fuzzy sets NB and PB becomes

$$\mathbf{h} = [\mathbf{B} | \mathbf{C}] = [\mathbf{B} | \mathbf{B}^r] \tag{19}$$

where $\mathbf{B}$ specifies the linguistic-hedge operators ranging from the first interval to the $\frac{n}{2}$-th interval of the fuzzy set NB while $\mathbf{B}^r$ specifies the linguistic-hedge operators ranging from the $(\frac{n}{2}+1)$-st interval to the n-th interval of the fuzzy set PB. Accordingly, the linguistic-hedge combination vectors we have to determine are $\mathbf{A}$ and $\mathbf{B}$ with the dimension of $\frac{n}{2}$.

Among all the various GAs, the simple GA (SGA) is the simplest one without loss of efficiency [4]. We proposed a modified version of the SGA, called modified simple-genetic-algorithm (MSGA), for increasing the linguistic-hedge combination variety while searching the optimal solution. Figure 7 shows the flow chart of the MSGA. The major difference between the MSGA and the SGA is that we remove ten individuals with lower fitness among the whole population and generate ten new individuals to fill the resulting vacancies in the population in the MSGA. This operation increases the variety of the combination of linguistic hedges and enhances the search ability. Before proceeding with this MSGA approach, there are two preliminaries to be finished.

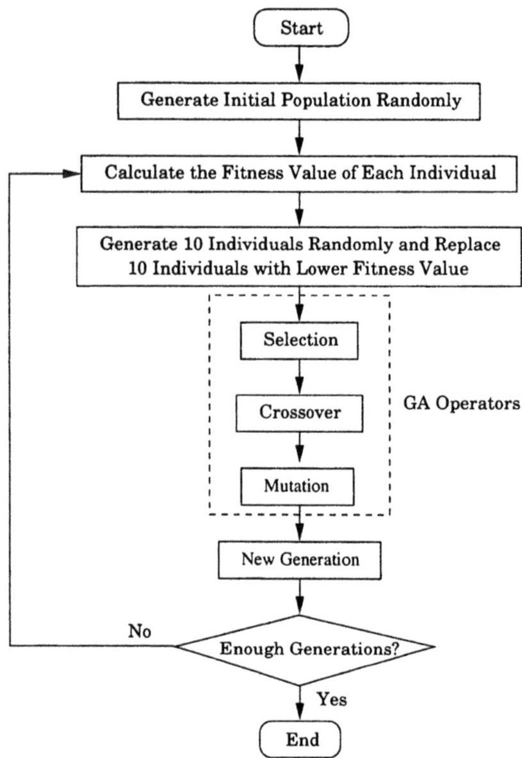

Fig. 7. Flow chart of modified simple-genetic-algorithm.

Two Preliminaries in MSGA

Definition of Suitable Coding One of the most attractive problems in MSGA is coding the solution space. For MSGA processing, more natural representations are more efficient and produce better solutions. Hence, the real-coded representation is used to manipulate the floating point linguistic-hedge operators. Each individual to be considered is encoded as a vector of floating point numbers, i.e.,

$$\mathbf{h} = [\mathbf{A}|\mathbf{A}^r] \quad \text{or} \quad \mathbf{h} = [\mathbf{B}|\mathbf{B}^r] \tag{20}$$

where

$$\mathbf{A} = [a_1 \cdots a_{\frac{n}{2}}], \tag{21}$$

$$\mathbf{B} = [b_1 \cdots b_{\frac{n}{2}}], \tag{22}$$

$$a_i, b_i \in S,\ i = 1, \cdots, \frac{n}{2}, \tag{23}$$

and

$$S = \{0.25, 0.5, 0.75, 1, 1.25, 1.5, 1.75, 2, 4\}. \tag{24}$$

This kind of encoding representation reduces the number of the possible individuals in the search space dramatically from 3×9^n to $2 \times 9^{\frac{n}{2}}$.

Choice of Fitness Function The second preliminary to be finished is choosing the problem-dependent fitness function. Different fitness functions promote different GA behaviors, which generate fitness values providing a performance measure of the problem considered. The general form of a fitness function consists of two functions and can be expressed as the composition of a scaling function $g(\mathbf{h})$ and an objective function $m(\mathbf{h})$, i.e.,

$$f(\mathbf{h}) = g \circ m(\mathbf{h}) \tag{25}$$

where $m(\mathbf{h})$ is the objective function returning a cost value, and $g(\mathbf{h})$ is the scaling function transferring the cost value to the fitness. We choose the fitness function using the power scaling function, which can be expressed as

$$f(\mathbf{h}) = \exp(-\sigma \cdot c(\mathbf{h})) \tag{26}$$

where $c(\mathbf{h})$ stands for the cost function which varies from problem to problem, and σ can be viewed as a discernment measure.

After deciding these two preliminaries, we should choose the genetic operators. This MSGA consists of three kinds of genetic operations which are *selection*, *crossover*, and *mutation*.

MSGA Operators

Selection Selection chooses the individuals in the population as parent individuals to create offspring for the next generation, whose purpose is to emphasize the fitter individuals in the population in hopes that their offspring will in turn have even higher fitness. In this work, the implementation method of fitness-proportionate selection is adopted. The selective probability $p_S(\mathbf{h}_i)$ of the i-th individual $\mathbf{h}_i$ is

$$p_S(\mathbf{h}_i) = f(\mathbf{h}_i)/\sum_{j=1}^{n} f(\mathbf{h}_j). \tag{27}$$

Crossover Instead of the single-point crossover, we adopt the two-point crossover to increase the candidate population variety of the linguistic-hedge combinations. For example, the parent individuals $\mathbf{h}_1$ and $\mathbf{h}_2$ given to be crossovered at the points k^+ and l^+ with the crossover probability p_c results in the new offspring $\mathbf{h}_1'$ and $\mathbf{h}_2'$ expressed as

$$\mathbf{h}_1' : h_{1i}' = \begin{cases} h_{2i} & k^+ < i < l^+ \\ h_{1i} & \text{otherwise} \end{cases} \tag{28}$$

and

$$\mathbf{h}_2' : h_{2i}' = \begin{cases} h_{1i} & k^+ < i < l^+ \\ h_{2i} & \text{otherwise} \end{cases} . \tag{29}$$

Mutation Each element in a linguistic-hedge combination string is a possible candidate for the mutated element that may be randomly replaced by all the linguistic-hedge operators according to the mutation probability p_m. As an illustration, the individual $\mathbf{h}_1'$ mutated in the k-th element and the l-th element results in the new offspring $\mathbf{h}_1''$ expressed as

$$\mathbf{h}_1'' : h_{1i}'' = \begin{cases} h_{1i}' & i \neq k, l \\ \text{elements selected from } S \text{ according to } p_m & \text{otherwise} \end{cases} . \tag{30}$$

In order to acquire a better performance, several parameters for MSGA should be set appropriately. We adopted the parameters suggested by De Jong [26] where the number of generations is 60, the population size is 100, the crossover rate is 0.6, and the mutation rate is 0.01.

4 Evaluations of the Linguistic-Hedge Fuzzy Logic Controller

We emphasize that the LHFLC with fewer rules can work better than a conventional FLC with more rules. To do this, two well-known nonlinear systems including the nonlinear plant model control system and the cart-pole balance system are used to verify the performance of the LHFLC. The number of rules chosen in this LHFLC is nine; the domain of each input variable is divided into 16 equal intervals. All simulations are performed with MATLAB [27].

4.1 Nonlinear Plant Model Control System

Problem Description The plant model of a non-BIBO nonlinear plant [24] can be expressed as

$$y(k+1) = 0.2y^2(k) + 0.2y(k-1) + 0.4\sin[0.5(y(k) + y(k-1))] \cdot \cos[0.5(y(k) + y(k-1))] + 1.2x(k) \quad (31)$$

where x is the input signal of the plant and y is its output signal. The reference model that the plant output will track is chosen as

$$r(k) = 1. \quad (32)$$

Figure 8 shows the plant control loop. We targets to determine the plant

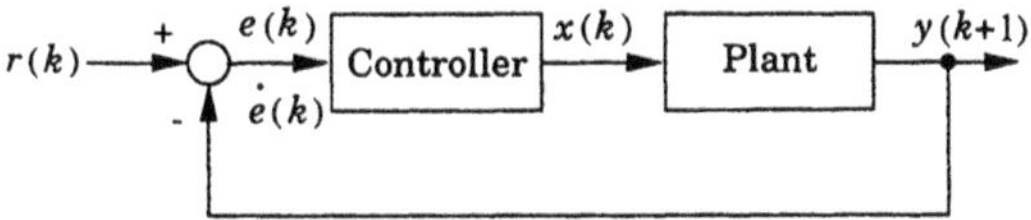

Fig. 8. Block diagram of nonlinear plant model control system.

input $x(k)$ such that

$$\lim_{k\to\infty} \mid r(k) - y(k) \mid < \epsilon \quad (33)$$

where ϵ is a suitably chosen constant. That is, the faster $y(k)$ tracks the reference signal $r(k)$, the better the controller will perform. In Fig. 8, the variables $e(k)$ and $\dot{e}(k)$ represent the error input and the change rate of error input of the controller, respectively, which are expressed as

$$e(k) = r(k) - y(k) \quad (34)$$

and

$$\dot{e}(k) = \frac{e(k) - e(k-1)}{\Delta t} \quad (35)$$

where the time step Δt is chosen as 0.1 seconds. Furthermore, $x(k)$ in (31) can be expressed as

$$x(k) = x(k-1) + \Delta x, \quad (36)$$

where Δx represents the increment of the plant input x at each iteration. Since the chosen plant model is nonlinear, it is difficult to handle it when the controller is determined by the classical control theory. The FLC is therefore a suitable choice to replace the role played by this controller. The input variables of the FLC are the error e and its change rate $\dot{e}$; the output variable is Δx. In order to stress the power of the proposed LHFLC, the performance of the system controlled by the conventional FLC with 7×7 rules and 3×3 rules is also concerned.

Simulation Results

- **Conventional FLC with 7×7 Rules**
  Figure 9 shows the rule table specifying the implication relationships between the input variables (e and $\dot{e}$) and the increment of the output variable Δx in the FLC. As shown in Fig. 10, the input variables e and $\dot{e}$ are characterized by seven fuzzy sets with the Gauss-like membership functions distributed in the interval $e \in [-0.6, 0.6]$ and $\dot{e} \in [-6, 6]$, respectively; the increment of the output variable Δx is characterized by

Δx ($\dot{e}$ \ e)	NB	NM	NS	ZE	PS	PM	PB
NB	NB	NB	NM	NM	NS	NS	ZE
NM	NB	NM	NM	NS	NS	ZE	PS
NS	NM	NM	NS	NS	ZE	PS	PS
ZE	NM	NS	NS	ZE	PS	PS	PM
PS	NS	NS	ZE	PS	PS	PM	PM
PM	NS	ZE	PS	PS	PM	PM	PB
PB	ZE	PS	PS	PM	PM	PB	PB

Fig. 9. Rule table of a size 7×7 for nonlinear plant model control system.

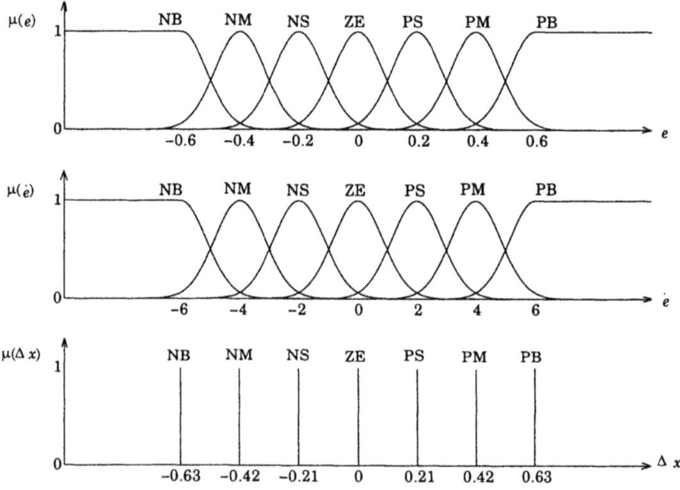

Fig. 10. Fuzzy membership functions for $e(k)$, $\dot{e}(k)$, and Δx of conventional FLC with 7×7 rules.

seven fuzzy singletons over the interval $\Delta x \in [-0.63, 0.63]$ with the support values (with membership equal to 1) located at -0.63, -0.42, -0.21, 0, 0.21, 0.42, and 0.63, respectively.

- **Conventional FLC with 3×3 Rules**
  In this case, the number of rules in the conventional FLC is reduced from

Δx	e: NB	e: ZE	e: PB
$\dot{e}$: NB	NB	NB	ZE
$\dot{e}$: ZE	NB	ZE	PB
$\dot{e}$: PB	ZE	PB	PB

Fig. 11. Rule table of a size 3×3 for nonlinear plant model control system.

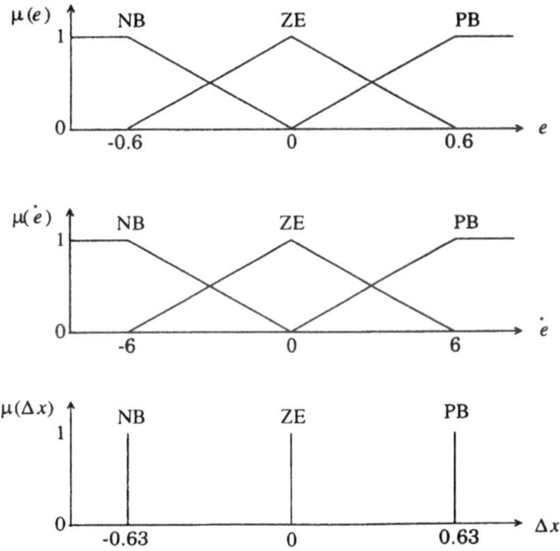

Fig. 12. Fuzzy membership functions for $e(k)$, $\dot{e}(k)$, and Δx of conventional FLC with 3×3 rules.

7×7 to 3×3 to run the same simulation. The rule table specifying the input/output relationships is shown in Fig. 11. Figure 12 shows the membership functions for the input variables (e and $\dot{e}$) scheduled by only three fuzzy sets with the simple shape membership functions linguistically labeled as NB, ZE, and PB distributed over the intervals $e \in [-0.6, 0.6]$ and $\dot{e} \in [-6, 6]$, respectively; the output variable Δx is characterized by three fuzzy singletons NB, ZE, and PB over the interval $\Delta x \in [-0.63, 0.63]$.

- **LHFLC with 3×3 Rules**
  The major difference between the LHFLC and the FLC with 3×3 rules is the inserted linguistic-hedge module. The linguistic-hedge combination is searched by MSGA and the fitness function is chosen as

$$f(\mathbf{h}) = \exp(-\sigma \cdot c_e(\mathbf{h})) \tag{37}$$

  where σ is selected as 0.2 and $c_e(\mathbf{h})$ is the cost function expressed as

$$c_e(\mathbf{h}) = \sum_{i=1}^{m} {e_i}^2(\mathbf{h}) \tag{38}$$

  in which m is the number of iterations during simulation. In the MSGA searching phase, the performance is measured according to its corresponding cost value. The lower the cost value, the better the linguistic-hedge combination searched. Hence, the variable $y(k)$ will track the reference signal $r(k)$ in the best manner as possible as it can. During this phase, the state $(e, \dot{e}) = (1, 0)$ is chosen as the initial state to search the optimal linguistic-hedge combination by MSGA. The simulation result indicates that the maximum average fitness value of 0.9 can be achieved at the 39-th generation. The resultant optimal linguistic-hedge combination vectors are

$$\mathbf{h} = [0.25 \quad 0.5 \quad 1 \quad 2 \quad 2 \quad 2 \quad 1 \quad 1 \quad 1 \quad 1 \quad 2 \quad 2 \quad 2 \quad 1 \quad 0.5 \quad 0.25] \tag{39}$$

  for the fuzzy set ZE and

$$\mathbf{h} = [0.5 \quad 2 \quad 2 \quad 0.25 \quad 4 \quad 1 \quad 2 \quad 0.25 \quad 0.25 \quad 2 \quad 1 \quad 4 \quad 0.25 \quad 2 \quad 2 \quad 0.5] \tag{40}$$

  for the fuzzy sets NB and PB.

The performance of the plant model controlled by this LHFLC is indicated by the solid line in Fig. 13. We define the settling time as the time for response to settle to within $\pm 0.1\%$ of the steady-state value. Therefore, the plant output $y(k)$ tracks the reference signal $r(k)$ with 0.3 s settling time in the proposed LHFLC system. In contrast, the plant output $y(k)$ tracks the reference signal $r(k)$ with 1.1 s settling time when it is controlled by the conventional FLC with 7×7 rules. Also, $y(k)$ tracks $r(k)$ with 2.7 s settling time when it is controlled by the conventional FLC with 3×3 rules. Clearly, the LHFLC which adopts the least number of rules (3×3 rules rather than 7×7 rules as those in the conventional FLC) and the simplest shape membership functions (the triangle-like membership functions rather than the Gauss-like ones as those in the conventional FLC) possesses the best performance. Table 2 summaries these results.

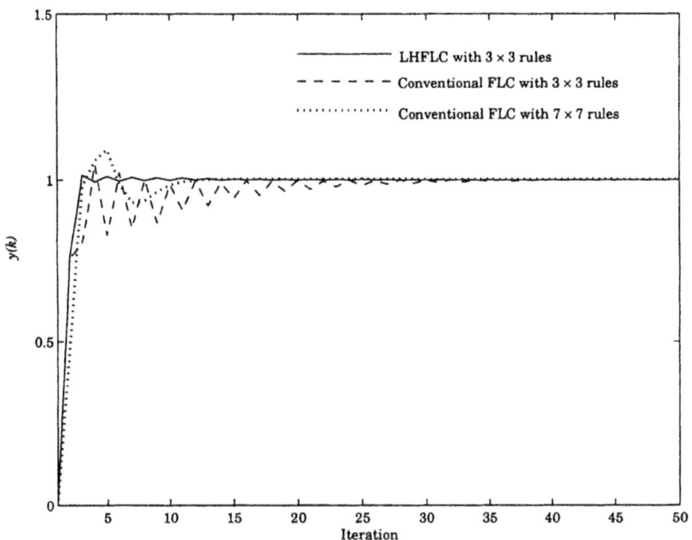

Fig. 13. Simulated transient response of nonlinear plant model control system.

Table 2. Performance factors of nonlinear plant model control system.

	Conventional FLC with 7×7 rules	Conventional FLC with 3×3 rules	LHFLC with 3×3 rules
Membership functions	Gauss-like	Triangle-like	Triangle-like
Settling times (s)	1.1	2.7	0.3

4.2 Cart-Pole Balance System

Problem Description The goal of a cart-pole balance system involves both vertically balancing a pole hinged to a motor-driven cart and causing the cart to be stopped at the specified position by applying forces on it. Figure 14 represents the cart-pole balance system, which can be described by the following nonlinear differential equations [28]

$$\ddot{\theta} = \frac{(M+m)g\sin\theta - \cos\theta[f + mL\dot{\theta}^2\sin\theta - \mu_c \mathrm{sgn}(\dot{x})] - \frac{\mu_p(M+m)\dot{\theta}}{mL}}{\frac{4}{3}(M+m)L - mL\cos^2\theta} \tag{41}$$

$$\ddot{x} = \frac{f + mL(\dot{\theta}^2\sin\theta - \ddot{\theta}\cos\theta)}{M+m} - \mu_c \mathrm{sgn}(\dot{x}) \tag{42}$$

where the related parameters are

$g = -9.8$ m/s^2, acceleration due to gravity;
$M = 1.1$ kg, mass of cart;

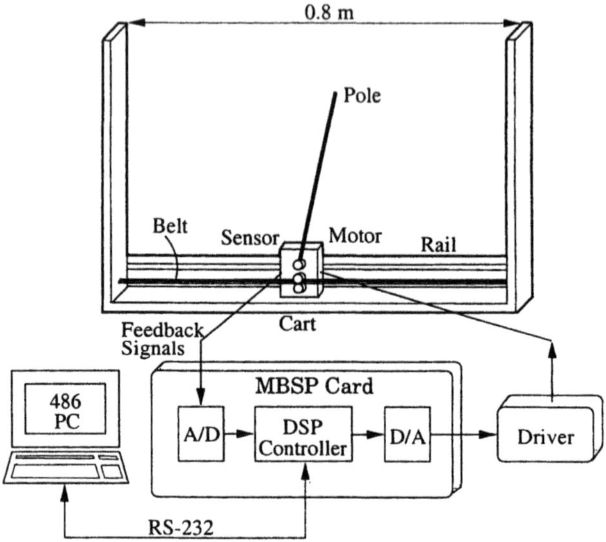

Fig. 14. Experimental setup of cart-pole balance system.

$m = 0.1$ kg, mass of pole;
$L = 0.5$ m, half-pole length;
$\mu_c = 0.1$, coefficient of friction of cart on track;
$\mu_p = 0.01$, coefficient of friction of pole on cart;
$\Delta t = 0.01$ s, the sample time;

the four state variables and the output variable are

x: position of the cart on the track (in m);
$\dot{x}$: cart velocity (in m/s);
θ: angle of the pole with the vertical (in rad);
$\dot{\theta}$: rate of change of the angle (in rad/s)
f: force (in N) applied to cart's center of mass.

We use the *switching-type* fuzzy sliding mode controller (FSMC) to control the cart-pole balance system as our platform. The switching-type FSMC proposed by Li [29] is a method based on FLC and sliding mode controller (SMC) [30]–[32], which achieves asymptotic stability of the system. The dynamics of the cart-pole balance system is divided into *approaching condition* and *departure condition.* Two different FSMCs to solve control problems for these two conditions should be designed, each characterized by the associated sliding surface chosen. The sliding surface of the cart-pole balance system for the approaching mode is designed as

$$s = e_\theta + 0.4e_{\dot{\theta}} - 0.3(e_x + 0.4e_{\ddot{\theta}}) = 0 \tag{43}$$

while that for the departure mode is chosen as

$$s = e_\theta + 0.4e_{\dot{\theta}} + 0.3(e_x + 0.4e_{\ddot{\theta}}) = 0 \tag{44}$$

in which the error vector $\mathbf{E} = [e_\theta \; e_{\dot{\theta}} \; e_x \; e_{\dot{x}}]^T$ is defined as the difference between the actual state vector $\mathbf{V} = [\theta \; \dot{\theta} \; x \; \dot{x}]^T$ and the desired state vector $\mathbf{V}_d = [\theta_d \; \dot{\theta}_d \; x_d \; \dot{x}_d]^T$. That is,

$$\mathbf{E} = \mathbf{V} - \mathbf{V}_d. \tag{45}$$

The input variables of the switching-type FSMC are s and its time derivative $\dot{s}$; the output variable is the force f applied to the cart. In the simulation as well as the experiment, the objective with which we are concerned is to control the pole being balanced at the position $x = 0.25$ m with the angle $\theta = 0$ rad.

Simulation Results

- **Switching-Type FSMC with 5×5 Rules**
  Figure 15 shows the rule table which specifies the implication relationships between the input variables (s and $\dot{s}$) and the output variable f in this controller. The five fuzzy sets with Gauss-like membership functions

$\dot{s}$ \ s (f)	NB	NS	ZE	PS	PB
NB	NB	NB	NM	NS	ZE
NS	NB	NM	NS	ZE	PS
ZE	NM	NS	ZE	PS	PM
PS	NS	ZE	PS	PM	PB
PB	ZE	PS	PM	PB	PB

Fig. 15. Rule table of a size 5×5 for cart-pole balance system.

  linguistically labeled as NB, NS, ZE, PS, and PB corresponding to the input variable s and $\dot{s}$ are designed over the intervals $s \in [-0.5, 0.5]$ and $\dot{s} \in [-0.05, 0.05]$, respectively, as shown in Fig. 16. The output variable f is characterized by seven fuzzy singletons designed over the interval $f \in [-100, 100]$.
- **Switching-Type FSMC with 3×3 Rules**
  To emphasize the capability of the LHFLC, the number of rules in the original switching-type FSMC is reduced from 5×5 to 3×3 to run the same simulation. In this case, the rule table is shown in Fig. 17. Each input variable is defined by three fuzzy sets with triangle-like membership functions distributed over the intervals $s \in [-0.5, 0.5]$ and $\dot{s} \in [-0.05, 0.05]$, respectively, as shown in Fig. 18. The output variable f is defined by three fuzzy singletons distributed over the interval $f \in [-100, 100]$.

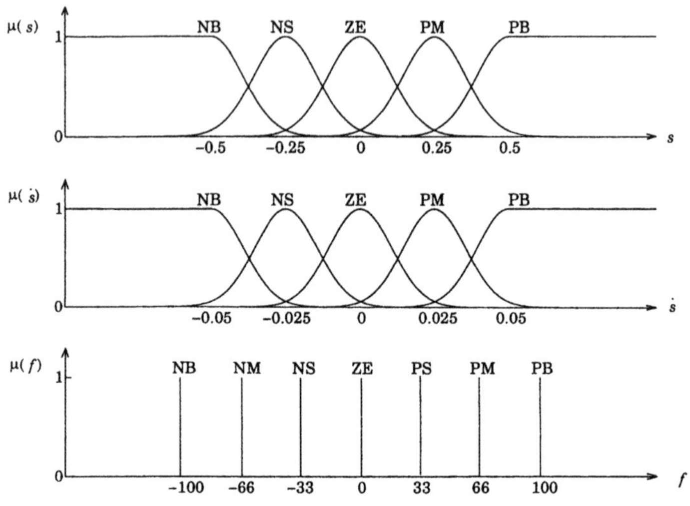

Fig. 16. Fuzzy membership functions for s, $\dot{s}$, and f of switching-type FSMC with 5×5 rules.

f \ s	NB	ZE	PB
$\dot{s}$ NB	NB	NB	ZE
$\dot{s}$ ZE	NB	ZE	PB
$\dot{s}$ PB	ZE	PB	PB

Fig. 17. Rule table of a size 3×3 for cart-pole balance system.

- **LHFLC with 3×3 Rules**

  In this case, the linguistic-hedge concept is applied to the design of switching-type FSMC in order to modify the membership functions of the input variables s and $\dot{s}$. The optimal linguistic-hedge combination is searched by MSGA according to the fitness function defined as

$$f(\mathbf{h}) = \exp(-\sigma(c_\theta(\mathbf{h}) + c_x(\mathbf{h}))) \tag{46}$$

where σ is selected as 0.1; $c_\theta(\mathbf{h})$ and $c_x(\mathbf{h})$ are the cost functions expressed as

$$c_\theta(\mathbf{h}) = \sum_{i=1}^{m} {e_{\theta_i}}^2(\mathbf{h}) \tag{47}$$

and

$$c_x(\mathbf{h}) = \sum_{i=1}^{m} {e_{x_i}}^2(\mathbf{h}) \tag{48}$$

in which m is the number of iterations during simulation. During the search process, the state $(x, \dot{x}, \theta, \dot{\theta}) = (0, 0, 0, 0)$ is chosen as the initial

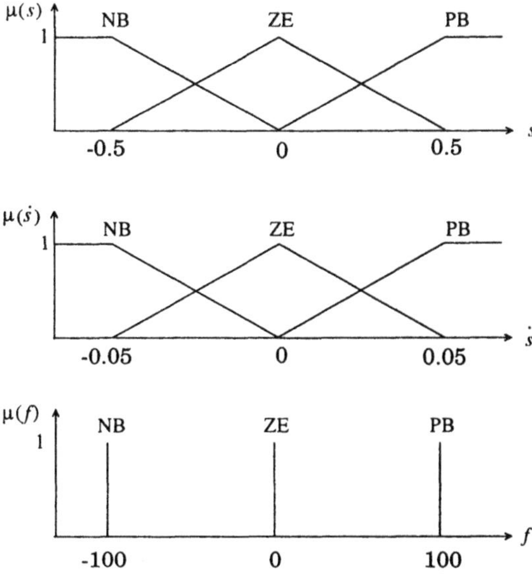

Fig. 18. Fuzzy membership functions for s, $\dot{s}$, and f of switching-type FSMC with 3×3 rules.

state to search the optimal linguistic-hedge combination by MSGA. The cost values returned by MSGA reflect the qualities of the linguistic-hedge combinations searched. The lower the cost value, the better the result searched. That is, we expect the pole angle θ to reach 0 rad and the cart to be located at the desired position $x = 0.25$ m as fast as possible. The simulation result indicates that the maximum average fitness value of 0.8 can be achieved at the 40-th generation. The resultant optimal linguistic-hedge combination vectors are

$$\mathbf{h} = [0.5 \quad 4 \quad 4 \quad 0.5 \quad 0.5 \quad 0.25 \quad 1 \quad 1 \quad 1 \quad 1 \quad 0.25 \quad 0.5 \quad 0.5 \quad 4 \quad 4 \quad 0.5] \tag{49}$$

for the fuzzy set ZE and

$$\mathbf{h} = [0.25 \quad 2 \quad 1 \quad 0.25 \quad 0.5 \quad 0.5 \quad 4 \quad 2 \quad 2 \quad 4 \quad 0.5 \quad 0.5 \quad 0.25 \quad 1 \quad 2 \quad 0.25] \tag{50}$$

for the fuzzy sets NB and PB.

Figure 19 shows the response of the cart-pole balance system controlled by the switching-type FSMC with either 5×5 or 3×3 rules and the LHFLC with associated $\mathbf{h}$. Obviously, either the pole angle response or the cart position response reveals that the LHFLC adopting the least number of rules and the simplest shape membership functions really enhances the performance of the FLC adopting 5×5 rules and Gauss-like membership functions or the FLC adopting 3×3 rules and triangle-like membership functions.

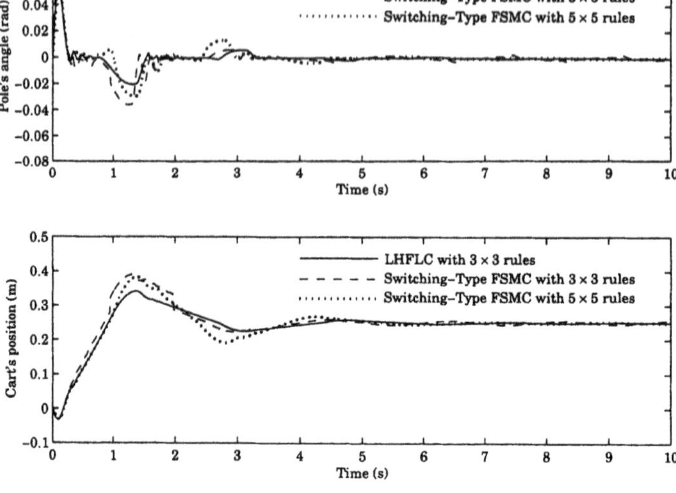

Fig. 19. Simulated transient response of the cart-pole balance system with initial conditions $(\theta, \dot{\theta}, x, \dot{x}) = (0, 0, 0, 0)$. (a) Response of pole angle. (b) Response of cart position.

Experimental Results To apply the proposed LHFLC to the real experimental system, the cart-pole balance system manufactured by Phimatic Enterprise Co. Ltd. is used to demonstrate this work. The arrangement of the whole experimental setup is illustrated in Fig. 14. The specifications of the related hardware are listed as follows.

Pole: 0.5 m length;
Drive force: DC motor (15 W);
Sensor: Photo encoder (500 pulse per rotation);
Micro-computer: 486 personal computer;
A/D, D/A & digital signal control card: MBSP card [33].

Initially, the LHFLC algorithm is programmed in C language. The control signals generated by the resultant execution code compiled from the C program are sent to the DSP controller of the MBSP card through the 25-pin RS-232 transmission line. The DSP controller processes the received signals to produce the control actions. These control actions are delivered to a dc motor via the D/A converter to apply the suitable force to the cart. The state of the cart-pole balance system is sensed by a photo encoder and fed back to the A/D converter of the MBSP card. This routine is run continuously until the system demand is met.

The responses of the pole angle and the cart position are recorded by computer and plotted in Fig. 20. Figure 21 exhibits the photograph of the cart-pole balance system controlled by the resultant LHFLC with an exposure of about 8 s.

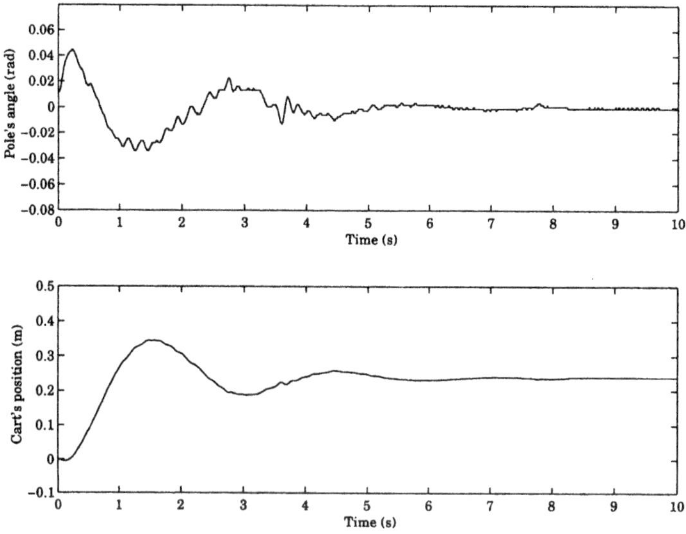

Fig. 20. Experimental result of the cart-pole balance system.

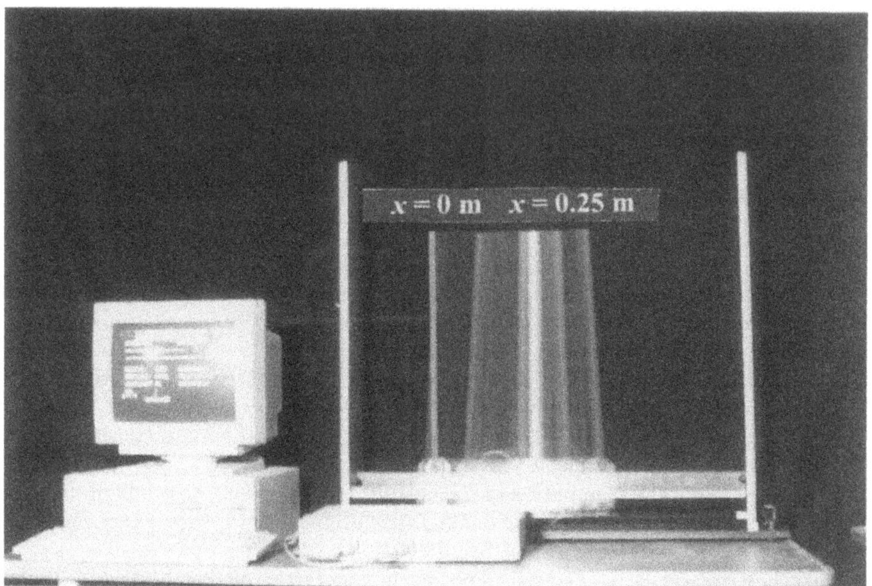

Fig. 21. Photograph of cart and pole during control with initial conditions $(\theta, \dot{\theta}, x, \dot{x}) = (0, 0, 0, 0)$.

5 Conclusion

In this chapter, we have investigated the concepts of linguistic hedges originated by Zadeh and explored how an linguistic-hedge operator affects the inference performance from the physical viewpoints. We have also introduced the LHFLC that combines the original FRBS and the concepts of linguistic hedges. In the LHFLC, the domains of input variables are partitioned into intervals in which the pieces of membership functions are dynamically adjusted by the linguistic hedges specified in the plant-dependent linguistic-hedge combination vectors according to the output signal fed back from the plant, which results into that we can employ fewer rules and simple-shape membership functions to achieve a better inference performance than that a conventional FLC does. The number of partitions of each input variable is 16. For a two-input-one-output LHFLC with each variable specified by three fuzzy sets with the membership functions of S-shape, $\wedge$-shape, and Z-shape, the number of inference rule is nine. In order to find a proper linguistic-hedge combination vector, we attached an MSGA module to the LHFLC. To verify the feasibility of this LHFLC, we have simulated the nonlinear plant model control system and the cart-pole balance system.

Due to the benefits mentioned such as characterizing the related variables by simple-shape membership functions and inferring control actions based on fewer rules, this architecture simplifies the complexity of the LHFLC design. From the architecture viewpoint, inserting a linguistic-hedge module allows us to use the simple triangle-like membership functions and a fewer number of rules instead of the more carefully designed membership functions and the large number of rules to reach the control goals. As a result, the membership function constructions and the rule developments become more simple works.

References

1. L. A. Zadeh. Fuzzy sets. *Information Control*, 8:338–353, 1965.
2. C. L. Karr. Design of an adaptive fuzzy logic controller using a genetic algorithm. In *Proceedings of the 4th International Conference on Genetic Algorithms*, pages 450–457, San Diego, CA, USA, 1991.
3. J. H. Holland. *Adaptation in natural and artificial systems*. Ann Arbor, MI: University of Michigan Press, 1975.
4. D. E. Goldberg. *Genetic algorithms in search, optimization and machine learning*. Reading, MA: Addison-Wesley, 1989.
5. R. Krishnapuram. Generation of membership functions via possibilistic clustering. In *Proceedings of the IEEE 5th International Conference Fuzzy Systems*, pages 902–908, Orlando, FL, USA, 1994.
6. C. J. Kim and B. D. Russell. Automatic generation of membership function and fuzzy rule using inductive reasoning. In *Proceedings of the IEEE 3rd International Conference on Industrial Fuzzy Control and Intelligence Systems*, pages 93–96, Houston, TX, USA, 1993.

7. L. X. Wang and J. M. Mendel. Generating fuzzy rules by learning from examples. *IEEE Transactions on Systems, Man, Cybernetics*, 22(6):1414–1427, 1992.
8. R. L. P. Chang and T. Pavlidis. Fuzzy decision tree algorithms. *IEEE Transactions on Systems, Man, Cybernetics*, 7(1):28–35, 1977.
9. E. T. Lee. Fuzzy tree automata and syntactic pattern recognition. *IEEE Transactions on Pattern Analysis Machine Intelligence*, 4(4):445–449, 1982.
10. S. Morishima and H. Harashima. Automatic rule extraction from statistical data and fuzzy tree search. *Systems and Computing in Japan*, 19(5):26–37, 1988.
11. I. B. Turksen and Y. Tian. Two-level tree search in fuzzy expert system. *IEEE Transactions on Systems, Man, Cybernetics*, 25(4):55–568, 1995.
12. B. D. Liu and C. Y. Huang. Design and implementation of the tree-based fuzzy logic controller. *IEEE Transactions on Systems, Man, Cybernetics*, 27(3):475–487, 1997.
13. L. A. Zadeh. Outline of a new approach to the analysis of complex systems and decision processes. *IEEE Transactions on Systems, Man, Cybernetics*, 3(1):28–44, 1973.
14. W. Banks. Mixing crisp and fuzzy logic in applications. In *WESCON'94, Idea/Microelectronics Conference Record*, pages 94–97, Anaheim, CA, USA, 1994.
15. B. Bouchon-Meunier. Linguistic hedges and fuzzy logic. In *Proceedings of the IEEE 1st International Conference on Fuzzy Systems*, pages 247–254, San Diego, CA, USA, 1992.
16. V. Novák. A horizon shifting model of linguistic hedges for approximate reasoning. In *Proceedings of the IEEE 5th International Conference on Fuzzy Systems*, pages 423–427, New Orleans, LA, USA, 1996.
17. C. Y. Huang, C. Y. Chen, and B. D. Liu. Current-mode linguistic hedge circuit for adaptive fuzzy logic controllers. *Electronic Letters*, 31(17):1517–1518, 1995.
18. C. Y. Chen, C. Y. Huang, and B. D. Liu. Current-mode fuzzy linguistic hedge circuit — contrast intensification. In *Proceedings of the IEEE International Symposium on Circuits and Systems*, pages 511-513, Atlanta, GA, USA, 1996.
19. C. Y. Chen, C. Y. Huang, B. D. Liu, and T. J. Su. A current-mode fuzzy linguistic hedge circuit — more or less. In *Proceedings of the IEEE 5th International Conference on Fuzzy Systems*, pages 1080–1085, New Orleans, LA, USA, 1996.
20. C. Y. Huang, C. Y. Chen, and B. D. Liu. Current-mode fuzzy linguistic hedge circuits. *Analog Integrated Circuits and Signal Processing*, 19(3):255–278, 1999.
21. C. Y. Chen, B. D. Liu, and J. Y. Tsao. Adaptive fuzzy logic controller blending the concepts of linguistic hedges and genetic algorithms. In *Proceedings of the IEEE 8th International Conference on Fuzzy Systems*, pages 1299–1304, Seoul, Korea, 1999.
22. B. D. Liu, C. Y. Chen, and J. Y. Tsao. Design of adaptive fuzzy logic controller based on linguistic-hedge concepts and genetic algorithms. *IEEE Transactions on Systems, Man, Cybernetics— Part B*, 31(1):32–53, 2001.
23. C. C. Lee. Fuzzy logic in control systems: fuzzy logic controller — Part I, II. *IEEE Transactions on Systems, Man, and Cybernetics*, 20(2):404–432, 1990.
24. C. C. Ku and K. Y. Lee. Diagonal recurrent neural networks for dynamic systems control. *IEEE Transactions on Neural Networks*, 6(1):144–156, 1995.
25. E. H. Mamdani. Application of fuzzy logic to approximate reasoning using linguistic synthesis. *IEEE Transactions on Computers*, 26(12):1182–1191, 1977.

26. K. A. De Jong. An analysis of the behavior of a class of genetic adaptive systems. Ph. D. dissertation, Department of Computer Science, Univcrsity of Michigan, Ann Arbor, 1975.
27. MATLAB, *Matrix Laboratory*, The Math Works, Inc., Natwick, MA, 1992.
28. P. H. Canno. *Dynamics of Physical Systems.* New York: McGraw-Hill, 1967.
29. T. S. Li and C. Y. Tsai. Parallel fuzzy sliding mode control of the cart-pole system. In *Proceedings of IEEE IECON'95*, pages 1468–1473, Orlando, FL, USA, 1995.
30. J. E. Slotine. *Applied nonlinear control.* Eaglewood Cliffs, NJ: Prentice-Hall, 1991.
31. J. Y. Hung, G. Senior, and J. C. Hung. Variable structure control: a survey. *IEEE Transactions on Industrial Electronics*, 40(1):2–22, 1993.
32. W. Gao and J. C. Hung. Variable structure control of nonlinear systems: a new approach. *IEEE Transactions on Industrial Electronics*, 40(1):45–55, 1993.
33. J. C. Huang. *MBSP manual.* Hsinchu, Taiwan, R.O.C.: Phimatic Enterprise Co., Ltd., 1995.

Automatic Construction of Fuzzy Rule-Based Systems: A trade-off between complexity and accuracy maintaining interpretability

Héctor Pomares, Ignacio Rojas, Jesús González

Department of Computer Architecture and Computer Technology,
University of Granada, E-18071, Granada, Spain
e-mails: {hpomares, irojas, jgonzalez}@atc.ugr.es

Abstract: The identification of a model is one of the key issues in the field of fuzzy system modeling and function approximation theory. An important characteristic that distinguishes fuzzy systems from other techniques in this area is their transparency and interpretability. Especially in the construction of a fuzzy system from a set of given training examples, little attention has been paid to the analysis of the trade-off between complexity and accuracy maintaining the interpretability of the final fuzzy system. In this chapter a systematic data based approach is proposed to determine fuzzy system structure and learning its parameters. In particular, two fundamental issues concerning fuzzy system modeling are addressed: fuzzy rule parameter optimization and the identification of system structure (i.e. the number of membership functions and fuzzy rules), taking always in mind the transparency of the obtained rule-base. This chapter presents a reliable method to obtain the structure of a complete rule-based fuzzy system for a specific approximation accuracy of the training data, i.e., it can decide which input variables must be taken into account in the fuzzy system and how many membership functions are needed in every selected input variable in order to reach the approximation target with the minimum number of parameters. The main criterion of comparison of all fuzzy systems obtained is their performance, measured as the accuracy of the model, versus complexity (number of parameters). Since determining which configuration has optimum characteristics is a clear example of fuzzy decision-taking, in this chapter the preferences of the human operator (end user) concerning the two conflicting objectives considered in evaluating the fuzzy system are translated into fuzzy rules.

1 Introduction

The automatic construction of fuzzy systems by means of a data set has traditionally focused on obtaining systems that feature a high degree of precision than on those that can be easily understood by a human operator. The transparency and interpretability of a fuzzy system depend on three main factors: the linguistic interpretability of the membership functions, the degree of complexity of the fuzzy system and the relevance of the fuzzy rules that comprise the system.

The problem of estimating an unknown function f from samples of the form $\{(\vec{x}^k; z^k); k=1,2,\ldots,K;$ with $z^k = f(\vec{x}^k) \in \mathbf{R}$, and $\vec{x}^k \in \mathbf{R}^N\}$ (i.e. function approximation from a finite number of data points), is one of the key issues in the field of fuzzy system modeling and function approximation theory [6, 7, 10, 13, 20, 23, 47]. The principal goal is to learn an unknown functional mapping between input and output vectors, using a set of known training samples [13]. Once this mapping is generated, it can be used for predicting the output values given new input vectors. Inputs and outputs can be continuous and/or categorical variables. This chapter is concerned with continuous output variables, thus considering regression or function approximation problems, as opposed to classification problems in which the output variable is categorical [7].

Recently, model-free systems, such as artificial neural networks or fuzzy systems, have been proposed to avoid the knowledge-acquisition bottleneck [14, 19, 20, 21, 26]. Fuzzy systems provide an attractive alternative to the "black boxes" characteristic of neural network models, because their behavior can be easily explained by a human being. In fact, the popularity and practicality of fuzzy systems derives from their ability to express relations that are either complex or not sufficiently understood, in terms of linguistic rules.

Historically, fuzzy rule bases have been constructed by knowledge acquisition from experts while the weights within neural nets have been learned from data [11, 41]. The most straightforward approach is to define rules and membership functions subjectively by studying a human operator. However, consulting an expert may be difficult and/or expensive; furthermore, translating the human operator's experience directly into the fuzzy linguistic values can be influenced by the intuition of the operators and designers, so that the fuzzy control rules may be incomplete or even contradict each other [51].

Many fuzzy systems that automatically derive fuzzy IF-THEN rules from numerical data have been proposed in the bibliography to overcome the problem of knowledge acquisition [1, 8, 10, 16, 17, 18, 20, 27, 30, 39]. The first relevant approach to tackle both tasks (i.e. structure and parameter identification) was proposed by T. Takagi and M. Sugeno in 1985 [46]. In this outstanding work, the consequents of the fuzzy rules were optimized using the least squares algorithm while the antecedents were calculated using the 'complex' method. Finally, they used a heuristic-combinational method to partition the input domain, obtaining more complex topologies to be subsequently optimized.

Since then, several authors have presented different approaches to fuzzy modeling, most of them based on clustering techniques [4]. In 1993, Sugeno and Yasukawa [42] published a paper on a FL-based approach to qualitative modeling, in which they proposed achieving structure identification by using a combination of fuzzy c-means clustering and the group method of data handling (GMDH). In their approach, the number of rules was calculated trying to minimize the variance in each output cluster and maximize the variance between clusters. The main advantage of their method is that they were able to separate structure identification from parameter identification. However, this approach may not be optimum since, as they noted, the two processes are mutually related and should not be separated. Identification methods based on clustering prove to be successful, especially in classification problems [4], but do not fully fit in approximation problems since clustering does not take into account the interpolation properties of the system. Similar techniques can be found in [3, 24, 29].

Analyzing the bibliography, we see that methods for automatically creating fuzzy systems from data, in which the structure of the fuzzy system and the parameters that define it are simultaneously optimized, have not been discussed in depth. The fact that much more effort has been dedicated to dealing with the problem of parameter adjustment than that of system identification is understandable since the latter is a very complex task for which it is very difficult to obtain reliable procedures. An additional difficulty is that, unlike methods for parameter adjustment, it is not possible to test system identification algorithms without using parameter adjustment algorithms. Thus, both problems must be tackled simultaneously.

This chapter presents a general and complete method to obtain the structure of a singleton fuzzy system automatically and simultaneously optimize the configuration of the membership functions (location and number) and the consequents of the rules of a fuzzy system. It takes into account the trade-off between complexity and accuracy, but at the same time seeking to achieve fuzzy systems that are both transparent and easily understood by a human operator.

The identification of the fuzzy system structure and the tuning of the parameters defining it are performed in conjunction. From an initially simple system, and by means of an overall analysis of the approximation error, the structure of such a system is modified in an autonomous way. The algorithm presented here obtains, automatically, various fuzzy system structures, each of which is optimized with respect to the parameters that define it (membership functions in the antecedent parts and consequents of the rules). The conclusions of the fuzzy rules are obtained optimally, i.e. for a particular configuration of membership functions, the consequents minimize the error between the desired output and that obtained by the fuzzy system. Simultaneously, the fuzzy sets defining the antecedent part of the rules are also tuned. As different structures of fuzzy systems with different degrees of complexity are able to approximate a given unknown function with different levels of accuracy, we have constructed an auxiliary fuzzy system to produce an index reflecting the compromise between the accuracy and the complexity of the rule set.

The fuzzy system that are automatically obtained should present the following fundamental characteristics:

- They must be easy to understand. Basically, this is achieved by optimizing the membership functions such that their linguistic interpretation is always transparent to the user.
- They must be accurate. For this purpose, which is fundamental when functional approximation is performed, it is necessary for the distribution or location of the membership functions in the input variables to be correct.

Moreover, the consequences of the fuzzy rules must be optimized. When this is so, error indices such as MSE are minimized.

- They must not be overly complex. The optimization of the structure of fuzzy systems is a very complicated matter that should be correctly carried out, since one of the elements preventing the system from being transparent and easy to understand is precisely an increase in the number of rules or membership functions of the fuzzy system. This chapter, while maintaining the importance of interpretability, describes how the structure of the fuzzy system can be modified automatically, developing from simple topologies to more complex fuzzy systems in which the density of the rules and membership functions (i.e. their number and position) is increased in an optimum fashion with the aim of improving the final accuracy achieved.

The organization of the rest of this work is as follows: the statement of the problem and the notation used is briefly described in Section 2, followed by the key issue of finding the optimum rules for a fixed configuration of membership functions in Section 3. The simultaneous optimization of the distribution of the membership functions and the consequents of the fuzzy rules is described in Section 4. Section 5 is concerned with how the structure of the fuzzy system is modified in an autonomous way (by recognizing where it is necessary to assign a larger number or density of rules). Because it is necessary to make a compromise between the accuracy and the complexity of the obtained system, Section 6 proposes an auxiliary fuzzy system to be responsible for selecting the most suitable structure. Finally, some comparative results are presented in Section 7 and conclusions are drawn in Section 8.

2 Problem Formulation

Let us consider a set D of desired input-output data pairs, belonging to an unknown function or system F. Each vector datum $(\vec{x}^k; z^k)$ can be expressed as $(x_1^k, x_2^k, \ldots, x_N^k; z^k)$, with $\vec{x}^k \; 0 \; \mathbf{R}^N$, $z^k \; 0 \; \mathbf{R}$ and k=1,2,...,K, being K the total number of samples. In this work, we will try to approximate these data using fuzzy systems comprising a set of IF-THEN fuzzy rules $R^*_{i_1 i_2 \ldots i_N}$ of the form:

$$IF\ x_1 \text{ is } X_1^{i_1} \ AND\ x_2 \text{ is } X_2^{i_2} \ AND \ldots AND\ x_N \text{ is } X_N^{i_N} \ THEN\ z = R_{i_1 i_2 \ldots i_N} \tag{1}$$

where $X_m^{i_m} \in \{X_m^1, X_m^2, \ldots, X_m^{n_m}\}$ with n_m being the number of membership functions of the input variable m. For the output variable, $R_{i_1 i_2 \ldots i_N}$ is the numeric consequent of the rule. In the literature there are many possibilities for the selection of the fuzzy operators that determine how to evaluate each individual rule and how to obtain a final conclusion from all the rules. There are also conflicting opinions about the best selection of these fuzzy primitives [15, 37, 50].

In our approach, the fuzzy inference method uses the product as T-norm and the centroid method with sum-product operator as the defuzzification strategy. The strength or α-level of rule $R^*_{i_1 i_2 \ldots i_N}$ is calculated by:

$$\alpha_{i_1 i_2 \ldots i_N}(\vec{x}) = \mu_{i_1 i_2 \ldots i_N}(\vec{x}) = \mu_{i_1 i_2 \ldots i_N}(x_1 x_2 \ldots x_N) = \prod_{m=1}^{N} \mu_{X_m^{i_m}}(x_m) \tag{2}$$

where the product operator is used to perform the conjunction of each of the individual membership functions within the premise and $\mu_{X_m^{i_m}}(x_m)$ is the membership degree of the m component of the vector $\vec{x}$ with respect to the linguistic function i_m. Using the above notation, the fuzzy function can be expressed as follows:

$$\widetilde{F}(\vec{x}^k;R,C) = \frac{\sum_{i_1=1}^{n_1} \sum_{i_2=1}^{n_2} \cdots \sum_{i_N=1}^{n_N} \left(R_{i_1 i_2 \ldots i_N} \cdot \prod_{m=1}^{N} \mu_{X_m^{i_m}}(x_m^k) \right)}{\sum_{i_1=1}^{n_1} \sum_{i_2=1}^{n_2} \cdots \sum_{i_N=1}^{n_N} \left(\prod_{m=1}^{N} \mu_{X_m^{i_m}}(x_m^k) \right)} \tag{3}$$

where an explicit statement is made of the dependency of the fuzzy output, not only on the input vector, but also on the matrix of rules *R* and on all the parameters that describe the membership functions *C*. Expression (3) is applicable for a complete rule base. If a rule is marked as not used, its degree of activation is considered to be zero.

The problem considered in this work may be stated in a precise way as that of finding a configuration C and generating a set of fuzzy rules $\Psi=\{R^*_p ; p=1,...,r\}$ from a data set D of K input-output pairs $(x_1^k, x_2^k, \ldots, x_N^k; z^k)$, such that the fuzzy system $\widetilde{F}(\vec{x}^k;R,C)$ correctly approximates the unknown function F using a trade-off between complexity and accuracy, trying to maintain the interpretation of the final approximator system obtained. As is common in all approximation schemes, the function to be minimized is the sum of squared errors:

$$J(R,C) = \sum_{k \in D} \left(F(\vec{x}^k) - \widetilde{F}(\vec{x}^k;R,C) \right)^2 \tag{4}$$

The index selected to determine the degree of accuracy of the obtained fuzzy approximation is the *Normalized Root-Mean-Square Error* (NRMSE) defined as:

$$NRMSE \equiv \sqrt{\frac{\overline{e^2}}{\sigma_z^2}} \tag{5}$$

where σ_z^2 is the variance of the output data, and $\overline{e^2}$ is the mean-square error between the obtained and the desired output. In this way, the NRMSE index describes the performance of the approximation, making it independent of scale factors or number of data.

Together with the information from the parameters comprising the system's rule set, it is first necessary to know how many membership functions each input variable has (the numbers n_m with m=1,...,N) and thus the total number of rules in the set.

In the proposed procedure we are reporting in the following sections, based on our papers [33, 34], from an initially simple system, the number of membership functions for each input variable, together with the number of rules comprising the system, is automatically determined without prior assignment by a human operator, thereby enabling the construction of a self-organized rule base. Fig. 1 gives a flowchart of the proposed method.

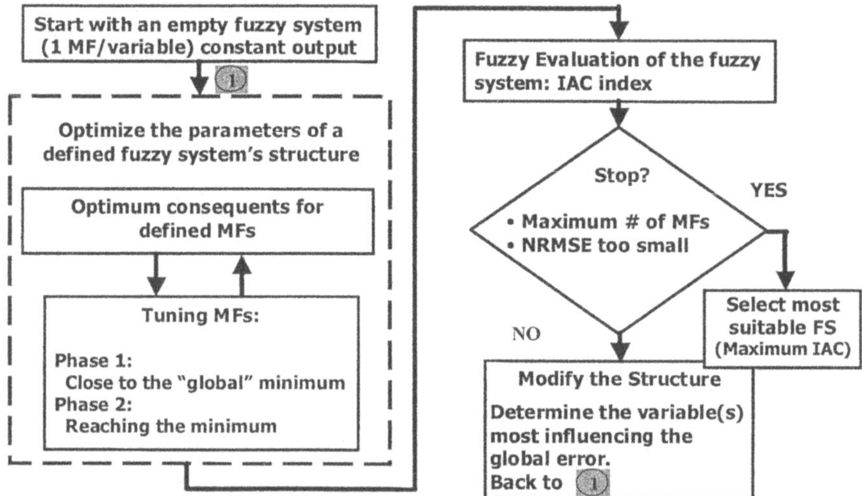

Fig. 1. General flowchart of the proposed algorithm

3 Finding the Optimum Rules for a Fixed Membership Function Configuration

The first case to examine is that in which the distribution and number of membership functions are fixed and, therefore, $\widetilde{F} = \widetilde{F}(\vec{x};R)$. Observing (3), $\widetilde{F}$ is not only a continuous and differentiable expression as a function of the consequent of the rules $R_{i_1 i_2 \ldots i_N}$, but this dependency is linear. This property allows us to find the optimum consequents of the rule-table as in [46] and subsequently in [43, 44]. The optimum fuzzy rule conclusion should cancel the first derivative from J, eq. (4) with respect to the consequent of the rules:

$$\frac{\partial J(R)}{\partial R_{j_1 j_2 \ldots j_N}} = -2 \cdot \sum_{k \in D} \left[\left(F(\vec{x}^k) - \widetilde{F}(\vec{x}^k;R,C) \right) \cdot \left(\frac{\partial \widetilde{F}(\vec{x}^k;R)}{\partial R_{j_1 j_2 \ldots j_N}} \right) \right] = 0 \qquad (6)$$

Substituting in the above expression and introducing the following notation:

$$S_{i_1 i_2 \ldots i_N, j_1 j_2 \ldots j_N} \equiv \sum_{k \in D} \frac{\prod_{m=1}^{N} \mu_{X_m^{i_m}}(x_m^k)}{\sum_{i_1=1}^{n_1} \sum_{i_2=1}^{n_2} \cdots \sum_{i_N=1}^{n_N} \left(\prod_{m=1}^{N} \mu_{X_m^{i_m}}(x_m^k) \right)} \cdot \frac{\prod_{m=1}^{N} \mu_{X_m^{j_m}}(x_m^k)}{\sum_{i_1=1}^{n_1} \sum_{i_2=1}^{n_2} \cdots \sum_{i_N=1}^{n_N} \left(\prod_{m=1}^{N} \mu_{X_m^{i_m}}(x_m^k) \right)}$$

$$S_{F, j_1 j_2 \ldots j_N} \equiv \sum_{k \in D} F(\vec{x}^k) \cdot \frac{\prod_{m=1}^{N} \mu_{X_m^{j_m}}(x_m^k)}{\sum_{i_1=1}^{n_1} \sum_{i_2=1}^{n_2} \cdots \sum_{i_N=1}^{n_N} \left(\prod_{m=1}^{N} \mu_{X_m^{i_m}}(x_m^k) \right)} \qquad (7)$$

we obtain the following system of linear equations:

$$\sum_{i_1=1}^{n_1} \sum_{i_2=1}^{n_2} \cdots \sum_{i_N=1}^{n_N} \left(R_{i_1 i_2 \ldots i_N} \cdot S_{i_1 i_2 \ldots i_N, j_1 j_2 \ldots j_N} \right) = S_{F, j_1 j_2 \ldots j_N} \qquad (8)$$

where j_1 ranges from 1 to n_1, j_2 from 1 to n_2, etc. This expression represents a system of $n_1 * n_2 * \ldots * n_N$ linear equations that coincides with the number of unknowns (the number of consequents of the rules). Therefore, if the determinant of the resulting matrix is not null, the solution is unique. The matrix is always two-dimensional and symmetrical (7). Moreover, it is a covariance-type matrix and the system can be very quickly solved using the Cholesky algorithm [35]. The obtained procedure is independent of the form and distribution of the selected membership functions (usually, triangular, gaussian, trapezoidal, etc.). Then, for a predefined membership function distribution, the optimum rule conclusion can be directly obtained.

4 Simultaneous Optimization of the Fuzzy Rule Base

In the previous section, the optimum consequents for a defined configuration of membership functions were obtained. However, the distribution of the membership functions (shape and location) has a strong influence on the performance of the systems, thus making it necessary to optimize them [5, 9, 20, 22, 28, 29, 31, 48]. To approach this issue, it is necessary to define a certain type of membership function.

Although the methodology is valid for any type of linguistic value, in order to maintain the *interpretability* of the final fuzzy system, the membership functions will be triangular functions with pair-wise overlap, i.e. each variable has a non zero membership value in at most two fuzzy sets. We will use a Triangular Partition (TP) [25, 40, 41, 44] where only the centers of the membership functions are stored, since the slopes of the triangles are calculated according to the centers of the surrounding membership functions (Fig. 2). Because the first and last center of the membership functions have been made to coincide with the domain of the function in each input variable, it is only necessary to store the value of the intermediate membership functions. This reduces the number of parameters that must be determined for the construction and subsequent optimization of a fuzzy system.

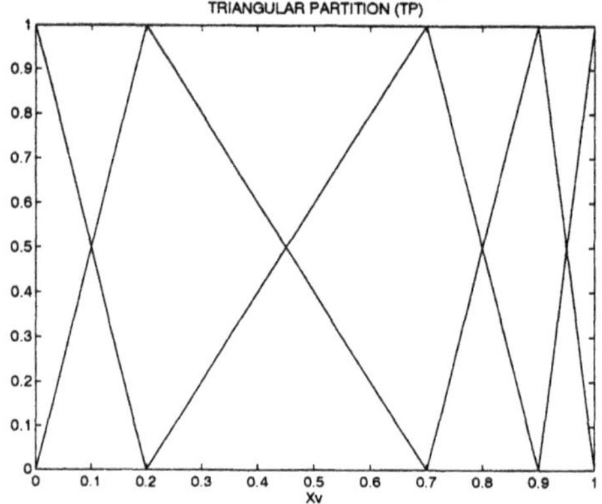

Fig. 2. Triangular Partition (TP)

Thanks to the membership function configurations selected (TP), it is straightforward for a human operator to *interpret* the final fuzzy system obtained. When the overlap is too high there are two main problems that arise for the implementation and understanding of fuzzy systems:

1. The number of activated rules increases exponentially, which becomes important when a hardware implementation is sought.
2. The membership functions lose their linguistic meaning when one input value activates more than two of them.

For each variable *m*, the membership degree of the m-component of the vector $\vec{x}$ with respect to the linguistic function i_m has the form:

$$\mu_{X_m^{i_m}}(x_m)=\begin{cases}\dfrac{x_m-c_m^{i_m-1}}{c_m^{i_m}-c_m^{i_m-1}} & if\ c_m^{i_m-1}\le x_m\le c_m^{i_m+1}\\[2ex] \dfrac{c_m^{i_m+1}-x_m}{c_m^{i_m+1}-c_m^{i_m}} & if\ c_m^{i_m}\le x_m\le c_m^{i_m+1}\\[2ex] 0 & otherwise\end{cases} \tag{9}$$

where $c_m^{i_m}$ represents the center of the i_m membership function of the input variable m. This decomposition of the domain of the input variables satisfies the following property [41]:

$$\sum_{i_m=1}^{n_m}\mu_{X_m^{i_m}}(x_m^k)=1\ \ ;\ \forall\vec{x}^k\in D \tag{10}$$

Using this property, it is apparent that the denominator of (3) is always equal to unity and thus, the fuzzy system considered can be expressed as:

$$\widetilde{F}(\vec{x};R,C)=\sum_{i_1=1}^{n_1}\sum_{i_2=1}^{n_2}\cdots\sum_{i_N=1}^{n_N}\left(R_{i_1i_2\ldots i_N}\cdot\prod_{m=1}^{N}\mu_{X_m^{i_m}}(x_m)\right) \tag{11}$$

Not only are few parameters required to define this type of distribution of membership functions, but also the interpretation of the system is straightforward for a human operator. In fact, the analysis of the internal states in the design of a fuzzy system (mainly the position of the linguistic values), with no restriction on the behavior of hybrid-fuzzy optimization techniques such as neuro-fuzzy networks and fuzzy-genetic algorithms [8, 32, 38, 45] shows that for each input vector many rules are activated and that the overlap is greater than two. Moreover, the position and domain of the membership functions do not usually match the characteristics of the problem being solved (the membership functions are sometimes located outside the domains of the input variables).

With the distribution of the membership functions selected and with the help of the direct method to determine the consequents of the rules, the number of parameters to be optimized for the fuzzy system $\widetilde{F}=\widetilde{F}(\vec{x};R,C)$ has been substantially reduced. For example, let us consider a fuzzy system with 3 input variables using 5 membership functions for each variable. If the rule-base is complete [11] a total of 134 parameters have to be optimized (5^3 consequents and 3·3 centers of membership functions). However, only the 9 centers of the linguistic values have to be optimized, because for a fixed configuration of the membership functions, the remaining parameters (the conclusion of the rules) are directly obtained. This improves the speed of convergence since it depends strongly with the number of parameters. Therefore, this section only refers to the optimization of the centers of the membership functions, it being implicit that each time there is a new configuration of the membership functions the optimum rules will be found using the method described in the previous section.

4.1 A prior stage to approach a near optimum solution

We wish to minimize function J (the error approximation) obtaining rules R and membership functions C. The dependency of $\widetilde{F} = \widetilde{F}(\vec{x};R,C)$ is not linear with respect to parameter C. Thus, using the gradient descent algorithm directly would lead the system to the first nearby local minimum relative to the initial conditions without taking into account whether this is the absolute minimum or not. How well it learns depends on how well we choose the initial set of parameters in C. It is therefore necessary to include a prior stage in our algorithm that allows us to approach a near optimum minimum in an efficient way, and without losing the linguistic interpretation inherent in the parameters defining the membership functions. The simplicity of the proposed system enables this task to be performed without difficulty. If the centers were to be located manually, without using an algorithm, they would be placed so that the errors were homogeneously distributed over the whole output range. An approximation that was very close to the original in one zone of the output range while greatly differing in another would not be considered a good approximation. Therefore, let us suppose that we can achieve near-optimum solutions by seeking the distribution of centers that produces equally distributed errors throughout all the areas defined. An overall analysis in the domain of the whole function is carried out to achieve the subsequent optimization of the overall function error J. The goal is for all the rules derived from the fuzzy system to be as precise as possible, such that the magnitude of the approximation error within each of the input spaces is similar. In order to obtain an accurate approximation of the unknown function with a homogeneous error distribution, the center of the i_m membership function of input variable m is associated with a "slope" $p_m^{i_m}$ of the form:

$$p_m^{i_m} = \frac{1}{\sigma_z^2}\left(\sum_{\substack{k \in data \\ x_m^k \in \left[c_m^{i_m-1}, c_m^{i_m}\right]}} e^2(\vec{x}^k) - \sum_{\substack{k \in data \\ x_m^k \in \left[c_m^{i_m}, c_m^{i_m+1}\right]}} e^2(\vec{x}^k) \right) \tag{12}$$

where e^2 and σ_z^2 are as defined in the previous section, and σ_z^2 is used to normalize the values of the slopes in order to make the whole process independent of scale factors. Therefore, each center $c_m^{i_m}$ has an associated parameter that represents the difference between the contribution of the preceding sector and the succeeding one to the value of the squared error. A positive value for such a slope means that the contribution to the left-hand sector is greater than that to the right and so the center must be moved to the left to counteract this effect. As the order of the centers cannot be allowed to vary, we perform the following movement:

$$\Delta c_m^{i_m} = \begin{cases} \dfrac{c_m^{i_m-1} - c_m^{i_m}}{b} \dfrac{p_m^{i_m}}{p_m^{i_m} + \dfrac{1}{T_m^{i_m}}} & if\, p_m^{i_m} \geq 0 \\ \dfrac{c_m^{i_m+1} - c_m^{i_m}}{b} \dfrac{\left|p_m^{i_m}\right|}{\left|p_m^{i_m}\right| + \dfrac{1}{T_m^{i_m}}} & if\, p_m^{i_m} < 0 \end{cases} \tag{13}$$

in which two new parameters are introduced: the first one is the active radius 'b', which is the maximum variation distance and is used to guarantee that the order of the membership function locations remains unchanged (a typical value is b=2, meaning that, at most, a center can be moved as far as the midpoint between it and its neighbor). The second parameter is the temperature $T_m^{i_m}$ of the center $c_m^{i_m}$, which indicates how far the center will be moved within the limits of possible movement. Thus, for very high temperatures the centers will move large distances, while at low temperatures these movements will be very small.

At first, temperatures are very high for all centers (typically 100-1000), though this is not a critical factor because, as the algorithm evolves, it adapts itself and adjusts the temperatures. The temperature of a center increases if the modification of the center in a certain number of iterations is in the same direction, while the temperature decreases when a change in direction occurs. The process finalizes when we obtain the configuration of centers that achieves an equidistant distribution of errors over the output surface (although in itself does not represent a good approximation). This procedure requires very few iterations to obtain a fuzzy system which, although in itself does not represent a good approximation, is assumed to be a good starting point to search for an approximation close to the optimum of the target function, using local-minimum search techniques.

4.2 Optimizing the position of the membership functions

Once we have a configuration assumed to be a near optimum one (as obtained above), then a method to find a local minimum is employed, thus reaching the desired minimum. One of the most commonly used methods to find local minima is the gradient descent. This method usually works well but in the present problem does not guarantee that the order of the membership function locations remains unchanged. Therefore, an alternative method also based on the gradient is proposed, which limits the movement of the membership function centers and prevents the order from being changed. The centers are optimized as follows:

$$\Delta c_m^{i_m} = -\eta_m^{i_m} \frac{\dfrac{\partial J(R,C)}{\partial c_m^{i_m}}}{\left|\dfrac{\partial J(R,C)}{\partial c_m^{i_m}}\right| + l_m^{i_m}} \tag{14}$$

where $l_m^{i_m}$ is a learning parameter and $\eta_m^{i_m}$ is given by:

$$\eta_m^{i_m} = \begin{cases} \dfrac{c_m^{i_m} - c_m^{i_m-1}}{b} & if\ \dfrac{\partial J(R,C)}{\partial c_m^{i_m}} \geq 0 \\ \dfrac{c_m^{i_m+1} - c_m^{i_m}}{b} & if\ \dfrac{\partial J(R,C)}{\partial c_m^{i_m}} < 0 \end{cases} \tag{15}$$

In graphic form, the movements of the centers are controlled by a saturation function as illustrated in Fig. 3. In the linear zone, the usual descent gradient (with no limitation) is employed, while for large movements of the centers, there are saturation regions to ensure the center does not move beyond a fraction b of the distance separating it from its neighbor. The learning parameters $l_m^{i_m}$ are progressively adapted to accelerate the search in a similar way to the parameter $1/T_m^{i_m}$ in equation (13). By this procedure, the center of each input membership function moves independently in the direction that reduces the squared error, without risking any alteration in the order of the centers and thus maintaining the linguistic interpretation of the fuzzy system obtained.

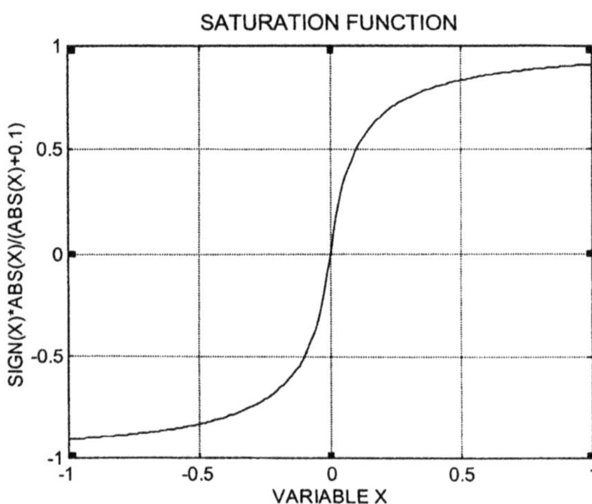

Fig. 3. Saturation function to control the movements of the membership function centers.

4.3 Example of this stage of the algorithm

To show the importance of the initial stage in the algorithm proposed, consider the function (Fig. 4):

$$f(x)=e^{-3x}\, sin(10\pi x) \tag{16}$$

defined in the interval [0,1]. We choose, at random, 100 possible initial configurations for each number of membership functions between 3 and 12 and perform a gradient descent for each of these to find the first relative minimum obtained from each of these initial configurations. Graphic results are given in Fig. 5a. It is apparent that even for one-dimensional functions there exist a large number of local minima and that these are more and more numerous as the number of membership functions to be optimized grows.

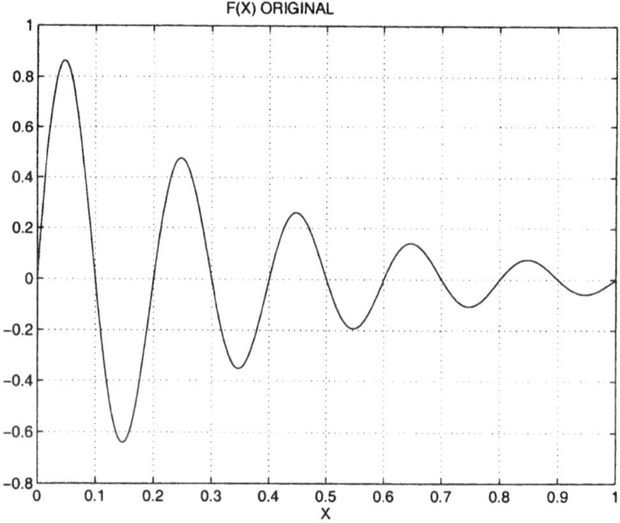

Fig. 4. Original function to be approximated

Table 1 presents the mean values obtained and the standard deviation for each distribution, together with the minimum values. In this table it is also shown the NRMSE obtained both by the complete algorithm and those derived from an equally-distributed configuration of membership functions. These data are presented in graphic form in Fig. 5b, where "*" represents the minima obtained by the complete algorithm, "x" signifies those derived from the equally-distributed initial configuration and a solid line shows the best values of each of the 100 samples for each configuration.

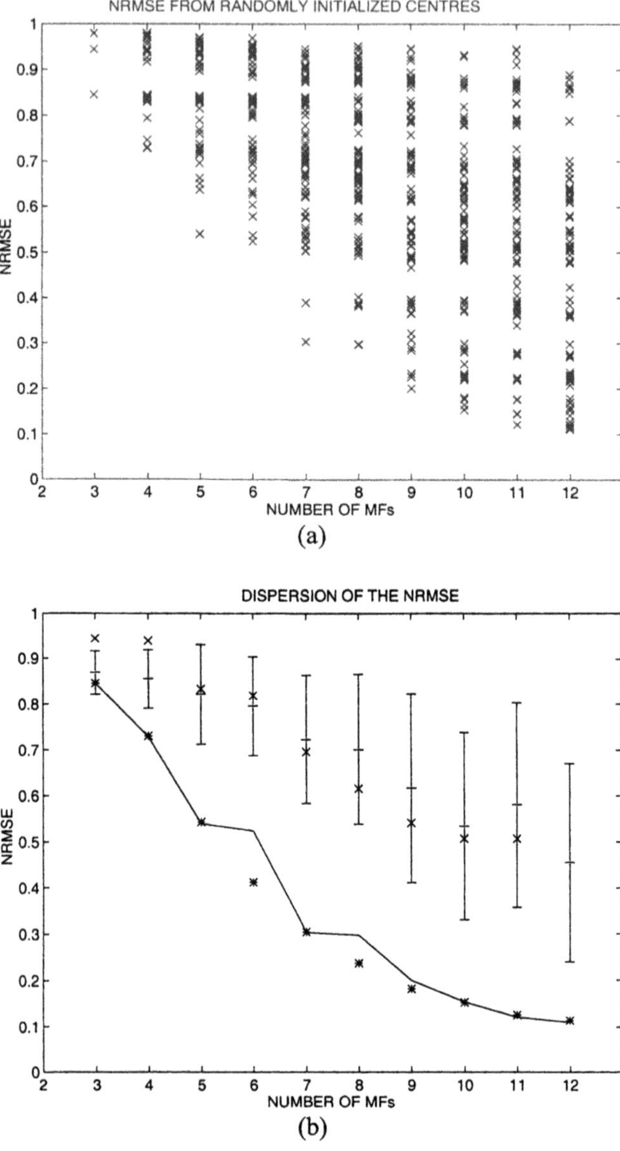

Fig. 5. (a) NRMSE obtained from 100 randomly configurations after gradient step. (b) Comparison with the proposed algorithm.

Table 1. Analysis of the evolution of NRMSE depending on the starting point where the gradient descent is initiated.

#MFs	NRMSE (average)	Dispersion	NRMSE (minimum)	NRMSE (equidistribution)	NRMSE (algorithm)
3	0.869	0.048	0.845	0.945	0.845
4	0.855	0.064	0.728	0.940	0.730
5	0.822	0.110	0.539	0.832	0.542
6	0.796	0.108	0.523	0.818	0.412
7	0.723	0.140	0.303	0.696	0.304
8	0.701	0.163	0.297	0.615	0.237
9	0.617	0.205	0.200	0.541	0.182
10	0.534	0.204	0.154	0.507	0.153
11	0.581	0.222	0.121	0.507	0.126
12	0.455	0.215	0.110	0.114	0.114

Several conclusions can be drawn from the above results:

- The use of membership functions that are equally-distributed throughout the input range is not a good solution. Except in the case of the 12 membership functions that, by chance, form a good starting point, the solution is practically equivalent to a random initialization.
- Neither by the use of a reasonably large number of random initial configurations is it possible to assert that one of them is a valid starting point to find an absolute minimum. In various cases, the approximation determined by the complete algorithm (using the prior stage) was found to be superior to the best of the random searches.
- By using the proposed initial stage we obtained an approximation that was always similar to the best of the random approximations, and on occasions, better. Nevertheless, there are configurations for which the best random approximation is slightly better than that obtained by the proposed algorithm. As stated in the previous section, the algorithm cannot guarantee to find the absolute minimum, although it always reaches a "good" one.

5 Alteration of the fuzzy system topology: system identification

At this point, we have finally obtained a near optimum configuration (location of the membership functions and rules) to approximate our target function for a fixed number of linguistic values. But an additional factor must now be examined: how to choose the number of membership functions for each input variable. In other words, if we want to improve the accuracy of the fuzzy system by inserting a new membership function in one of the available input variables, in which one should we place it?

Let us suppose that we want to approximate the function given by

$$F(x_1,x_2) = 1.3356 \cdot \begin{Bmatrix} 1.5(1-x_1) + e^{(2x_1-1)} \, \boldsymbol{sin}\left(3\pi(x_1-0.6)^2\right) + \\ e^{3x_2-0.5} \, \boldsymbol{sin}\left(4\pi(x_2-0.9)^2\right) \end{Bmatrix} \tag{17}$$

with $x_1, x_2 \in [0,1]$ (see Fig. 6a) using about 25 rules for its approximation. Within this range, we could use a configuration of 6 membership functions defined for x_1 and 4 for x_2, i.e., a 6x4 configuration, obtaining an approximation error (NRMSE) of 0.364 (Fig. 6b). Alternatively, we could try to approximate the function using a 5x5 configuration whose NRMSE equals to 0.254 (see Fig. 6c), which is better than the previous one but worse than the 4x6 configuration, with which an NRMSE of 0.104 could be obtained (Fig. 6d).

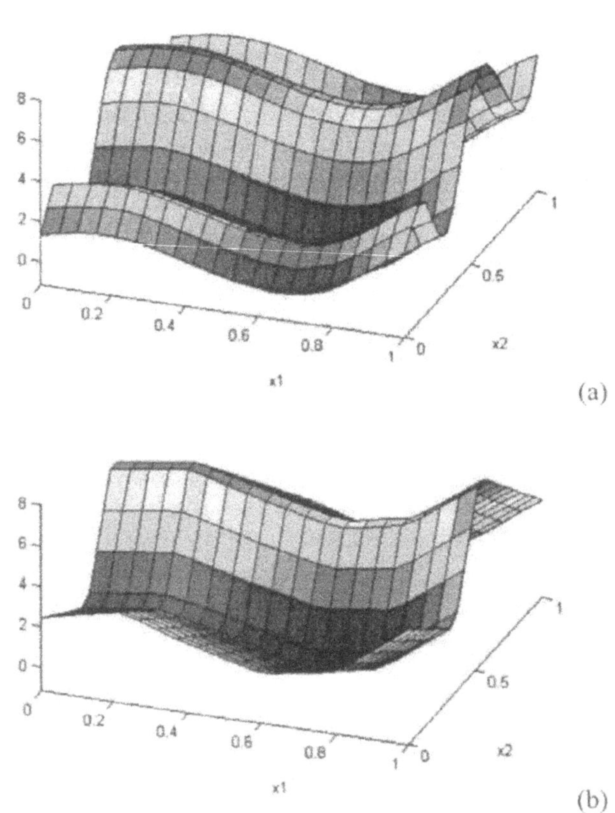

(a)

(b)

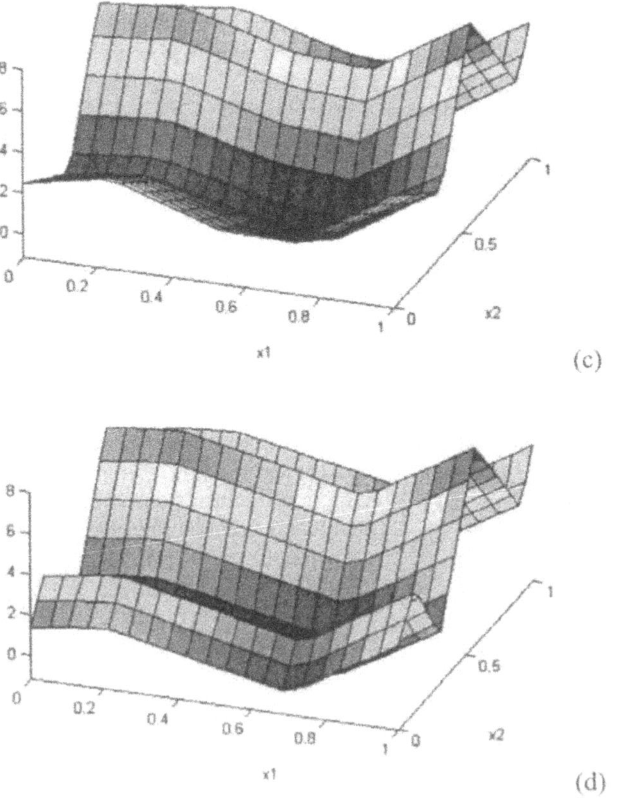

Fig. 6. a) Original function b) Approximation using a 6x4 configuration (NRMSE = 0.384) c) Approximation using a 5x5 configuration (NRMSE = 0.254) d) Approximation using a 6x4 configuration (NRMSE = 0.104)

In all these examples, complexities of the fuzzy systems are similar but this is not the case when looking at their accuracy in approximating our target function. As it is apparent, the determination of the fuzzy system structure is *fundamental* for a correct approximation accuracy. This is the issue we are going to tackle in this section.

The main goal of this part of the algorithm is to [42]:

- Select the input variables from a set of input candidates that are truly significant within the approximation requested.
- Find an optimum partition of the input space and, consequently, the optimum number of rules that are necessary to attain a certain approximation accuracy.

This problem is very complex. For the first point, the only method described in the literature uses a combinatorial tree, i.e., basically it constructs a whole optimized fuzzy system with every possible combination of input candidates (which is very tedious and time consuming). For the second one, clustering techniques have been commonly used but the condition for determining the number of clusters does not generally depend on the approximation error, is very arbitrary and does not take into account the interpolative properties of the system (and so, it is better suited for classification problems).

When we have I/O data, the input candidates are the input variables used in the I/O samples. It may occur that not all of them are significant in the model or, at least, are not necessary to attain a certain approximation accuracy. In this work, points 1 and 2 are tackled jointly using the same methodology. The procedure is capable of selecting which of the input candidates should increase its number of membership functions in order to diminish the approximation error more rapidly. If the selected input candidate has only 1 membership function defined, the decision is equivalent to considering this input candidate as one of the input variables in the fuzzy system. Therefore, the whole problem of structure determination is solved.

To approach this issue, the error distribution for each variable is examined. Let us consider, in a first approach, the ideal (continuous) case:

Let us start by inspecting input variable x_v. For this purpose, we split the domain of the remaining variables into infinitesimal intervals of width *dx*

$$\begin{aligned}&[x_1, x_1 + dx_1] \times \cdots \times [x_{v-1}, x_{v-1} + dx_{v-1}] \times [D_v^-, D_v^+] \times \\ &[x_{v+1}, x_{v+1} + dx_{v+1}] \times \cdots \times [x_N, x_N + dx_N]\end{aligned} \tag{18}$$

where D_v^-, D_v^+ are the lower and upper limits of the domain of variable v, and we consider each one of them individually as one-dimensional subfunctions of variable x_v using the MFs already defined in that variable. As this is a fixed configuration, and the problem is linear, the approximation process can be done in a straightforward manner, and a one-dimensional function is obtained:

$$F^v_{x_1,\ldots,x_{v-1},x_{v+1}\ldots,x_N}(x_v) \tag{19}$$

this approximates the error surface in the region considered using the MFs of variable x_v and without the influence of the ones defined in the other variables. As stated above, if function (19) is capable of accurately approximating the error surface in its definition domain, we conclude that, as far as this region is concerned, there is no need to insert a new membership function into variable x_v. Conversely, if this approximation is poor then there is a high degree of responsibility of variable x_v in the error existing in that region.

By adding up all the approximation errors with respect to the error surface for each infinitesimal region, we can compute an index reflecting the degree of responsibility of variable x_v in the existing global error:

$$J_\nu = \int_{X_1} dx_1 \cdots \int_{X_{\nu-1}} dx_{\nu-1} \int_{X_{\nu+1}} dx_{\nu+1} \cdots \int_{X_N} dx_N \left[\int_{X_\nu} dx_\nu \left(e(\vec{x}) - F^{\nu}_{x_1,\ldots,x_{\nu-1},x_{\nu+1},\ldots,x_N}(x_\nu) \right)^2 \right] \tag{20}$$

where $e(\vec{x})$ is the error surface, i.e., the error between the original function and the one approximated by the current fuzzy system.

Due to the fact that we will always have a discrete set of I/O data, the above expressions must be translated into a more tractable (discrete) form. To accomplish this, the infinitesimal intervals must be discretized. Thus, expression (18) now reads

$$[x_1, x_1 + \Delta x_1] \times \cdots \times [x_{\nu-1}, x_{\nu-1} + \Delta x_{\nu-1}] \times [D_\nu^-, D_\nu^+] \times [x_{\nu+1}, x_{\nu+1} + \Delta x_{\nu+1}] \times \cdots \times [x_N, x_N + \Delta x_N] \tag{21}$$

and we consider all the points of the error surface within this region as having the same input values in all variables except variable x_ν thus obtaining a one-dimensional function of x_ν. Obviously, the smaller the width of the region the better the similarity between the continuous and discrete cases.

Finally, we compute the approximation index (20) (but now replacing integral signs by summation signs) for each input variable and compare; the input variable with highest J_ν is the one selected to increment the number of membership functions which will be homogeneously distributed as the starting point for the next stage of the algorithm. If we define

$$J_{max} = \max_{\nu}(J_\nu) \tag{22}$$

then the set of input variables selected can be given by:

$$\{x_\nu / J_\nu \geq \rho \cdot J_{max}, \nu = 1 \ldots N\} \tag{23}$$

with ρ being a number in the range [0,1]. If we chose $\rho = 0.1$, then we would also include those variables whose index J_ν was within 10% of the maximum value.

For instance, in the previous example, if we split the input domain of each variable into 20 regions, the evolution of the normalized root-mean-square error with the number of fuzzy rules is presented in Table 2. The values of the J indices are also shown in this table together with the NRMSE obtained in case of selecting the reverse decision. From the table it can be seen that the algorithm always select the best choice in each case and, what is not less important, the algorithm is capable of starting with a initial fuzzy system with no rules at all.

Table 2. Evolution of NRMSE with the number of MFs, obtained automatically by the algorithm starting from an initial fuzzy system with no rules. Running time of the whole process: 6 min. 10 sec. on a Pentium II, 450MHz

Config.	NRMSE	J_1	J_2	NRMSE reverse decision
1x1	1.000	0.380	0.919	
1x2	0.939	0.380	0.850	2x1: 1.000
1x3	0.783	0.379	0.677	2x2: 0.939
1x4	0.544	0.380	0.375	2x3: 0.783
2x5	0.461	0.376	0.245	
3x5	0.278	0.114	0.245	2x6: 0.394
3x6	0.146	0.113	0.090	4x5: 0.257
4x6	0.104	0.052	0.089	3x7: 0.135
4x7	0.088	0.052	0.068	5x6: 0.098
4x8	0.076	0.052	0.067	5x7: 0.076
4x9	0.062	0.052	0.032	5x8: 0.075
5x9	0.041	0.027	0.031	4x10: 0.062

For example, it can be seen that for an approximation accuracy of 0.5, the first of the input variables could be considered as a perturbation noise. Nevertheless, if we want to obtain more accurate approximations, this variable should start to be taken into account and, as the algorithm evolves, we obtain fuzzy systems more precise but also more complex. Besides, as expected, the configuration reached with the complexity of about 25 rules is the 4x6 configuration which, as previously noted, is much better than the 6x4 or the 5x5 configurations.

The proposed method has two main advantages (apart from its validity to accomplish its task):

1. This method is independent of the number of membership functions defined in each variable. It can also deal with the case when only one membership function is defined in an input variable (which is equivalent to not taking this variable into account in the fuzzy system). In this case, every one dimensional function is approximated by a constant when analyzing this variable.
2. The type of membership function defined in each input variable is not a problem for the method since it approximates all one-dimensional subfunctions using the MFs already defined in the analyzed variable (which can be triangular, gaussian, trapezoidal, etc.).

6 Fuzzy Evaluation of the Configurations Obtained by the Algorithm

To evaluate the fuzzy system obtained, we have used the error approximation criterion, but to take into account the parsimony principle, that is, the number of parameters to be optimized in the system, a new term to describe the complexity of the derived fuzzy system must be added. Therefore, once the principal algorithm has ended, it is of great relevance to determine which configuration of membership functions (and thus of rules) should be selected to approximate the function, from the possible configurations provided by the algorithm. If we intend the data to be accurately approximated by the fuzzy rules, from the set of obtained configurations the one providing the smallest mean squared error should be taken as the solution. However, this configuration will generally need a high number of rules and membership functions for the input variables. If, on the contrary, we want the fuzzy system to be the simplest possible, even when the approximation error is moderately high, we will choose a solution where the total number of rules is small. In general, the best model will be the one giving the smallest error with the lowest number of parameters (i.e. the one presenting least structural complexity). There must exist a compromise between the accuracy we are looking for and the consequent complexity of the system.

One solution to this problem can be found by using the Minimum Description Length [36]. The description length of a model is defined as the sum of the quantity of data needed to describe it and the quantity needed to describe the error between this model and reality. The problem of selecting from different fuzzy systems the one presenting the best results, while simultaneously taking into account the goals of accuracy and simplicity, also arises in optimization problems where more than one solution can be obtained, i.e. multi-objective optimization or Pareto optimization [12].

Since determining which configuration has optimum characteristics is a clear example of fuzzy decision-taking, in this work the preferences of the human operator (end user) concerning the two objectives considered in evaluating the fuzzy system are translated into fuzzy rules. Thus, to approach this problem, instead of using classical indices such as the Akaike Information Criterion and the Minimum Description Length [2, 36], we propose a fuzzy system responsible for selecting the most adequate structure from the different solutions provided by the algorithm, taking into account the compromise between the accuracy of the approximation of the fuzzy system and its complexity (measured as the number of rules). The output of this fuzzy system, defined in the interval [0,10] in this work, is the so-called *Index of Accuracy of the approximation versus system Complexity* (IAC). In this way, the human operator expresses the degree of compromise between the accuracy of the system and its complexity by means of fuzzy rules of the following type:

IF Root Mean Squared Error is Small AND Number of Parameters is Big THEN IAC is Medium

These depend on the particular specifications of the problem to be resolved. It must be pointed out that one given configuration may seem suitable for certain applications and not for others. Therefore, it is the final user that must provide the fuzzy rules to construct the IAC index. A simple example of rules to define this index is presented in Table 3. Since this rules are merely an example, the output of the IAC system will be considered here as an impartial information of which configurations seem to have a better trade-off between accuracy and complexity. Therefore the best configuration will be the one that maximizes the IAC index.

Table 3. Possible rules to obtain the IAC index (Z = Zero, VS = Very Small, S = Small, M = Medium, B = Big, VB = Very Big.).

NRMSE \ #Parms	Z	S	M	B
Z	VB	VB	B	M
S	VB	B	M	S
M	M	S	VS	Z
B	VS	Z	Z	Z

Fig. 7 reflects the IAC index obtained for the different configurations obtained in the example presented in the previous section.

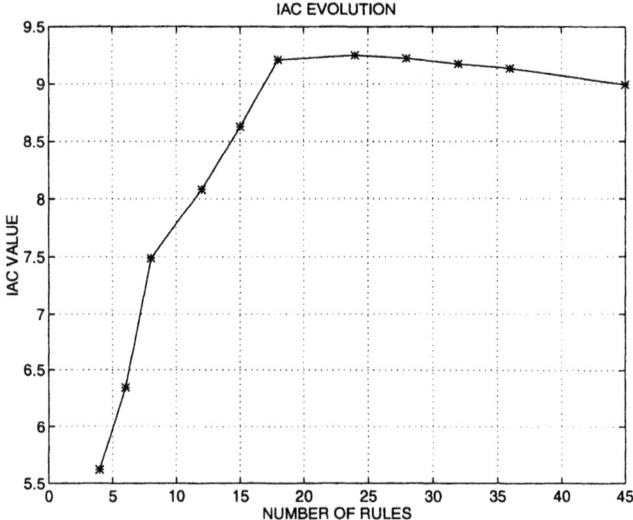

Fig. 7. Evolution of the IAC index for the different fuzzy systems obtained automatically by the algorithm

According to the figure, the IAC index considers the 4x6 structure as the most successful one considering the trade-off between accuracy and complexity for this example using the rules of Table 3.

7 Some Comparisons with Other Approaches

Finally, the results of the proposed self-generating fuzzy rule-table algorithm are compared with other methods for the direct synthesis of fuzzy systems using triangular membership functions reported in the bibliography. Table 4 presents various functions of 1 and 2-input variables used by different authors as examples to test their proposed algorithms, together with the respective error indices. In each of the functions presented, the authors introduce alternative methodologies which produce different error indices. T. Sudkamp et al. [41] examine different algorithms for constructing fuzzy rules from input-output training data, and propose a theory of completion and interpolation for extending the training information to the entire rule base. They analyze the learning algorithm proposed by Wang et al. [47] and introduce modifications to improve it. In order to compare the behavior of the proposed algorithm, not only concerning the approximation error obtained but also its structural complexity (number of rules), different configurations of the fuzzy systems obtained by our methodology are presented. On the right of the table, the error indices produced by our methodology for the test data set are presented for different structures of fuzzy systems *(in bold type the structure with highest IAC)*. It is important to note that not only is the number of parameters needed to define the fuzzy system (i.e. number of rules and membership functions) considerably reduced compared with other approaches, but also that the performance of the fuzzy system obtained is greatly improved.

8 Conclusion

In this chapter, we have proposed a new learning method to automatically obtain the structure of a fuzzy system and derive fuzzy rules and membership functions from a given set of training data, taking into account the trade-off between complexity and accuracy, but at the same time seeking to achieve fuzzy systems that are both transparent and easily understood by a human operator.

The algorithm presented here is able to recognize where it is necessary to assign a larger number of rules and to increase the density of membership functions in a specific input variable for an accurate approximation of the target function. The proposed approach to function approximation automatically determines the structure and parameters of the fuzzy system, while knowledge acquisition or training is very fast.

The approximation accuracy of the fuzzy system derived by the presented method has been empirically compared with the results of other fuzzy-rule generating methodologies. These comparisons with previous studies reveal that the proposed methodology not only obtains better approximation results, but also provides fuzzy systems that, in terms of the number of rules and membership functions, are less structurally complex; therefore fuzzy systems can be more easily interpreted and understood.

Table 4. Comparison of the proposed algorithm with other methods for direct synthesis of fuzzy systems

Function	Proposed in the bibliography					Our Approach			
	Algorithm		N° of Rules	Performance index: Ave. Error	Performance index: Max. Error	N° of Rules	NRMSE	Ave. Error	Max. Error
$F(x)=x^3$	T.Sudkamp & R.J.Hammell [41]	Region growing	15	0.044	0.193	4	0.080	0.026	0.075
			25	0.021	0.083				
		Weighted average	15	0.044	0.193				
			25	0.021	0.083	**6**	**0.032**	**0.011**	**0.032**
	L. Wang & J.M. Mendel [47]		15	0.029	0.079				
			25	0.018	0.088				
	Improvement of L.Wang [41]		15	0.079	0.071	8	0.017	0.006	0.016
			25	0.088	0.050				
$F(x)=\sin(2\pi\cdot x)$	T.Sudkamp & R.J.Hammell [41]	Region growing	15	0.131	0.408	**6**	**0.112**	**0.068**	**0.217**
			25	0.052	0.167				
		Weighted average	15	0.180	1.051				
			25	0.082	0.759	8	0.0823	0.047	0.162
	L. Wang & J.M. Mendel [47]		15	0.060	0.159				
			25	0.026	0.092				
	Improvement of L.Wang [41]		15	0.071	0.179	10	0.0237	0.026	0.078
			25	0.031	0.083				
$F(x_1,x_2)=x_1^2+x_2^2-1$	T.Sudkamp & R.J.Hammell [41]	Region growing	15x15	0.038	0.286	5x5	0.0617	0.0225	0.080
			25x25	0.035	0.250				
		Weighted average	15x15	0.060	1.286				
			25x25	0.148	1.333	**6x6**	**0.0394**	**0.0144**	**0.053**
	L. Wang & J.M. Mendel [47]		5x5	0.107	0.5				
			15x15	0.034	0.143				
	Improvement of L.Wang [41]		5x5	0.154	0.4	7x7	0.0273	0.0099	0.035
			15x15	0.03	0.143				

References

[1] S.Abe and M.S.Lan. Fuzzy rules extraction directly from numerical data for function approximation. *IEEE Trans. Syst. Man and Cyber.*, 25(1):119-129, 1995.

[2] H. Akaike. *Information theory and extension of the maximum likelihood principle.* Akademia Kiado, Budapest, pages 267-810, 1973.
[3] S.Barada and H.Singh. Generating optimal adaptive fuzzy-neural models of dynamical systems with applications to control. *IEEE Trans. Syst. Man and Cyber., Part C*, 28(3):371-391, 1998.
[4] J.C.Bezdek. *Pattern Recognition with Fuzzy Objective Function Algorithms.* Plenum Press, New York, 1981.
[5] J.R.Boston. Effects of membership function parameters on the performance of a fuzzy signal detector. *IEEE Trans. Fuzzy Systems*, 5(2):249-255, 1997.
[6] D.S.Chen and R.C.Jain. A robust back propagation learning algorithm for function approximation. *IEEE Trans. Neural Networks*, 5(3):467-479, 1994.
[7] V.Cherkassky, D.Gehring, and F.Mulier. Comparison of adaptive methods for function estimation from samples. *IEEE Trans. Neural Networks*, 7(4):969-984, 1996.
[8] C.K.Chiang, H.Y.Chung, and J-J.Lin. A self-learning fuzzy logic controller using genetic algorithms with reinforcements. *IEEE Trans. Fuzzy Systems*, 5(3):460-467, 1997.
[9] M.R.Cinvalar and H.J.Trussell. Constructing membership functions using statistical data. *Fuzzy Sets and Systems*, 18:1-14, 1986.
[10] J.A.Dickerson and B.Kosko. Fuzzy Function Approximation with Ellipsoidal Rules. *IEEE Trans. Syst. Man and Cyber.- Part B*, 26(4):542-560, 1996.
[11] D.Driankov, H.Hellendoorn, and M.Reinfrank. *An introduction to fuzzy control.* Springer-Verlag, 1993.
[12] C.M. Fonseca and P.J. Fleming. Multiobjective optimization and multiple constraint handling with evolutionary algorithms -- part I: A unified formulation. *IEEE Trans. Syst. Man and Cyber.- Part A*, 28(2):26-37, 1998.
[13] J.H.Friedman. An overview of predictive learning and function approximation. In *From Statistics to Neural Networks: Theory and Pattern Recognition Applications*, volume 136. V.Cherkassky, J.H.Friedman, and H.Wechsler, Eds. Berlin: Springer-Verlang, NATO ASI Series F, 1994.
[14] M.Funabashi and A.Maeda. Fuzzy and neural hybrid expert systems: synergetic AI. *IEEE Expert*, pages 32-40, 1995.
[15] M.M.Gupta and J.Qi. Design of fuzzy logic controllers based on generalized T-operators. *Fuzzy Sets and Systems*, 40:473-489, 1991.
[16] C.M.Higgins and R.M.Goodman. Fuzzy rule-based networks for control. IEEE Trans. Fuzzy Systems, 2(1):82-88, 1994.
[17] A.Homaifar and E.McCormick. Simultaneous design of membership functions and rule sets for fuzzy controllers using genetic algorithms. *IEEE Trans. Fuzzy Systems*, 3(2):129-139, 1995.
[18] T.Hong and C.Lee. Induction of fuzzy rules and membership functions from training examples. *Fuzzy Sets and Systems*, 84:33-37, 1996.
[19] H.Ishibuchi, R.Fujioka, and H.Tanaka. Neural Networks that learn from Fuzzy If-Then Rules. *IEEE Trans. Fuzzy Systems*, 1(2):85-96, 1993.
[20] J.S.R.Jang, C.T.Sun, and E.Mizutani. *Neuro-Fuzzy and soft computing.* Prentice Hall, ISBN 0-13-261066-3, 1997.
[21] J.S.R.Jang and C.T.Sun. Neuro-fuzzy modelling and control. *Proc. IEEE*, 83(3):378-405, 1995.
[22] J.S.R.Jang. Anfis: Adaptive-network-based fuzzy inference systems. *IEEE Trans. Syst. Man and Cyber.*, 23:665-685, 1993.
[23] T-Y.Kwok and D-Y.Yeung. Objective functions for training new hidden units in con-

structive neural networks. *IEEE Trans. Neural Networks*, 8(5):1131-1148, 1997.
[24] R.Langari and L.Wang. Complex systems modelling via fuzzy logic. *IEEE Trans. Syst., Man, Cybern., Part B*, 26:100-106, 1996.
[25] C.C.Lee. Fuzzy logic in control systems: fuzzy logic controller - Part I,II. *IEEE Trans. Syst. Man and Cyber.*, 20(2):404-435, 1990.
[26] W.Li. A method for design of a hybrid neuro-fuzzy control system based on behavior modeling. *IEEE Trans. Fuzzy Systems*, 5(1):128-134, 1997.
[27] S.Liu and S.Hu. A method of generating control rule model and its application. *Fuzzy Sets and Systems*, 52:33-37, 1992.
[28] A.Lotfi, H.C.Andersen, and A.C.Tsoi. Matrix formulation of fuzzy rule-based systems. *IEEE Trans. Syst. Man and Cyber.- Part B*, 26(2):332-339, 1996.
[29] A.Lotfi and A.C.Tsoi. Importance of membership functions: a comparative study on different learning methods for fuzzy inference systems. *Proc. Third IEEE Int. Conf. Fuzzy Systems*, Orlando (FL), pages 1791-1796, 1994.
[30] V.S.Manorajan, A.S.Lazaro, D.Edwards, and A.Athalye. A systematic approach to obtaining fuzzy sets for control systems. *IEEE Trans. Syst. Man and Cyber.*, 25(1):206-213, 1995.
[31] H.Nomura, I.Hayashi, and N.Wakami. A learning method of fuzzy inference rules by descent method. *Proc. of IEEE Int. Conf. Fuzzy Systems*, pages 203-210, 1992.
[32] D.Park, A.Kandel, and G.Langholz. Genetic-Based new fuzzy reasoning models with application to fuzzy control. *IEEE Trans. Syst. Man and Cyber.*, 24(1):39-47, 1994.
[33] H.Pomares, I.Rojas, J.González, and A.Prieto. Structure Identification in Complete Rule-Based Fuzzy Systems. *IEEE Trans. Fuzzy Systems*, to appear.
[34] H.Pomares, I.Rojas, J.Ortega, J.Gonzalez, and A.Prieto. A Systematic Approach to a Self-generating Fuzzy Rule-table for Function Approximation. *IEEE Trans Syst., Man and Cyber. Part B*, 30(3):431-447, 2000.
[35] W.H.Press, B.P.Flannery, S.A.Teukolsky, and W.T.Vetterling. *Numerical Recipes in C.* Cambridge University Press, ISBN: 0 521 43720 2, 1993.
[36] J.Rissanen. Modelling by shortest data description. *Automatica* 14:464-471, 1978.
[37] I.Rojas, J.Ortega, F.J.Pelayo, and A.Prieto. Statistical analysis of the main parameters in the fuzzy inference process. *Fuzzy Sets and Systems*, 102:157-173, 1999.
[38] I.Rojas, J.J.Merelo, J.L.Bernier, and A.Prieto. A new approach to fuzzy controller designing and coding via genetic algorithms. *IEEE Int. Conf. Fuzzy Systems*, Barcelona, pages 1505-1510, 1997.
[39] R.Rovatti and R.Guerrieri. Fuzzy sets of rules for system identification. *IEEE Trans. Fuzzy Systems*, 4(2):89-102, 1996.
[40] E.H.Ruspini. A new approach to Clustering. *Info Control*, 15:22-32, 1969.
[41] T.Sudkamp and R.J.Hammell. Interpolation, Completion and Learning Fuzzy Rules. *IEEE Trans. Syst. Man and Cyber.*, 24(2):332-342, 1994.
[42] M.Sugeno and T.Yasukawa. A fuzzy-logic based approach to qualitative modelling. *IEEE Trans. Fuzzy Syst.*, 1(1), 1993.
[43] M.Sugeno and G.Kang. Structure identification of fuzzy model. *Fuzzy Sets and Systems*, 28:15-33, 1988.
[44] M.Sugeno and K.Tanaka. Successive identification of a fuzzy model and its application to prediction of complex systems. *Fuzzy Sets and Systems*, 41:315-334, 1991.
[45] H.Surmann, A.Kanstein, and K.Goser. Self-organizing and genetic algorithms for an automatic design of fuzzy control and decision systems. *Proc. EUFIT 93*, pages 1097-1104, 1993.

[46] T.Takagi and M.Sugeno. Fuzzy identification of systems and its applications to modelling and control. *IEEE Trans. Syst. Man and Cyber.*, 15:116-132, 1985.
[47] L.X.Wang and J.M.Mendel. Generating fuzzy rules by learning from examples. *IEEE Trans. Syst. Man and Cyber.*, 22(6):1414-1427, 1992.
[48] L.-X.Wang and J.M.Mendel. Fuzzy basis functions, universal approximation, and orthogonal least square learning. *IEEE Trans. Neural Networks*, 3:807-814, 1992.
[49] R.R.Yager and D.P.Filev. Unified structure and parameter identification of fuzzy models. *IEEE Trans. Syst. Man, Cybern.*, 23:1198-1205, 1993.
[50] R.R.Yager. Quantifier guided aggregation using OWA operators. *Int. J. Int. Syst.*, 11:49-73, 1996.
[51] Z.Zhang, M.Mizumoto. On rule self-generating for fuzzy control. Int. J. Int. Syst., vol.9, :1047-1057, 1994.

Using Individually Tested Rules for the Data–based Generation of Interpretable Rule Bases with High Accuracy

Timo Slawinski[1], Peter Krause[1], and Harro Kiendl[1]

Department of Control Engineering, University of Dortmund,
Otto-Hahn-Str. 4, 44221 Dortmund, Germany.
E-mail: timo@slawinskis.de, {krause, kiendl}@esr.e-technik.uni-dortmund.de

Abstract. Fuzzy–modeling is predominately applied, if transparent and interpretable results are desired. However, especially for data–based approaches it turns out that interpretability and high accuracy are often partly contradictory modeling objectives. Considering the interpretability the complexity of the final rule base, i.e. the number and structure of the rules as well as the number and shape of the membership functions, is a critical factor. Obviously, a high accuracy often comes along with a high complexity of the rule base. Furthermore, many of the common approaches are based on optimizing the rule base as a whole only with respect to the modeling error. Therefore it cannot be guaranteed that each individual rule is locally reasonable. Alternative approaches, based on optimizing or testing individual rules, generally do not reach the aimed accuracy and can lead to large and redundant rule bases. We overcome these drawbacks by new concepts presented in this contribution: They allow to improve the accuracy and to reduce the size of rule bases, consisting of individually tested rules. The applicability of the approaches is demonstrated for various tasks of different complexity. It turns out, that with the new concepts a data–based generation of small interpretable rule bases with high accuracy is possible even for high dimensional problems.

Keywords: *Fuzzy Modeling, data–based rule generation, Fuzzy–ROSA method, feedback of modeling quality, rule reduction, conversion in TSK fuzzy models*

1 Introduction

The use of fuzzy modeling is often motivated by the interpretability of the resulting model. Fuzzy models consist of qualitative *if–then* rules, which corresponds to the way human knowledge is usually presented. The main advantages of interpretability are a higher acceptance by the experts and an improved possibility to adapt the model by hand. Furthermore, in the case of data–based fuzzy modeling interpretability allows additional insight into the considered system. This is especially important for complex systems and if only few prior knowledge is available.

The applicability of different methods for data–based rule generation has been demonstrated impressively in numerous publications. However, many of the authors consider only the obtained accuracy and do not discuss the interpretability. From our point of view the interpretability depends mainly on two factors: the complexity of the final rule base and the relevance of the individual rules.

Essential for the complexity are the number and the structure of the rules. In praxis fuzzy systems, designed by experts through knowledge extraction, usually do not consist of more than about hundred rules. Accordingly, rule bases with a higher number of rules have to be considered as very complex. In this context it is favorable to use generalizing rules, which do not consider all linguistic variables in the premise. Such a generalizing rule which is composed of fewer expressions than the total number of input variables has the advantage to covers several linguistic input situations. In particular, for high dimensional problems with a large number of input variables, rule bases consisting of generalizing rules turn out to be much smaller than rule bases consisting of specialized or complete rules only. Considering the conclusion part of the fuzzy rules, a linguistic expression (*Mamdani rules*) is better interpretable than a function (*Takagi–Sugeno–Kang rules*). However, latter are more flexible and supply usually a higher accuracy. Also important for the complexity of the final rule base is the total number and shape of the membership functions. A high accuracy is often obtained by providing individual membership functions for each rule. In addition it is not always guaranteed, that the membership functions are convex and normalized. In both cases the complexity increases and the interpretability degrades.

Common approaches for data–based fuzzy modeling are often based on optimizing the rule base as a whole by simply minimizing the modeling error. This approach does not guarantee, that each individual rule is locally reasonable, i.e. relevant. In addition, this approach may lead to a high number of free parameters, e.g. structure and selection of the rules. Due to the curse of dimensionality few of the proposed methods can be applied to high dimensional problems, with more than four or five input variables. To overcome these drawbacks, our approach is based on testing individual fuzzy rules. Only those rules are added to the rule base, which have passed a statistical (relevance) test. Consequently, each accepted rule is reasonable (relevant) in sense of the applied test. Furthermore the problem of finding a good rule base is reduced to the problem of finding good rules. However, the modeling quality and the interaction of the rules is not directly considered in the rule generation process. This may lead to a lower accuracy and to large and redundant rule bases.

In consideration of these difficulties, we present here three concepts for the feedback of the modeling quality in rule generation process of individually tested rules. The general purpose is the generation of small interpretable rule bases with high accuracy even for high dimensional problems. Firstly, we introduce a new iterative search concept. The basic idea is to generate specialized rules only in subspaces with poor modeling quality. Especially for high dimensional problems this specific search often allows to reach the achievable accuracy with much less rules. Secondly, we describe an optimizing conflict reduction based on a genetic algorithm. After the rule search this reduction is applied in order to find the optimal subset of relevant rules. Due to the inconsistencies and redundancies in the initial rule base it is mostly possible to improve the accuracy and simultaneously reduce the number of rules. Furthermore, here the initial rule base consists only of individually tested rules. Consequently the search space is usually much smaller and the final rule base is better interpretable than in other approaches. The last presented concept is especially developed for approximation tasks. By a conversion of the generated rule base in a more flexible Takagi–Sugeno–Kang (TSK) fuzzy model the accuracy can be significantly improved. It has to be pointed out that the conversion preserves the interpretability. Accordingly, the resulting TSK fuzzy model can be interpreted by using the corresponding relevant Mamdani fuzzy rules.

The paper is organized as follows. In Sect. 2 a brief introduction in the data–based generation of individual tested rules with the Fuzzy–ROSA method is given. Afterwards the concepts for a feedback of the modeling quality in the rule generation process are motivated. The new specific search on subspace with poor modeling quality is introduced in Sect. 3. In Sect. 4 the optimizing conflict reduction is described and in Sect. 5 the conversion in interpretable TSK fuzzy models is presented. The results obtained by applying the concepts on several benchmark problems are discussed and compared in Sect. 6. Finally we present our conclusions and give an outlook in Sect. 7.

2 Concepts for Improving the Accuracy

In this Section we present concepts for a feedback of modeling quality in the generation process of individually tested rules. The general aim is to reach a higher accuracy without significant loss of interpretability or if possible to improve both. The concepts were originally developed for the Fuzzy–ROSA method. However, under certain conditions the concepts can also be used for other approaches, based on testing individual fuzzy rules.

2.1 Data–based Generation of Individually Tested Rules with the Fuzzy–ROSA Method (FRM)

The basic idea of the Fuzzy–ROSA[1] method (FRM) is to apply a relevance test to single fuzzy rules to assess their ability to describe a relevant aspect of the system under consideration [1–4]. This reduces the problem of finding good rule bases to the problem of finding relevant rules. As a result the FRM can handle complex applications but does not reach the maximal possible accuracy in general. Moreover, since each rule with a high relevance is supposed

[1] **R**ule–**O**riented **S**tatistical **A**nalysis

to express an important aspect of the system, such rules are meaningful by themselves, which lead to more transparent and comprehensible rule bases. Also important for the interpretability of the final rule base are the number and shape of the membership functions as well as the chosen structure of rules. Furthermore both are factors, which determine the resulting structure of the search space.

Structure of the Search Space In the FRM the membership functions for the linguistic values of the input and output variables must be defined prior to the rule generation process. By this the complexity, i.e. the number of free parameters, is further reduced. In order to enforce the interpretability of the generated rules normalized triangular or trapezoidal membership functions are predominately used in the FRM[2]. For the same reason the FRM only considers fuzzy rules of the Mamdani type [5]:

$$\text{IF } P \text{ THEN } C \quad \text{with} \quad P = e_1 \wedge \cdots \wedge e_c \quad \text{and} \quad C = e_Y \, . \tag{1}$$

The premise P consists of a conjunction[3] of c linguistic expressions e referring to the V input variables $X_1 \ldots X_V$ and the conclusion C is a linguistic expression referring to the output variable Y. The number of linguistic expressions is called combination depth. In the FRM the size of the search space can be restricted by a maximum admitted combination depth c_{max}, which is usually chosen to be smaller than the number V of input variables. The premise of a rule can also be interpreted as a linguistic input situation. If $c = c_{max} = V$, i.e. each linguistic variable is considered, we speak of a complete input situation and rule, respectively. Otherwise, if $c < c_{max} \leq V$, we speak of an incomplete or generalizing input situation and rule, respectively. After the project definition, i.e. the defining of the membership function and the choosing of the maximum combination depth c_{max}, the corresponding search space, i.e. the set of all potential rules (hypotheses) [2,6], is specified.

Rule Generation Process The basic task of the rule generation process is to set up the potential rules (hypotheses) and apply a statistical relevance test to the hypotheses. Only rules that pass this test are accepted as relevant fuzzy rules. To meet different modeling goals, several rule test and rating strategies have been developed [4,2]. According to the original relvance test [1,3], a rule is relevant if the constrained probability $p(C|P)$ of its conclusion C for a given premise P exceeds the unconstrained probability $p(P)$, where

[2] In contrast to the more general view in [3,5] we do not consider here negated statements in the conclusion (negative rules).

[3] To enforce the interpretability combinations of linguistic expressions, that refer to the same linguistic variable, are usually prohibit (combination restriction). For simplification negated statements in the premise are not explicit considered in the following (complement restriction).

these probabilities are estimated based on the given data sets. In order to consider the statistical credibility, confidence intervals are calculated to a given confidence level and used instead of the estimated probabilities. As a result of the test, each rule is given a rating index, which can be interpreted as a certainty factor. The rule base is generated by iteratively collecting relevant rules. Depending on the size of the search space, i.e. total number of hypotheses, a *complete search*, an *evolutionary search* or a combination of both is applied. The complete search guarantees, that all relevant rules are found. If a complete search would lead to not acceptable computing times, we usually apply the evolutionary search described in [7].

Rule Reduction and Analysis The final steps of the FRM is the rule reduction and the analysis of the rule base. In order to obtain small and transparent rule bases two types of subsequent rule reductions are available [8]. *Online reduction* aims at preventing redundant rules. It is applied when a new rule is found and added to the rule base. Rules are considered as redundant in the rule base if the latter contains a more general rule with a higher rating index and the same conclusion. The following *offline reduction* strategies can be executed after the rule generation process: The *situation/structure-based conflict reduction* deletes all rules that do not belong in any complete/generalizing input situation to the best rules (with the highest rating indices). This guarantees that the reduced rule base covers as many input situations as it did before. In contrast, the *data-based rule reduction* removes rules, which are mainly based on data points, that also supports better rules. This can be done very efficiently, but the coverage is no longer be conserved in general. Finally, an analysis of the obtained rule base allows to rate the modeling process and to provide feedback that can be used to modify the problem formulation. However, the overall performance of the generated rule base does not affect directly the rule generation process so far.

2.2 Feedback of Modeling Quality in the Generation Process

In this Section we introduce the concepts for a feedback of modeling quality in the rule generation process of the FRM. On the left of Fig. 1 the FRM is displayed in its initial form. Depending on certain parameters, the rule search sets up and tests individual rules. All relevant rules are collected in the rule base. The overall performance of the rule base, generated so far, can be analyzed and assessed in respect of several indicators for the modeling quality. The following indicators are widely used.

Accuracy: An appropriate measure for the accuracy of the rule base is the modeling error. Here we use the mean absolute/square error (MAE/MSE) for approximation tasks and the relative classification error (RCE) for classification tasks.

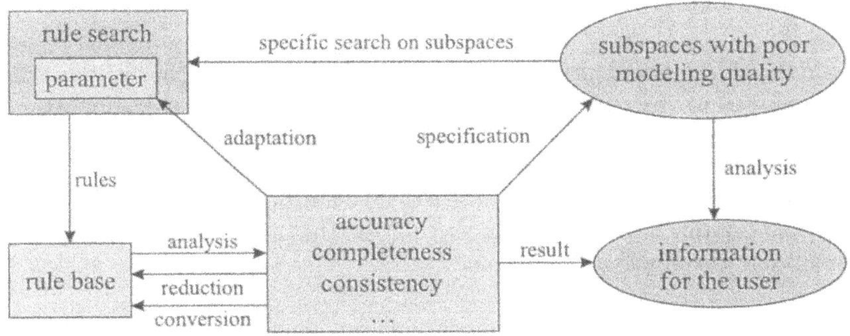

Fig. 1. Concepts for the feedback of modeling quality.

Completeness: Often a complete coverage, i.e. an activation of at least one rule in each input situation, is desired. The coverage rate (number of input situation covered divided by the total number of input situations) is a suitable measure for the completeness of a rule base.

Consistency: Especially in data–based approaches the obtained rule bases can produce multimodal output membership functions. Sometimes this ambiguities are caused by inconsistencies and therefore are hard to interpret. Approaches for the quantification of the consistency of a rule base are presented for example in [9].

Beside the information for the user these indicators can be used for optimization. An example is the *optimizing conflict reduction*, presented in Sect. 4. There a genetic algorithm is used to reduce simultaneously the modeling error and number of rules. Optionally, the coverage with respect to the input situations supported by the learning data can be considered. However, it is quite evident, that so far these indicators only provide a numerical value for the accuracy, completeness or consistency, respectively. Therefore no direct information how to improve the modeling quality is given.

A valuable information concerns the subspace with a poor modeling quality, e.g. those input situations which are not covered by any rule. Latter would allow us to generate exclusively rules in the not covered subspace. Due to this restriction of the search space such a specific generation is generally more efficient and additionally smaller rule bases are obtained. In order to profit from these advantages we have developed and implemented the specific search on subspaces with poor modeling quality (Sect. 3). An essential prerequisite was the design of efficient algorithms and suitable data structures for the specification of the subspace with poor modeling quality.

Especially for approximation tasks a further approach for improving the accuracy was developed. By means of a transformation the Mamdani rules are converted into Takagi–Sugeno–Kang (TSK) rules. As described in Sect. 5 the function in the conclusion of the TSK–rules is identified with respect to the modeling error. In contrast to other TSK–approaches the initial Mamdani rules, which are less complex, can be used for an interpretation of the rule base.

It should be noted that a further possibility for a feedback of the overall performance of the rule base in the generation process is the adaptation of the strategy parameters of the rule search. In the hybrid evolutionary search concept (HESC) presented in [7] a fuzzy system is used for a dynamically adaptation of the strategy parameters of an evolutionary algorithm. Different from the approaches discussed in this paper, the main aim of the HESC was to improve the efficiency of search.

3 Specific Search on Subspaces (SSS) with Poor Modeling Quality

The specific search on subspaces (SSS) described in this section can be divided in two main steps. Firstly, efficient data structures and algorithms are developed for the specification of subspaces with poor modeling quality (Sect. 3.1). Secondly, a rule search is designed for setting up rules covering the specified subspace (Sect. 3.2). It should be noted that this Section is addressed to those, who are interested in the details of the algorithms and data structures. Otherwise it is advised to skip this Section.

3.1 Specification of Subspaces with Poor Modeling Quality

In the following we present two representations of the subspace with poor modeling quality. They were originally developed for the specification of the not covered subspace [10], but as described below they can also be used for the specification of subspaces with poor accuracy.

Representation by Set of Sets (SoS) Considering only premises with a combinations depth $c = 1$, the not covered subspace, i.e. the set $\mathcal{I}_\mathcal{E}$ of complete input situations I which are not covered by any rule can be represented as follows.

Definition 1. (Representation by a set of linguistic expressions) A set of linguistic expressions $\mathcal{E} = \{e_1, \ldots, e_n\}$ represents the set $\mathcal{I}_\mathcal{E}$ of composeable input situations, i.e. an input situation I is in $\mathcal{I}_\mathcal{E}$ iff each linguistic expression in I is in $\mathcal{E}$. Then we also speak[4] of $\mathcal{E}$ is more general ($\geq_g$) than I.

[4] In this sense a set of linguistic expressions $\mathcal{E}_1$ is more general than another set of linguistic expressions $\mathcal{E}_2$, iff $\forall e \in \mathcal{E}_2\ \exists e' \in \mathcal{E}_1 : e = e'$. In contrast the definition of $\geq_g$ for premises/input situations is $P_1 \geq_g P_2 \Leftrightarrow \forall e \in P_1 \exists e' \in P_2 : e = e'$. We use the function *Fill*(P) after Def. 4 for constructing the corresponding $\mathcal{E}_P$ for a given premise P.

Thereby a complete input situation $I = e_1 \wedge \cdots \wedge e_V$ after Eq. 1 is interpreted as set of linguistic expression $\{e_1, \ldots, e_V\}$. If $\mathcal{E} = \mathcal{E}_0$, i.e. it contains all defined linguistic expressions, then the set $\mathcal{I}_0$ of all possible input situations is represented. If for any input variable no linguistic expression is in $\mathcal{I}_\mathcal{E}$ the empty set $\{\emptyset\}$ is represented, i.e. each input situation is covered. We consider now a rule base consisting only of one rule with the premise $P = e_i$. The corresponding not covered subspace $\mathcal{I}_\mathcal{E}$ is represented by $\mathcal{E} = \mathcal{E}_0 \setminus \{e_i\}$. Accordingly for any additional rule in the rule base only the linguistic expression considered its premise must be removed from $\mathcal{E}$. With the objective to permit also rules with a combination depth $c_{max} > 1$ the representation is extended as follows.

Definition 2. (Representation by a set of sets of linguistic expressions) A set $\tilde{\mathcal{E}}_{SoS}$ of sets $\mathcal{E}_a$ of linguistic expressions with $a \in \{1, \ldots, |\tilde{\mathcal{E}}_{SoS}|\}$ represents the set $\mathcal{I}_{SoS}$ of input situations I with $\exists a \in \{1, \ldots, |\tilde{\mathcal{E}}_{SoS}|\} : I \in \mathcal{I}_{\mathcal{E}_a}$ after Def. 1.

According to this definition $\{\mathcal{E}_0\}$ is a suitable representation for the set $\mathcal{I}_0$ of all possible input situations.

Representation by Set of Pairs (SoP) We introduce the idea of this representation by means of a simple example. For this we consider an input space with V input variables and S linguistic values for each input variable. For a given premise[5] $P = e_{11} \wedge e_{21}$ an input situation I is element of the not covered subspace, iff the following condition holds:

$$(e_{11} \in I \Rightarrow \exists j \in \{2, \ldots, S\} : e_{2j} \in I) \wedge (e_{21} \in I \Rightarrow \exists k \in \{2, \ldots, S\} : e_{1k} \in I) .$$

For simplification we use the notation $(I \leq_g \mathcal{E}_B \Rightarrow I \leq_g \mathcal{E}_F)$ for the implication in the following[6]. This condition is a conjunction consisting here of two and in general of c implications. In the antecedent and consequent of each implication an input situation is compared to a set of linguistic expressions. This leads to the following definition.

Definition 3. (Representation by set of pairs) A set $\mathcal{T}_{SoP}$ of pairs of $(\mathcal{E}_B, \mathcal{E}_F)$ with $\mathcal{E}_B, \mathcal{E}_F \subseteq \mathcal{E}_0$ represents the subset $\mathcal{I}_{SoP} \subseteq \mathcal{I}_0$ of input situations, whose elements I hold the following condition:

$$\forall (\mathcal{E}_B, \mathcal{E}_F) \in \mathcal{T}_{SoP} : (I \leq_g \mathcal{E}_B \Rightarrow I \leq_g \mathcal{E}_F) .$$

According to this definition $\{(\mathcal{E}_0, \mathcal{E}_0)\}$ is a suitable representation for the set $\mathcal{I}_0$ of all possible input situations.

[5] The first index of e_{ij} indicates the linguistic variable and the second index the linguistic value.

[6] We use the function $fill(P)$ with P chosen as $\{e_{11}\}$ and $\{e_{22} \ldots, e_{2S}\}$ for the construction of $\mathcal{E}_B$ and $\mathcal{E}_F$, respectively.

Specification of the not Covered Subspace For n given premises P the not covered subspace $\mathcal{I}$ can be specified after the following scheme:

$$\begin{aligned} \mathcal{I} &= \mathcal{I}_0 \,, \\ \mathcal{I} &= \mathcal{I} \setminus \mathcal{I}(P_1) \,, \\ &\vdots \\ \mathcal{I} &= \mathcal{I} \setminus \mathcal{I}(P_n) \,. \end{aligned} \tag{2}$$

Thereby $\mathcal{I}(P)$ denotes the set of input situations covered by the premise P. For the representation $\tilde{\mathcal{E}}_{SoS}$ after Def. 2 a corresponding algorithm (Alg. 1) can be found in Sect. 8. It can be shown, that in this approach for $c_{max} > 1$ the computational effort for the specification increases exponentially with number of rules. In contrast to this the representation $\mathcal{T}_{SoP}$ after Def. 3 has the advantage, that in combination with Alg. 3 the computational effort increases only polynomiell with the number of rules. If as result $\mathcal{T}_{SoP} = \emptyset$ is obtained, each input situation is covered, i.e. $\mathcal{I}_{SoP} = \emptyset$. However, even if $\mathcal{T}_{SoP} \neq \emptyset$ an empty set $\mathcal{I}_{SoP}$ can be represented, because it cannot be guaranteed that for each pair $(\mathcal{E}_B, \mathcal{E}_F) \in \mathcal{T}_{SoP}$ an input situation $I \in \mathcal{I}_{SoP}$ with $I \leq_g \mathcal{E}_B$ really exists[7]. To avoid this a subsequent reduction (analysis) of $\mathcal{T}_{SoP}$ must be applied [10]. In the worst case the computational effort of this reduction increases again exponentially with the number of rules. Even if this approach (without reduction) leads only to a *sufficient* criterion for completeness, it turns out that the completeness can be determined for most of the practical applications considered in [10]. Additionally, the representation $\mathcal{T}_{SoP}$ may be used in any case without the reduction for the specific search on subspaces (Sect. 3.2).

Specification of Subspace with Poor Accuracy In our approach the representation $\tilde{\mathcal{E}}_{SoS}$ after Def. 2 is used for the specification of the subspace with poor accuracy. Thereby the set $\tilde{\mathcal{E}}_{SoS}$ is composed of sets $\mathcal{E}_a$ with a high modeling error ϵ_a. Firstly, all possible premises P_a with certain combination depth are generated. Secondly the corresponding sets $\mathcal{E}_a$ are constructed by using the function *Fill*(P_a) after Def. 4. Finally, the associated error ϵ_a is calculated for the D learning data points $(\boldsymbol{x}_q, y_q)$ as follows:

$$\epsilon_a := \frac{\sum_{q=1}^{D} \mu_{P_a}(\boldsymbol{x}_q) \cdot ||y_q - \hat{y}(\boldsymbol{x}_q)||}{\sum_{q=1}^{D} \mu_{P_a}(\boldsymbol{x}_q)} \,. \tag{3}$$

The defuzzified output value of the fuzzy system which is based on the considered rule base for a given input vector $\boldsymbol{x}$ is denoted by $\hat{y}(\boldsymbol{x})$. For the mean absolute or square error (MAE/MSE) the norm $|y - \hat{y}|$ or quadratic norm

[7] This may be caused by certain combinations of implications in Def. 3, which cannot be fulfilled simultaneously.

$(y-\hat{y})^2$ is used for the evaluation $||\cdot||$ of the difference between y and $\hat{y}$. In classification task we use the relative classification error RCE, with $||\cdot|| = 1$ if $y \neq \hat{y}$ and $||\cdot|| = 0$ else. The subspace $\mathcal{I}_{SoS}$ with poor accuracy is then represented by $\tilde{\mathcal{E}}_{SoS}$ consisting of a chooseable number of sets $\mathcal{E}_a$ with the highest modeling error ϵ_a. Another possible approach, not investigated further, is to choose all sets $\mathcal{E}_a$ with a modeling error ϵ_a higher than a problem specific threshold ϵ_{max}.

3.2 Specific Search on Subspaces (SSS)

In our approach an iterative search concept is applied for the specific search. Starting with a low combination depth a preliminary rule base is generated. Then the subspace with poor modeling quality is specified. In the final step this information is used to generate only those rules with a higher combination depth, which are covering partly the subspace with poor modeling quality.

Set of Rules Covering the Subspace The set $\mathcal{P}$ of premises, which are covering at least one input situation I of the subspace with poor modeling quality $\mathcal{I}$ can be defined as

$$\mathcal{P} := \{P | \mathcal{I}(P) \cap \mathcal{I} \neq \emptyset\} . \tag{4}$$

The basic idea for the construction of $\mathcal{P}$ is to start with generalizing premises $P \in \mathcal{P}$, i.e. in the simplest case with a combination depth $c = 1$. To each generalizing premise (P with $c < V$) in the set $\mathcal{P}$ after Eq. 4 we can add linguistic expressions in such a way that the resulting premise P' is also in $\mathcal{P}$. The set $\mathcal{E}^+$ of permitted linguistic expression for a given premise P is defined as

$$\mathcal{E}^+(P) := \{e_{ij} | (P \cup \{e_{ij}\}) \in \mathcal{P} \wedge \nexists e_{lk} \in P \text{ with } i = l\} . \tag{5}$$

For the representation $\tilde{\mathcal{E}}_{SoS}$ after Def. 2 the set $\mathcal{E}^+(P)$ of permitted linguistic expressions can be constructed as follows. Firstly, the set union of the linguistic expressions of those elements $\mathcal{E}_a \in \tilde{\mathcal{E}}_{SoS}$, which are more general ($\mathcal{E}_a \geq_g P$) than the premise P in the sense of Def. 2, is determined. Secondly, all linguistic expressions are removed, that refer to variables which are considered in the premise P. The construction of $\mathcal{E}^+(P)$ for the representation $\mathcal{T}_{SoP}$ is realized[8] by means of step 3–7 of Alg. 3. Subsequently, also all linguistic expressions must be removed, which refer to variables which are considered in the premise P. For reasons discussed above, it occurs that in this way a superset of $\mathcal{E}^+(P)$ is generated, if the representation $\mathcal{T}_{SoP}$ is used without reduction. It should be noted, that also approaches were developed, for generating rules, which are covering many or exclusive input situations of the subspace with poor modeling quality [10].

[8] For the execution the settings $P_{cur} = P$ and $\mathcal{E}_F^P = \mathcal{E}_0$ must be choosen.

For the complete rule search on subspaces each premise P is be interpreted as a node of a tree search, comparable to the tree–oriented search concept presented in [2]. The corresponding set of branches, starting from this node is given by $\mathcal{E}^+(P)$. In order to guarantee that each node is considered only one time we add only expressions $e_{ij} \in \mathcal{E}^+(P)$ to the premise with regard to the following restriction

$$\forall e_{kl} \in P : k < i \ . \tag{6}$$

Starting from a initial set of generalizing premises or from the node $P = \emptyset$ the branches are built up until the chosen maximum combination depth $c_{max} \leq V$ is reached.

Iterative Search Concept In our approach we use the iterative search concept depicted in Fig. 2 for the specific search on subspaces. Starting with

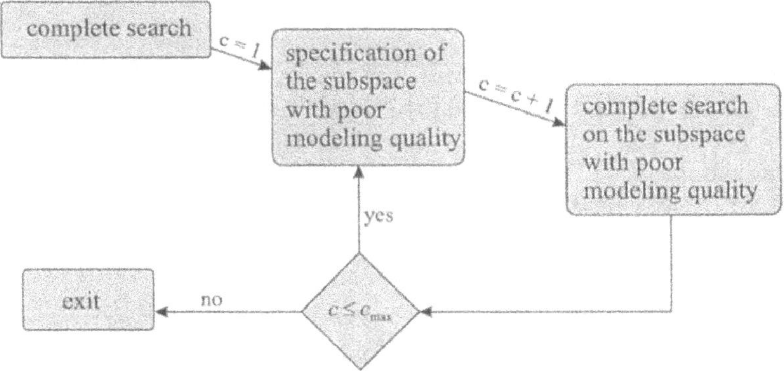

Fig. 2. Specific search on subspaces.

an initial rule set with a combination depth $c = 1$ the subspace of poor modeling is specified as described above. Then the complete search on subspaces is applied to generate exclusively rules, which are covering (partly) the subspaces with poor modeling quality. This two iteration steps are repeated until the chosen maximum combination depth c_{max} is reached. In Sect. 6 we demonstrate, that with the specific search on subspace with poor accuracy a compact rule bases with a high accuracy can be obtained. Due to space limitations we do not present here results for the specific search on the not covered subspaces. As shown in [10] this approach is useful for an efficient generation of a complete rule base.

4 Optimizing Conflict Reduction (OCR)

The aim of optimizing conflict reduction [11] is to reduce the number of rules in consideration of the modeling or classification error. In contrast to other GA–based approaches, in this reduction step of the FRM only the set of all relevant rules and not the set of all possible rules have to be considered. However, to find the optimal subset of relevant rules is still a complex optimization problem. Assuming a rule base with R rules and a minimum number R_{min} of rules after reduction, there are

$$\sum_{\varphi=1}^{R-R_{min}} \binom{R}{\varphi} \tag{7}$$

different reduction possibilities in the search space. For 100 rules and a minimum number of 20 rules, the search space consists of 10^{30} possibilities. For 1000 rules and a minimum number of 200 rules, the search space consists of 10^{301} possibilities.

Because of the high complexity and the large number of local optima, a genetic algorithm has been designed for optimization.

4.1 Fitness Function

For the definition of the fitness function three aspects are considered:

1. Small modeling and classification errors are desired.
2. The number of rules should be reduced as much as possible.
3. It is often desirable that the number of covered data sets is preserved.

Since the first aspects are partly competitive, the user have to determine a weighting factor $w \in [0,1]$ in order to define a cost function. This weighting factor is used for the evaluation of the fitness function[9]:

$$f(\mathbf{a}_i) = \begin{cases} w \cdot \frac{\epsilon(\mathbf{a}_i)^2}{\epsilon(\mathbf{a}_0)^2} + \frac{(1-w)\cdot R(\mathbf{a}_i)}{R(\mathbf{a}_0)} & \text{if } \epsilon(\mathbf{a}_0) \neq 0 \\ w \cdot \epsilon(\mathbf{a}_i)^2 + \frac{(1-w)\cdot R(\mathbf{a}_i)}{R(\mathbf{a}_0)} & \text{if } \epsilon(\mathbf{a}_0) = 0 \end{cases} . \tag{8}$$

Here $\mathbf{a}_0$ is the initial individual representing the initial rule base and $\mathbf{a}_i$ is the actual individual under investigation that represents a possible, reduced rule base. Further on $\epsilon(\mathbf{a}_0)$ is the modeling error of the initial individual and $\epsilon(\mathbf{a}_i)$ is the modeling error of the actual individual. $R(\mathbf{a}_0)$ is the number of rules of the initial individual and $R(\mathbf{a}_i)$ is the number of rules of the actual individual. Depending on the application, the function ϵ can vary as described in section 2.2.

[9] Good individuals have low fitness values.

Small values for the weighting factor w lead predominantly to rule reduction, high values of w lead predominantly to a reduction of the modeling or classification error. The optimum for $w = 0$ is the empty rule base, the optimum for $w = 1$ is a rule base with the smallest modeling or classification error, independent of the number of rules.

In contrast to the similar approach in [12], the number of rules plays here an important role in the evaluation of the fitness values. In [12] a genetic algorithm is used to generate individual rules where the fitness of an individual depends only on the rating of the rule. Here, each individual represents a subset of rules of the rule base. In order to rate an individual, the input/output behavior has to be determined. For small initial rule bases (less than 50 rules), the number of rules is less important. However, for large rule bases, there is a substantial difference between 100 rules or 2000 rules. In such a case, a slightly higher modeling error is accepted in favor of a further rule reduction.

To consider the third aspect, the fitness $f(\mathbf{a}_i)$ is multiplied by a penalty factor $p(\mathbf{a}_i)$ in the form of

$$p(\mathbf{a}_i) = \frac{D - D_0}{D - D_i}, \tag{9}$$

where D is the number of data points, D_0 is the number of data points that are not covered by the initial rule base, and D_i is the number of data points that are not covered by the actual rule base.

4.2 Genetic Operators

One individual in a population consists of r bits, with r being the number of rules of the initial rule base. Each bit represents the state of a rule. A rule can have the state activated or deactivated. The genetic operators are standard operators [13]. The recombination is realized by a uniform crossover operator with fitness proportional selection. The n–point–crossover operator leads to worse results as the algorithm converges too fast. For the mutation operator, a number $z \in [0, 1]$ is selected randomly for each bit. For $z \leq p_m$, the bit is inverted. As selection operator, a tournament selection with two descendants is chosen. The actual best individual of the evolution process is kept outside of the populations. The initialization of the starting population has turned out to be of strategic importance [11]. The standard initialization[10] has proven to be not favorable for complex rule bases. Instead the number z is chosen by the user in such a way, that only few rules are activated.

[10] For each individual, a number $z \in [0; 1]$ is randomly selected. For each bit, b_j, a second number $z_b \in [0; 1]$ is randomly selected. If $z_b \leq z$ the bit is activated

5 Conversion in Interpretable Takagi–Sugeno–Kang Fuzzy Models

The FRM aims at generating rule bases consisting of Mamdani fuzzy rules. The advantage of such rules is, that they can be easily interpreted and comprehended. This advantage is often lost if a Takagi–Sugeno fuzzy model is directly generated. The reasons can be that for each of the rules individually membership functions are provided and that these membership functions are strongly overlapping. Another reason may be that the conclusions become hard to be interpreted because of a high number of parameters which come with a high dimension. On the other side, the advantage of Takagi–Sugeno fuzzy models is that they allow in general to reach a higher accuracy. Therefore the FRM is extended for the generation of Takagi–Sugeno rules as follows: Starting with a generated Mamdani rule base this rule base is converted into a Takagi–Sugeno rule base. Essential is that the conversion is so designed that it increases the model accuracy but preserves the interpretability of the rules.

In contrast to a Mamdani rule (1) the conclusion of a Takagi–Sugeno rule consists of a function $f(\mathbf{x})$ depending on the values of the input variables $\{x_1, \ldots x_V\}$:

$$\text{IF } P \text{ THEN } f(\mathbf{x}) \,. \tag{10}$$

Most commonly a linear combination of the input values is used as a conclusion:

$$f(\mathbf{x}) = a_0 + \sum_{i=1}^{V} a_i \cdot x_i \,. \tag{11}$$

The value of the output of the fuzzy model is calculated by

$$y_D = \frac{\sum_{j=1}^{R} p_j \cdot f_j(\mathbf{x})}{\sum_{j=1}^{R} p_j} \,, \tag{12}$$

where p_j is the fulfillment of the premise of rule j and $f_j(\mathbf{x})$ is the function of the conclusion of rule j.

5.1 Motivation

In order to clarify the here proposed method we consider a SISO system with the input variable X and the output variable Y. The data points describing the system are distributed as shown in Fig. 3. For this system the relevant rule

$$\text{IF } X = \textit{medium} \text{ THEN } Y = \textit{medium} \tag{13}$$

is found.

If only rule (13) is active, the resulting output value of the model is for $X = \textit{medium}$ constant as shown in Fig. 3 by the solid line. As the visible

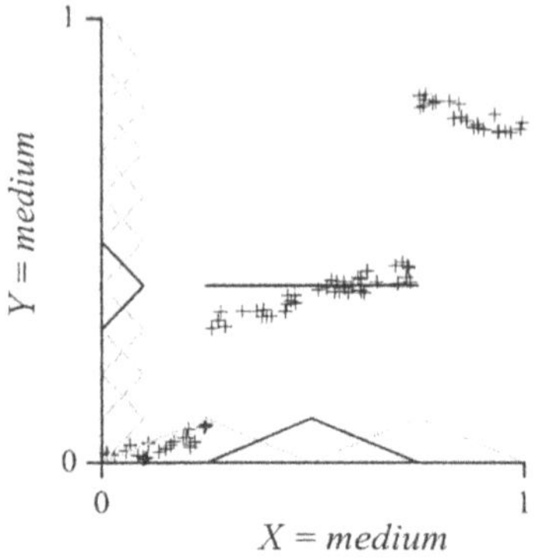

Fig. 3. Example of a SISO system. The solid line shows the output value of the fuzzy model using a Mamdani fuzzy model.

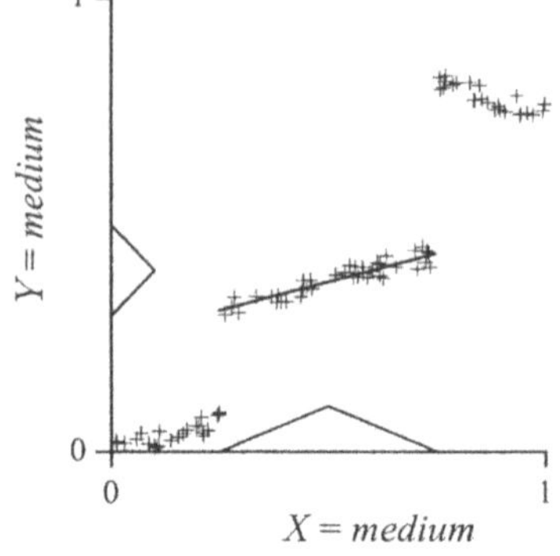

Fig. 4. Example of a SISO system. The solid line shows the output value of the fuzzy model using a Takagi–Sugeno fuzzy model with rule (14).

trend in the data cannot be modeled adequately by the linguistic value of the Mamdani rule the resulting modeling error becomes relative high. If instead the rule

$$\text{IF } X = \textit{medium} \text{ THEN } 0.25 + 0.125 \cdot x \tag{14}$$

with a function as conclusion is used the modeling error is drastically reduced as depicted in Fig. 4. Here, the solid line shows the new output values which reflect the true behavior of the system much better.

Thus, to make use of advantage of the Takagi–Sugeno rule, we transform the Mamdani rule by replacing the linguistic expression through a function. In general, an arbitrary function with some free parameters can be used to replace the linguistic expression of the conclusion. In order to identify the free parameters it is necessary to choose a suitable set of data points for the identification process.

The set of all data points $\mathcal{D}_L$ is given by

$$\mathcal{D}_L = \{d_1, \dots d_D\} \quad \text{mit} \quad d_i = \{\mathbf{x}_i^T \, , \, y_i\} \, . \tag{15}$$

However, only those data points which fulfill the premise of the considered rule are involved. Thus, we need only to consider the subset

$$\mathcal{D}_P = \{d_i \mid d_i \in \mathcal{D}_L \wedge \mu_P(\mathbf{x}_i) > 0\} \subseteq \mathcal{D}_L \, . \tag{16}$$

From this, we form the subset

$$\mathcal{D}_{P \wedge C} = \{d_i \mid d_i \in \mathcal{D}_P \wedge \mu_C(y_i) > 0\} \subseteq \mathcal{D}_P \, , \tag{17}$$

which contains all data points fulfilling the premise and the conclusion. In order to maintain interpretability and to consider the conclusion of the Mamdani rule it is nearby to introduce the condition

$$\mu_P(\mathbf{x}) > 0 \Rightarrow \mu_C(f(\mathbf{x})) > 0 \,. \tag{18}$$

This Condition holds if a suitable function, such as a linear combination of the input variables, is used and the identification of the parameters is based on the set $\mathcal{D}_{P \wedge C}$ [14,15].

On the other hand there are some reasons to drop condition (18):

- In some cases, for example using the FRM with certain test and rating strategies, the number of elements of $\mathcal{D}_{P \wedge C}$ might be significantly smaller than the number of elements of $\mathcal{D}_P$. Then, if only the data points of $\mathcal{D}_{P \wedge C}$ are used for identification, the improvement counts only towards a small number of data points. But, the modeling quality of the data points of $\mathcal{D}_P \setminus \mathcal{D}_{P \wedge C}$ will probably decrease. This would lead to an overall decrease of the modeling quality.
- The determination of a sensible function which fulfills condition (18) is generally difficult. One reason for this is, that an equally distribution of the data points fulfilling the premise is unusual (due to the often limited amount of data points or to the shape and the location of the membership functions).
- The condition (18) is not a necessary condition for maintaining transparency and interpretability. The original Mamdani rule base can still be used to illustrate system correlations while the transformed Takagi–Sugeno rule base can be interpreted as a specification of the former rule base.

Because of this reasons it seems to be reasonable to restrict the data points for identification to the set $\mathcal{D}_{P \wedge C}$. In addition, the results have shown, that the violation of condition (18) is not problematic. Therefore, we use the set $\mathcal{D}_P$ for the identification of the parameters of the functions generally.

5.2 Generalizing Rules

Some methods for the generation of fuzzy models, like the FRM, support the use of generalizing rules. In the premise of a generalizing rule not all input variables are addressed. Thus, it is one possibility to identify only those parameters of a function which are connected to the variables used in a given premise. This would simplify the identification, especially when the number of input variables is large. On the other hand it often turned out empirically, that identifying all parameters in any case is favorable, because by this additional trends of variables not in the premise can be adequately considered to improve the modeling quality. In [16] it is shown that both approaches lead to an improvement of the model quality, but the approach to identify always all parameters is in every case significantly better. This compensates the advantage of a faster identification by far, so that only the second approach should be used.

5.3 Transformation of the Conclusion

As described above, all Mamdani rules

$$\text{IF } P \text{ THEN } C \tag{19}$$

of a rule base are to be transformed into Takagi–Sugeno rules

$$\text{IF } P \text{ THEN } f(\mathbf{x}) \,. \tag{20}$$

In principal, there are many possibilities to choose the function $f(\mathbf{x})$. Here, we look only at two approaches — a linear combination and a multilinear combination of the input variables. Another approach — using a polynomial for each dimension — is described in [16]. So far we have not considered the more general approach of using a multidimensional polynomial because of the high number of free parameters.

Linear Combination The use of a linear combination

$$f(\mathbf{x}) = a_0 + \sum_{i=1}^{V} a_i \cdot x_i \tag{21}$$

of the input variables is the simplest approach for the function in (20). For the identification of the $V + 1$ parameters we use a least squares algorithm. If the required $V + 1$ linear independent data points are not available the constant value

$$f(\mathbf{x}) = \frac{1}{D_P} \sum_{l=1}^{D_P} y_l \tag{22}$$

is chosen for $f(\mathbf{x})$. This value is the mean value of the D_P output values of the set $\mathcal{D}_P$.

Multilinear Combination The multilinear function approach

$$f(\mathbf{x}) = a_0 + \sum_{i=1}^{2^V-1} a_i \prod_{k=0}^{V-1} x_{k+1}^{I_k(i)} \tag{23}$$

is an extension of the above linear approach. It has the property, that the function is linear in the direction of each coordinate, but considers nonlinear dependencies between the input variables. Herein $I_0(i)$ and $I_k(i)$ are given by

$$I_0(i) = \begin{cases} 1\,, & \text{if } i \text{ odd} \\ 0\,, & \text{if } i \text{ even} \end{cases} \quad \text{for } k = 0$$

$$I_k(i) = \begin{cases} 1\,, & \text{if } \left(i - \sum_{l=0}^{k-1} I_l \cdot 2^l\right)/2^k \quad \text{odd} \\ 0\,, & \text{if } \left(i - \sum_{l=0}^{k-1} I_l \cdot 2^l\right)/2^k \quad \text{even} \end{cases} \quad \text{for } k > 0 \quad . \tag{24}$$

For the identification of the 2^V parameters a least squares algorithm is used. If the necessary 2^V linear independent data points are not available, function (22) is used instead.

6 Results and Comparison

In this Section the different concepts for the feedback of the modeling quality in the rule generation process of the FRM are compared. The aim here is to assess and discuss the different concepts with respect to accuracy and interpretability and not to compare them to other methods in literature. Such a detailed comparison can be found in Appendix 8.1. For the comparison four well known benchmark problems were chosen[11]. As described in [2] the WINE and SAT problem are examples for a small and a complex classification problem, respectively. The MACKEY and KIN problem are a small and a complex approximation problem [2].

The modeling results are assessed by the obtained accuracy, i. e. the modeling error on validation data, and the interpretability, i. e. the complexity of the final rule base (number of rules). As references we use the best results obtained with the FRM yet as well as the results obtained by the systematic approach FRM_{sys} presented in [10]. The systematic approach was originally developed to support non experts by applying the FRM. It is based on some heuristic procedures for the project definition and a deterministic scheme for the adjustment of the strategy parameters of the FRM. Then the rule generation is performed with a maximum combination depth of $c_{max} = 2$. It turns out that in this way satisfying results can be obtained in an acceptable time.

As shown in Tab. 1 the systematic approach FRM_{sys} can lead to large rule base, especially if it is applied to the more complex problems SAT and KIN. In order to reduce the number of rules the optimizing conflict reduction (OCR) presented in Sect. 4 is performed in a subsequent step. For the easier problems WINE and MACKEY the OCR lead to a significant decrease

[11] The corresponding data sets are available at the *UCI Repository of Machine Learning Databases* and *Data for Evaluating Learning in Valid Experiments*:
http://www.ics.uci.edu/~mlearn/MLRepository.html
http://www.cs.utoronto.ca/~delve/data/datasets.html

Table 1. Modeling results (error (RCE/MAE) on validation data) for the different benchmark problems. The number of rules is indicated in brackets. For the best result FRM_{best} corresponding literature is cited.

	FRM_{sys}		FRM_{sys}^{OCR}		FRM_{SSS}		FRM_{SSS}^{OCR}		FRM_{best}		
WINE	11.2	(105)	10.7	(15)	11.3	(83)	9.5	(17)	6.2	(131)	[10]
SAT	18.2	(2683)	18.2	(1044)	17.9	(574)	16.0	(99)	12.7	(204)	[11]
MACKEY	0.07	(59)	0.05	(20)	0.07	(46)	0.05	(21)	0.01	(92)	[15]
KIN	0.22	(1530)	0.12	(457)	0.24	(308)	0.17	(127)	0.16	(309)	[10]

of the number of rules and additionally to a slight improvement of the accuracy. With 15 and 20 rules with a maximum combination depth $c_{max} = 2$, the rule bases are very good to interpret. In contrast, the rule bases of the more complex problems SAT and KIN are still large after applying the OCR. Considering the accuracy the results are different: On the one hand the modeling error is drastically reduced by the OCR for the KIN problem, on the other hand there is no improvement obtained for the SAT problem. To summarize, with the FRM_{sys}^{OCR} approach a satisfying compromise between accuracy and interpretability is found for WINE and MACKEY. For KIN a high accuracy is obtained, but with nearly 500 rules, the rule base is still complex and consequently hard to interpret. Finally, the rule base for the SAT problem is neither interpretable nor lead to a high accuracy.

We consider now the specific search on the subspaces with poor accuracy presented in Sec. 3.2. In order to allow a fair comparison, the iterative search concept depicted in Fig. 2 is also restricted to a maximum combination depth $c_{max} = 2$. The results for this approach FRM_{SSS} are shown in Tab. 1. It has to be pointed out that for the SAT and KIN problem the number of rules is drastically smaller than in the FRM_{sys} and even smaller than in the FRM_{sys}^{OCR} approach. Nevertheless the accuracy is still comparable to the FRM_{sys} approach. This results can be significantly improved by also applying the OCR. In contrast to the FRM_{sys}^{OCR} approach described above, the FRM_{SSS}^{OCR} approach leads for SAT and KIN to much smaller rule bases with only about 100 rules. From our point of view such a size of a rule base is still interpretable. It should be noted, that in principle it would be possible to get the same small rule base with the FRM_{sys}^{OCR} approach. This is due to the fact, that the initial rule base obtained by FRM_{sys} approach is a superset of the FRM_{SSS} rule base. However, it seems that the high number of rules in FRM_{sys} rule base makes it too difficult for the OCR to find a small subset with an acceptable accuracy. This explanation is congruent with the results obtained for the easier problems WINE and MACKEY. Due to the smaller initial rule bases for the OCR the final rule bases of the FRM_{sys}^{OCR} and the FRM_{SSS}^{OCR} approach are comparable.

Altogether the results show, that with a feedback of the modeling quality, it is possible to improve drastically the interpretability without impairing the accuracy. Rather especially for the more complex application the accuracy can be improved simultaneously. Depending on the complexity of the considered problem different approaches are recommended: For easier problems with a small initial rule base ($\ll 1000$) it is sufficient to apply only the OCR. For more complex problems it is promising to generate firstly a smaller initial rule base with the specific search on subspaces with poor modeling quality (SSS). Then in a subsequent step the OCR is applied. A comparison with the best results FRM_{best} obtained with the Fuzzy–ROSA method yet lead to the conclusion, that a further improvement of the accuracy is possible at the expense of interpretability.

So far only fuzzy models of the Mamdani type are considered. As described in Sec. 5 exclusively for approximation tasks a conversion in TSK fuzzy models can lead to a much higher accuracy without loss of interpretability. The modeling results for the conversion of the rule bases described above are shown in Tab. 2. Thereby a multilinear function is used for MACKEY and a linear function is used for KIN. It should be noted that the OCR is applied to the converted fuzzy models FRM_{sys} and FRM_{SSS} of Tab. 2 respectively[12]. Therefore the number of rules of the OCR approaches FRM_{sys}^{OCR} and FRM_{SSS}^{OCR} in Tab. 2 are different to the approaches in Tab. 1. To summarize, the conversion in TSK fuzzy models leads in all considered approaches in Tab 2 to the highest accuracy obtained with the FRM yet. More important also the size of the rule base is almost the smallest. Similar to the results discussed above for the easier MACKEY problem with the FRM_{sys}^{OCR} and the FRM_{SSS}^{OCR} approach comparable rule bases are obtained. For the more complex problem KIN also the FRM_{SSS}^{OCR} approach lead to the best results.

Table 2. Results (error (MAE) on validation data) after the conversion in Takagi–Sugeno–Kang fuzzy models. The number of rules after the OCR is indicated in brackets.

	FRM_{sys}	FRM_{sys}^{OCR}	FRM_{SSS}	FRM_{SSS}^{OCR}
MACKEY	0.009	0.006 (20)	0.011	0.007 (17)
KIN	0.19	0.12 (359)	0.15	0.10 (62)

[12] This is due to the fact, that a direct conversion of the FRM_{sys}^{OCR} and FRM_{SSS}^{OCR} of Tab. 1 does not reach the same accuracy.

7 Concluding Remarks

In this paper we present several concepts for a feedback of the modeling quality in the rule generation process of individually tested rules. A genetic algorithm is used in the optimizing conflict reduction (OCR) in order to reduce simultaneously the number of rules and the modeling error. It turns out that for not to complex problems the OCR lead to very easily interpretable rule bases with a good accuracy. For more complex problems the complete search may lead to large rule bases, which are not well manageable by the OCR. In general smaller initial rule bases can be obtained with the specific search on subsets with poor accuracy. In combination with the OCR this approach leads also for more complex problems to a good compromise between accuracy and interpretability. A further improvement of the accuracy is possible for approximation tasks by a conversion in TSK fuzzy models.

8 Appendix

8.1 Results for the Benchmark Problems

WINE: Differing from the results in Table 1 and 2, which are based on a two times cross validation, we present here a ten times cross validation in order to guarantee a fair comparison with the results[13] in [17]: $FRM_{sys} \hat{=} FRM_{best}$: $R = 141$, $\epsilon_{learn} = 3.2\%$, $\epsilon_{vali} = 6.2\%$ and $FRM_{sys+OCR}$: $R = 35$, $\epsilon_{learn} = 1.8\%$, $\epsilon_{vali} = 6.7\%$. Thereby ϵ_{vali} and ϵ_{learn} denote the relative classification error (RCE) on validation and learning data in percent, respectively and R is the number of rules. For each input variable five equidistant trapezoidal membership functions are defined. The number of classes (singletons) for the output variable is three.

Fig. 5. WINE: relative classification error (RCE) on validation data ϵ_{vali} for the approaches ∇ FRM_{sys}, $\diamond$ $FRM_{sys+OCR}$, $\star$ FRM_{best} and $\times$ literature [17].

SAT: Our results for this classification task obtained by a two times cross validation are as follows: FRM_{sys}: $R = 2683$, $\epsilon_{learn} = 18.2\%$, $\epsilon_{vali} = 18.8\%$, $FRM_{sys+OCR}$: $R = 1044$, $\epsilon_{learn} = 15.4\%$, $\epsilon_{vali} = 18.2\%$ and FRM_{best}:

[13] In [17] the data set were split randomly in ten parts. In ten applications of the learning method nine parts were used as learning and the left part as validation data sets. The modeling result is obtained by averaging over the ten learning cycles.

$R = 204$, $\epsilon_{learn} = 11.0\%$, $\epsilon_{vali} = 12.7\%$. Thereby ϵ_{vali} and ϵ_{learn} denote the relative classification error (RCE) on validation and learning data in percent, respectively and R is the number of rules. For each input variable nine equidistant trapezoidal membership functions are defined. The number of classes (singletons) for the output variable is six.

Fig. 6. SAT: relative classification error (RCE) on validation data ϵ_{vali} for the approaches ∇ FRM$_{sys}$, $\diamond$ FRM$_{sys+OCR}$, $\star$ FRM$_{best}$ and $\times$ literature [18].

MACKEY: The task for the chaotic Mackey–Glass time series is to predict the next value based on the recent values[14]. For this approximation task we performed a two times cross validation, that leads to the following results: FRM_{sys}: $R = 59$, $\epsilon_{learn} = 0.07/0.007$, $\epsilon_{vali} = 0.07/0.008$, $FRM_{sys+OCR}$: $R = 20$, $\epsilon_{learn} = 0.05/0.004$, $\epsilon_{vali} = 0.05/0.005$ and $FRM^{OCR}_{Sys+TSK} \hat{=} FRM_{best}$: $R = 20$, $\epsilon_{learn} = 0.006/0.00007$, $\epsilon_{vali} = 0.007/0.00012$. Thereby ϵ_{vali} and ϵ_{learn} denote the mean absolute/square error (MAE/MSE) on validation and learning data, respectively and R is the number of rules. For each input variable four equidistant triangular membership functions are defined. The number of triangular, equidistant membership functions for the output variable is nine.

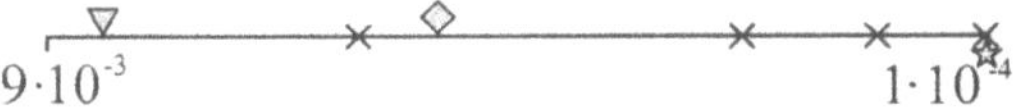

Fig. 7. MACKEY: mean square error (MSE) on validation data ϵ_{vali} for the approaches ∇ FRM_{sys}, $\diamond$ $FRM_{sys+OCR}$, $\star$ FRM_{best} and $\times$ literature [15].

KIN: The following results[15] are obtained for this approximation task: FRM_{sys}: $R = 1530$, $\epsilon_{learn} = 0.21/0.07$, $\epsilon_{vali} = 0.22/0.07$, $FRM_{sys+OCR}$: $R = 457$, $\epsilon_{learn} = 0.11/0.02$, $\epsilon_{vali} = 0.12/0.02$ and $FRM^{OCR}_{SSS+TSK} \hat{=} FRM_{best}$: $R = 62$, $\epsilon_{learn} = 0.09/0.01$, $\epsilon_{vali} = 0.10/0.01$. Thereby ϵ_{vali} and ϵ_{learn} denote the mean absolute/square error (MAE/MSE) on validation and learning data, respectively and R is the number of rules. For the input variables five up to seven equidistant trapezoidal membership functions or occasionally singletons are defined. The number of the trapezoidal, equidistant membership functions for the output variable is nine.

[14] In our experiments we use the last four past values

[15] Our results are based on the *kin32fm* data sets.

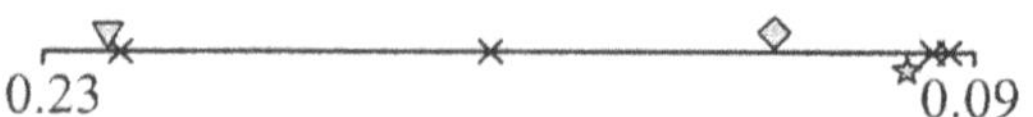

Fig. 8. KIN: mean absolute error (MAE) on validation data ϵ_{vali} for the approaches $\triangledown$ FRM_{sys}, $\diamond$ $FRM_{sys+OCR}$, $\star$ FRM_{best} and $\times$ literature [15].

8.2 Algorithms

Algorithm 1 Construction of a representation $\tilde{\mathcal{E}}_{SoS}$ after Def. 2 for the not covered subspace $\mathcal{I}_{SoS}$.

Require: n premises P_q, $\tilde{\mathcal{E}}_{SoS} = \{\mathcal{E}_0\}$ and q=1
Ensure: $\tilde{\mathcal{E}}_{SoS}$
1: **while** $(\tilde{\mathcal{E}}_{SoS} \neq \emptyset) \wedge (q \leq n)$ **do**
2: $\tilde{\mathcal{E}}'_{SoS} = \emptyset$
3: **for all** $\mathcal{E}_a \in \tilde{\mathcal{E}}_{SoS}$ **do**
4: **for all** $e_{i_p j_p} \in P_q$ **do**
5: $\mathcal{E}_{a,p} = \mathcal{E}_a \backslash \{e_{i_p j_p}\}$
6: **if** $\nexists\, l \in \{1, \ldots, V\} : \forall\, e_{ij} \in \mathcal{E}_{a,p} : l \neq i$ **then**
7: $\tilde{\mathcal{E}}'_{SoS} = \tilde{\mathcal{E}}'_{SoS} \cup \mathcal{E}_{a,p}$
8: **end if**
9: **end for**
10: **end for**
11: $\tilde{\mathcal{E}}_{SoS} = \tilde{\mathcal{E}}'_{SoS}$, $q = q + 1$
12: **end while**
13: **if** $\tilde{\mathcal{E}}_{SoS} = \emptyset$ **then**
14: return: rule base complete
15: **else**
16: return: the not covered subspace is represented by $\tilde{\mathcal{E}}_{SoS}$
17: **end if**

Definition 4. (Function *Fill*(P)) For a given premise P the function *Fill*(P) calculates the set $\mathcal{E}_P$ of linguistic expressions, which contains all linguistic expressions of P and for each input variable X_i not considered in P all S_i defined linguistic expressions $\{e_{i1}, \ldots, e_{iS_i}\}$.

Algorithm 2 *Insert*$((\mathcal{E}_B, \mathcal{E}_F), \mathcal{T}_{SoP})$.

Require: Pair $(\mathcal{E}_B, \mathcal{E}_F)$ and set of pairs $\mathcal{T}_{SoP}$
Ensure: $\mathcal{T}_{SoP}$
1: **for all** $(\mathcal{E}_B^q, \mathcal{E}_F^q) \in \mathcal{T}_{SoP}$ **do**
2: **if** $\mathcal{E}_B \geq_g \mathcal{E}_B^q$ **then**
3: **if** $\exists\, i \in \{1,\dots,V\} \,: \forall\, j \in \{1,\dots,S_i\} : e_{ij} \notin (\mathcal{E}_F \cap \mathcal{E}_F^q)$ **then**
4: $\mathcal{T}_{SoP}$ without $\mathcal{E}_B^q$ {call to Alg. 3}
5: **else**
6: $\mathcal{T}_{SoP} = (\mathcal{T}_{SoP} \backslash \{(\mathcal{E}_B^q, \mathcal{E}_F^q)\}) \cup (\mathcal{E}_B^q, (\mathcal{E}_F^q \cap \mathcal{E}_F))$
7: **end if**
8: **end if**
9: **end for**
10: $\mathcal{T}_{SoP} = \mathcal{T}_{SoP} \cup \{(\mathcal{E}_B, \mathcal{E}_F)\}$

Algorithm 3 Calculate $\mathcal{T}'_{SoP} = \mathcal{T}_{SoP}$ without P, i.e. the representation after Def. 3 of $\mathcal{T}'_{SoP}$, which represents all input situations of $\mathcal{T}_{SoP}$ except of those, which are covered by the premise P.

Require: premise P and set of pairs $\mathcal{T}_{SoP}$
Ensure: $\mathcal{T}_{SoP}$
1: **for all** $e_{ij} \in P$ **do**
2: $P_{cur} = P \backslash \{e_{ij}\}, \mathcal{E}_F^P = \mathcal{E}_0$
3: **for all** $(\mathcal{E}_B, \mathcal{E}_F) \in \mathcal{T}_{SoP}$ **do**
4: **if** $\mathcal{E}_B \geq_g P_{cur}$ **then**
5: $\mathcal{E}_F^P = \mathcal{E}_F^P \cap \mathcal{E}_F$
6: **end if**
7: **end for**
8: $\mathcal{E}_F^{cur} = \mathcal{E}_F^P \backslash \{e_{ij}\}$
9: **if** $\nexists\, i \in \{1,\dots,V\} \,: \forall\, j \in \{1,\dots,S_i\} : e_{ij} \notin \mathcal{E}_F^{cur}$ **then**
10: $\mathcal{T}_{SoP} = Insert((Fill(P_{cur}), \mathcal{E}_F^{cur}), \mathcal{T}_{SoP})$
11: **else**
12: **if** $P_{cur} = \emptyset$ **then**
13: $\mathcal{T}_{SoP} = \emptyset$, exit {$\mathcal{T}_{SoP}$ represents $\emptyset$}
14: **else**
15: $\mathcal{T}_{SoP}$ ohne P_{cur} {recursive call}
16: **end if**
17: **end if**
18: **end for**

References

1. A. Krone and H. Kiendl. Automatic generation of positive and negative rules for two–way fuzzy controllers. In *Proceedings of the Second European Congress on Intelligent Techniques and Soft Computing (EUFIT '94)*, volume 1, pages 438–447, Aachen, 1994. Verlag Mainz.
2. T. Slawinski, A. Krone, P. Krause, and H. Kiendl. The fuzzy–rosa method: A statistically motivated fuzzy approach for data–based generation of small interpretable rule bases in high–dimensional search spaces. In M. Last, A. Kandel,

and H. Bunke, editors, *Data Mining and Computational Intelligence*, pages 141–166. Physica-Verlag, Heidelberg, 2001.

3. A. Krone and H. Taeger. Data–based fuzzy rule test for fuzzy modelling. *Fuzzy Sets and Systems*, 126:343–358, 2001.
4. H. Jessen. Test and rating strategies for automatic fuzzy rule generation and application to load prediction. In M. Mohammadian, editor, *New Frontiers in Computational Intelligence and its Applications*, pages 11–21. IOS Press, Amsterdam, Niederlande, 2000.
5. H. Kiendl. Decision analysis by advanced fuzzy systems. In J. Kacprzyk L. Zadeh, editor, *Computing with Words in Information/Intelligent Systems*, pages 223–242. Physica-Verlag, Heidelberg, 1999.
6. A. Krone, T. Slawinski, and P. Krause. Search Space Structuring as a Key to Cope with Different Problem Sizes in the Field of Fuzzy Modeling. In M Jamshidi, M. Fathi, and T. Furuhashi, editors, *Soft Computing, Multimedia, and Image Processing (Proceedings of the Fourth World Automation Congress (WAC '00), Maui, USA)*, volume 11, pages 346–354, Albuquerque, NM, USA, 2000. TSI Press.
7. T. Slawinski, A. Krone, U. Hammel, D. Wiesmann, and P. Krause. A hybrid evolutionary search concept for data–based generation of relevant fuzzy rules in high dimensional spaces. In *Proceedings of IEEE International Conference on Fuzzy Systems (FUZZ-IEEE '99) Seoul, Korea, 1999*, volume 3, pages 1432–1437, Piscataway, NJ, 1999. IEEE Press.
8. A. Krone. Advanced rule reduction concepts for optimizing efficiency of knowledge extraction. In *Proceedings of the Fourth European Congress on Intelligent Techniques and Soft Computing (EUFIT '96)*, volume 2, pages 919–923, Aachen, 1996. Verlag Mainz.
9. W. Pedrycz. *Fuzzy Control and Fuzzy Systems.* Research Studies LTD, Taunton, Großbritanien, 1993.
10. T. Slawinski. *Analyse und effiziente Generierung von relevanten Fuzzy Regeln in hochdimensionalen Suchräumen.* bei der Universität Dortmund eingereichte Dissertation, 1 edition, 2001.
11. A. Krone, P. Krause, and T. Slawinski. A New Rule Reduction Method for Finding Interpretable and Small Rule Bases in High Dimensional Search Spaces. In *Proceedings of the Ninth IEEE International Conference on Fuzzy Systems, (FUZZ-IEEE '00), San Antonio, USA*, volume 2, pages 696–699, Piscataway, NJ, 2000. IEEE Press.
12. O. Cordon, A. Gonzalez, F. Herrera, and R. Perez. Encouraging cooperation in the genetic iterative rule learning approach for qualitative modeling. In *Computing with Words in Information/Intelligent Systems*, pages 95–117. Physica–Verlag, Heidelberg, 1999.
13. Th. Bäck, U. Hammel, and H.-P. Schwefel. Evolutionary computation: Comments on the history and current state. *IEEE Transactions on Evolutionary Computation*, 1:1–15, 1997.
14. P. Krause and T. Slawinski. Das Fuzzy–ROSA–Verfahren: Von der regelorientierten statistischen Analyse zur datenbasierten Generierung von interpretierbaren Takagi–Sugeno–Systemen. *at – Automatisierungstechnik*, 9(49):391–399, 2001.
15. P. Krause. Generierung von Takagi–Sugeno–Fuzzy–Systemen aus relevanten Fuzzy–Regeln. In *Tagungsband des 10. Workshops Fuzzy Control des GMA–*

FA 5.22, Dortmund, pages 84–97, Karlsruhe, 2000. VDI/VDE GMA-FA 5.22, Wissenschaftliche Berichte Forschungszentrum Karlsruhe (FZKA 6509).

16. P. Krause. *Datenbasierte Generierung von transparenten und genauen Fuzzy-Modellen für mehrdeutige Daten und komplexe Systeme.* bei der Universität Dortmund eingereichte Dissertation, 1 edition, 2001.
17. R. Holve. Investigation of automatic rule generation for hierarchical fuzzy systems. In *Proceedings of the Seventh IEEE International Conference on Fuzzy Systems (FUZZ-IEEE '98), Anchorage, USA, 1998*, volume 2, pages 973–978, Piscataway, NJ, 1998. IEEE Press.
18. D. Michie, D.J. Spiegelhalter, and C.C. Taylor. *Machine learning, Neural and Statistical Classification.* Ellis Horwood, Hemel Hempstead, Großbritannien, 1994.

SECTION 4

EXTENDING THE MODEL STRUCTURE TO IMPROVE THE ACCURACY

A description of several characteristics for improving the accuracy and interpretability of inductive linguistic rule learning algorithms *

Eugenio Aguirre, Antonio González, and Raúl Pérez

Departamento de Ciencias de la Computación e Inteligencia Artificial. E.T.S. de Ingeniería Informática. Universidad de Granada 18071-Granada, Spain

Abstract. The learning algorithms can be an useful tool for helping to the humans to understand the behavior of phenomena from a set of samples. In particular, those algorithms that represent the knowledge obtained by linguistic fuzzy rules are appropriate for this task. However, it is not sufficient that the knowledge representation is close to the humans comprehension. Furthermore, it is necessary that the knowledge is expressed as simple as possible.

In this chapter, we classify some techniques, models and tools for improving the knowledge obtained by inductive linguistic rule learning algorithms from three different points of view: those that increase the knowledge interpretability, those that increase the knowledge accuracy keeping its interpretability and those that simultaneously increase the accuracy and interpretability of the knowledge.

In this study, we have considered fuzzy rules expressed by the Disjunctive Normal Form (DNF).

1 Introduction

One of the possible applications of learning algorithms consists of obtaining information about the behavior of different systems. In some cases, we can obtain a set of samples of the system considering some input parameters and watching the values returned on some output parameters. This set of examples provides us a certain knowledge about the behavior of the system.

Usually, for obtaining a most elaborated description of the system, the example set is used as the main component of an inductive learning algorithm. The goal of these algorithms consists of finding a theory that generalizes the example set and permits to describe some relations between input and output parameters.

Two important elements appear in the formulation of the inductive rule: the representation of the hypothesis or theory and the background knowledge used. In the literature we can find several learning algorithms that propose different combination of these elements [12,39,44]. The choice of these elements determines the bias of the knowledge obtained by the learning algorithm. The features of the knowledge to be learned have to be established at the initial stages of the development of the inductive learning algorithm.

* This work has been supported by the CICYT under Project TAP99-0535-C02-01

Thus, when the interpretability of the knowledge is a desirable feature, we must select an adequate representation model. The rules are representation models of the information that the humans can understand and frequently are used in theirs communications. Among the different rule models, those that include linguistic labels for the description of the concepts present two advantages:

a) They permit to establish flexible partitions on the domain of the attributes involved in the system, where each partition describes the semantic of a label and it has an interpretable meaning in the human language.
b) They work with vagueness and imprecision in the description of the concepts. It is an important feature of the communication among persons since it permits to reduce the information needed in order to understand the concepts.

The learning algorithms that use a linguistic rule model whose main objective is to obtain theories with a good interpretability are members of an area of the Fuzzy Modeling called Linguistic Fuzzy Modeling (LFM). In opposite of these, the learning algorithms that are member of the Precise Fuzzy Modeling (PFM) have as main objective to obtain theories with a good accuracy.

This classification of the fuzzy modeling is caused for the impossibility, in general, of obtaining the highest degrees of interpretability and accuracy simultaneously. Therefore, considering that the quality of the knowledge learned depends of its accuracy and its interpretability, the learning algorithms of the LFM try to find an acceptable balance between these parameters, being the interpretability the main goal.

In this work, we show several proposals for improving the behavior of learning algorithms of the LFM classified in three aspects:

a) proposals that permit to increase the interpretability keeping the accuracy,
b) proposals that permit to increase the accuracy keeping the interpretability and
c) proposals that permit to increase the interpretability and the accuracy simultaneously.

Before to develop a learning algorithm whose objective is obtaining interpretable knowledge, we find a problem: while the accuracy is a measurable parameter and among the researchers in the area there is a certain consensus about how it must be calculated (combining consistence and completeness, error percentage, success percentage, etc), the interpretability is a subjective and complex concept that has not a simple formulation yet.

However, we can establish some general criteria for determining what rule bases are more understandable than others, taking into account criteria as the simplicity of the individual rules, hypothesis with a reduced number of rules and variables and the simplicity of the rule and the reasoning model. In our case, we consider the following criteria

- number of rules,
- number of relevant variables involved in the rule base and
- number of variables in the antecedent of the rules.

Moreover, since we have used DNF rules, the number and grouping of values assigned to a variable will be also considered.

Therefore, this work analyzes different alternatives for improving the interpretability of the learning algorithms of the LFM following the previous general criteria and it takes as referential point the works developed for the construction of algorithms based on the genetic iterative model, such as [23], considering and relating other proposals that can be found in the literature.

The chapter is organized as follows. Section 2 introduces the DNF rule model, studies how this rule model is appropriate for obtaining interpretable knowledge versus the most common rule models used in the fuzzy modeling and shows how the architecture of the learning algorithm can improve the interpretability of the rule base from three points of view: reducing the number of rules, obtaining short descriptions and selecting relevant features. Section 3 shows how to improve the accuracy of linguistic fuzzy models taking into account the process of selecting rules and by the inclusion of the semantic hedges. Section 4 describes a refinement algorithm of theories as a process for increasing the accuracy and interpretability of the knowledge simultaneously and Section 5 points out some conclusions.

2 Increasing the interpretability in LFM

In the introduction of this chapter, we associate the degree of interpretability of a knowledge base represented by rules to the satisfaction of some criteria of the rule base. These criteria are related with the number of rules and the simplicity of each rule. But, the initial component of this study is the model of rule used by the learning algorithm. Next subsection analyzes the DNF rule model.

2.1 The DNF rule model

The language known as Disjunctive Normal Form (DNF) was proposed by Michalski [39] and it is a simple language for representing complex concepts by a disjunctive rule set. The antecedent of the rule is composed of the conjunction of one or more variables, while the consequent determines the concept that must be assigned when the antecedent is activated. Furthermore, internal disjunctions of values in each variable are permitted in the antecedent of the rule. We can consider the DNF rules as an extension of the Mamdani-type rules [38].

An example of DNF rule is the following:

IF

femur_length is (*medium* or *big-medium* or *big*) and
head_diameter is (*medium* or *big*) and
foetus_sex is *male*

THEN

foetus_weight is *normal*

where the domain of ***femur_length*** is {*small*, *small-medium*, *medium*, *big-medium*, *big* }, the domain of ***head_diameter*** is {*small*, *medium*, *big*} and the domain of ***foetus_sex*** is {*male*, *female*, *unknown*}. The domain of the consequent variable ***foetus_weight*** is {*low*, *normal*, *high*}.

Using this language it is possible to determine when a variable is relevant for describing a class. If a variable takes the whole values of its domain, then this variable is not needed for describing the rule and it can be eliminated from the description of the rule [1], for example, in the following rule:

IF

femur_length is (*medium* or *big-medium* or *big*) and
head_diameter is (*medium* or *big*) and
foetus_sex is (*male* or *female* or *unknown*)

THEN

foetus_weight is *normal*

the ***foetus_sex*** variable can be considered irrelevant and the rule can be expressed as:

IF

femur_length is (*medium* or *big-medium* or *big*) and
head_diameter is (*medium* or *big*)

THEN

foetus_weight is *normal*

So, the learning algorithms that use the DNF rule model for representing the learned knowledge, can detect, in a natural way, the irrelevant variables and therefore, can reduce the number of variables involved in the description of a rule and obtains more understandable rules. Furthermore, the DNF rules

[1] This property is really true when we use crisp values. In the fuzzy case, in [21], we obtained a similar result using a mathematical model that assign the convex hull of the membership functions to adjacent fuzzy sets

may be a good tool for developing feature selection mechanisms as we will later show.

The possibility of including a disjunction of values in the antecedent variables offers another two advantages under the point of view of the interpretability:

- When a variable takes all values of its domain except one, we can reinterpret the assignment to the variable by the NOT operator. In the rule of the last example, we can see that the ***head_diameter*** variable takes two of the tree values of its domain. We can reinterpret this rule in the following way:

  IF

  femur_length is (*medium* or *big-medium* or *big*) and
  head_diameter is **not** *small*

  THEN

  foetus_weight is *normal*

  In general, we can transform the assignment for describing that values are not taken for the variable using the NOT operator applied on all values. In the above example, the ***femur_length*** variable can be transformed as:

  femur_length is **not** (*small* or *medium-small*)

  being its interpretation that the variable does not take the value *small* AND does not take the value *medium-small*. Obviously, using the NOT operator, the number of values involved in the assignment of the variable is always less or equal than to the number of value defined in the variable domain and therefore, the assignments are simpler and more understandable.
- When the domain associated to a variable is defined on an ordered referential set, and the values of the domain can be ordered, in some cases we can introduce three connectives for improving the interpretability of the assignment without modifying its semantic. These three relational connectives are: **less or equal than** ($\leq$), **greater or equal than** ($\geq$) and **between**. So, if the disjunction is composed by a sequence of adjacent values of the domain of the variable, we can apply one of previous connectives, the first of the them, when the sequence begins in the less value, the second, when the sequence finishes in the last value and the third otherwise. Therefore, in our example, the first variable of the antecedent takes the last three values on its ordered domain. We can reformulate the rule in the following way:

  IF

  femur_length is greater or equal than *medium* and

head_diameter is **not** *small*
THEN
foetus_weight is *normal*

These features show that the DNF rules have a high degree of flexibility for describing the knowledge in a simple and interpretable way.

2.2 Iterative strategy for learning rules

An important element for obtaining a reduced number of rules is the architecture of the learning algorithm. In the literature, we can find several proposal for the problem of learning multiple concepts [9,12,15,26,44]. Trying to classify these proposal, we have considered the decomposition of the problem as an important aspect. So, we consider the following classification:

a) Proposal without decomposition problem: The algorithms based on this category consider indivisible the learning problem and they try to extract a knowledge that offers a global solution to this problem. Inside of this category, we can consider the inductive learning algorithms based on neural networks or algorithms based on Pittsburgh approach.
b) Proposal with decomposition problem: In this category we consider those algorithms that split the learning problem in a subset of simpler problems. The goal solution is obtained by the aggregation of the solution of each subproblem.

On the second type, we can find in the machine learning literature several strategies, such as the learning algorithm based on decision trees [9,44]. However, we are interested in a special strategy used by a wide number of learning algorithms called the **iterative strategy**. The iterative strategy consists of transforming the problem of learning m concepts in m problems where each one of them has to learn only one concept. Basically, this strategy fixes a class that must be learned and taking into account the examples set and by generalization and specification mechanisms obtain a description of the class. Therefore, a partial solution of the problem is composed by one or more descriptions for a concept. The global solution is obtained by the aggregation of the partial solutions.

The main differences among the learning algorithms based on this strategy is found in the way in which they obtain the partial solutions. We can distinguish two features that must be defined in the learning algorithm:

1. a criterion for evaluating if a certain set of descriptions can be considered as a solution for representing a certain concept, and
2. a search algorithm for finding this set of descriptions.

In general, the algorithms based on this strategy take into account the consistency and completeness conditions [39] for establishing the criterion of evaluation of a rule set. These two conditions must be satisfied for the global solution, but they can be reformulated for their application on partial solution verifying the following property: if all partial solutions satisfy the consistency and completeness conditions, then the global solutions composed by the aggregation of the partial solutions satisfy these conditions.

The problem of obtaining a partial solution can be decomposed again in several subproblems as was shown in [18]. Basically, this alternative has the following structure:

0. Let $X_1, X_2, \ldots, X_n$ be the predictive variables included for learning the behavior of a determined system. Let Y the consequent variable defined on a discrete domain composed by m concepts. Furthermore, let E be an example set that describe the behavior of the system in the past. Let GLOBAL and PARTIAL be two rule sets empty at the beginning.
1. Select a B concept that is not learned yet.
2. While PARTIAL is not a partial solution do:
   2.1. Find the best antecedent D for describing the B concept.
   2.2. PARTIAL $\leftarrow$ PARTIAL $\cup$ D.
   2.3. Penalize the description of PARTIAL with the intention that they cannot be selected again.
3. GLOBAL $\leftarrow$ GLOBAL $\cup$ PARTIAL.
4. PARTIAL $\leftarrow \emptyset$.
5. If there are more concept for learning, go to step 1. Otherwise, GLOBAL is returned as global solution of the learning problem.

Using the previous process we do not need to know a priori the number of necessary rules in order to describe a concept. We only need to establish a criterion for determining if a description set is a partial solution and a criterion to define the concept of best antecedent.

Normally, above criteria are based on the extension of the concepts of consistency and completeness. The consistency condition is easy to extend since if the partial solution satisfies this condition, each particular rule of the partial solution verifies the condition too. However, the completeness condition cannot be extended in this sense. On the other hand, we know that the aggregation of the rules obtained for a concept must verify the completeness condition. Thus, we can consider that the completeness condition establishes a criterion for determining when a set of learned rules is sufficient for describing a particular concept.

The consistency and completeness conditions define strong restrictions on the knowledge that must be learned. In a great number of real cases, it is very difficult to keep these conditions because the presence of noise or mistakes in the examples of the training set. Furthermore, this problem increases when we work with fuzzy rules. For this reason, in [21] a new formulation for these concepts was proposed.

This formulation is based on the following idea: in crisp models, the concepts of completeness and consistency are clear, however when we work with fuzzy examples or/and fuzzy rules, the concepts above must be reformulated in order to adapt them to the special characteristics of the linguistic labels. Thus, in [21] we proposed a adaptation of these concepts on fuzzy models called the degree of completeness and the degree of soft consistency for a rule, where both definitions use the concept of number of positive and negative examples as defined in [21,23].

Definition 1. The degree of completeness of a rule R (with a consequent Y=B) is defined as

$$\Lambda(R) = \frac{n^+(R)}{n_B}$$

where $n^+(R)$ is the number of positive examples of the rule R and n_B is the number of examples in the training set of the class B.

The soft consistency degree [21] is based on the possibility of admitting some noise in the rules. The definition is based on two main ideas: to extend the crisp consistency definition to fuzzy rules and moreover to make gradual this measure. Thus, in order to define the soft consistency degree we use the following sets:

$$\Delta^k = \{R \mid n^-(R) < k\, n^+(R)\}$$

with $k \in (0, 1]$ and

$$\Delta^0 = \{R \mid n^-(R) = 0\}$$

which represents in the general case the set of rules having a number of negative examples strictly less than a percentage (depending on k) of the positive examples and the complete consistent rules with k=0.

Definition 2. The degree to which a rule R with $n^+(R) > 0$ satisfies the soft consistency condition is

$$\Gamma_{k_1 k_2}(R) = \begin{cases} 1 & \text{if R} \in \Delta^{k_1} \\ \frac{k_2 n^+(R) - n^-(R)}{n^+(R)(k_2 - k_1)} & \text{if R} \notin \Delta^{k_1} \text{ and R} \in \Delta^{k_2} \\ 0 & \text{otherwise} \end{cases}$$

where $k_1, k_2 \in [0, 1]$ and $k_1 < k_2$, and $n^-(R)$, $n^+(R)$ are the number of negative and positive examples to the rule R.

This definition uses two parameters, k_1 is a lower bound of the noise threshold and k_2 is an upper bound of the noise threshold. The above formula gives a degree 1 to rules in Δ^{k_1}, that is, rules having a admissible number of negative examples (measured as a percentage, in k_1, of the number of positive examples). It gives a degree 0 to rules out of Δ^{k_2}, that is, rules having an excessive number of negative examples (measures as a percentage, in k_2, of the number of positive examples). Since, if $k_1 < k_2$ then $\Delta^{k_1} \subseteq \Delta^{k_2}$, a linear variation is assigned to rules between both extremes. In practical experimentation we have obtained good results using the extreme values $k_1 = 0$ and $k_2 = 1$.

The algorithms based on this general structure for learning concepts and that use as mechanism for searching the best rule in each iteration a genetic algorithm, are called iterative genetic algorithms [20]. This kind of algorithms is an alternative to the two classical approaches for learning using genetic algorithm, Michigan and Pittsburgh, and it presents two main advantages: the search space is reduced by the decomposition of the problem and it permits to control the quality of each particular rule.

From the point of view of the interpretability, this second feature is very important since for defining the valuation about the goodness of the rule it can be included criteria for improving the interpretative quality of the rules. In [10] criteria were proposed for evaluating the concepts of best rule in this sense. So, the evaluation function is composed by three criteria:

Criterion 1: A combination between the degree of consistency and the degree of completeness of the rule. Both measures are a relaxation of the strictness classical consistency and completeness conditions applied to individual rules. The rule with maximum value for this criterion is simultaneously a consistent rule covering a great number of examples of the concepts that must be learned from the training set. Therefore, the algorithm tends toward knowledge bases with a reduced number of consistent rules.

Criterion 2: Number of variables in the antecedent of the rule. An elemental component of the learning algorithm for applying this criterion is that the rule model allows to eliminate variables in the description of the rule (the DNF rules allows us to do it). Minimizing this criterion, we obtain rules with a low number of variables involved in its antecedent, that is, the antecedent is composed only by the variables that permit to discriminate among the class that is being learned and the rest of the class. Therefore, the criterion favors feature selection.

Criterion 3: Configuration of the assigned values to the antecedent variables. When we use DNF rules, we prefer those assignment to the variables that permit to apply the operator **not**, **less or equal than**, **greater or equal than** and **between** (described in the Section 2.2) for obtaining simpler descriptions. This criterion is called simplicity in the assignment.

Combining these three criteria by a lexicographical evaluation function, we obtain a reduced set of rules with simple assignment on a reduced number of variables in the antecedent of each rule.

The results obtained using this strategy and this criteria in a learning algorithm are good [10]. However, when there is a high number of variables involved in the learning problem can be observed that still appear unnecessary variables in the description of the rule. The reason of this behavior is the difficulty of the algorithm for considering a variable irrelevant for a concept. Therefore, it is necessary to define a feature selection process in the learning algorithm. In the next two subsections, we briefly describe some feature selection mechanisms and we study a general model of feature selection for the iterative genetic model algorithms.

2.3 Feature selection

The objective of the inductive learning algorithms consists of establishing a hypothesis for describing the behavior of the examples in the training set. Normally, the classical machine learning algorithms have associated inductive criteria for trying to find a knowledge of the system by non complex structures, being the *Occam's Razor* philosophy one of the most used criteria.

The inclusion of these criteria can be considered as a trial for defining an ideal performance of the algorithm in the sense of selecting the feature set that produces the best behavior of the system. In general, find the minimal structure is a hard-NP problem. Furthermore, the algorithms that take into account this problem, use heuristic functions for navigate through the space of feasible hypothesis.

The selection of relevant features and the elimination of the irrelevant variables is one of the main problems of the machine learning algorithms because the detection of the irrelevant variables provokes a reduction on the search space and allow to obtain more understandable rules.

A classification of the feature selection models taking into account the strategy used for evaluating the feature subsets is proposed in [6]. It distinguishes among the following three classes:

a) **Filter model**: These algorithms are called filter methods, because they filter out irrelevant variables before induction occurs. The preprocessing step uses general characteristics of the training set to select some features and exclude others. Thus, filtering methods are independent of the induction algorithm that will use their output, and they can be combined with any method (figure 1).
   The simplest filtering scheme is to evaluate each feature individually based on its correlation with the target function and then to select the k features with the highest value.
   We can find several proposed of filter methods and next we describe some of them:

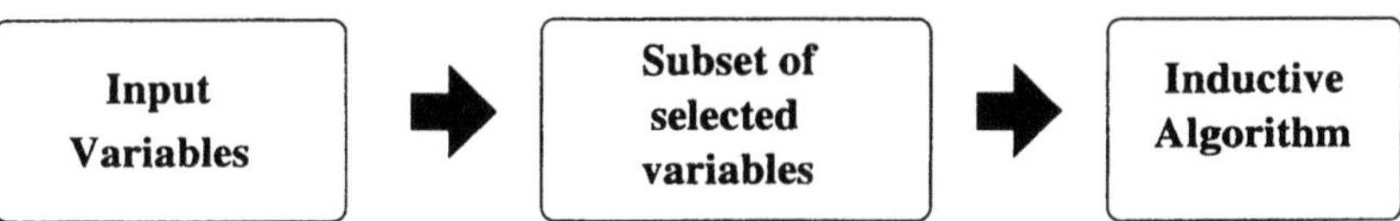

Fig. 1. Filter model

- RELIEF [27,30,47]: This is an statistical method for feature selection and it is based on a distance measurement and the assignment of weights to the variables for determining its relevance in the system. The general idea of the method consists of considering a variable as relevant if it can be used for distinguishing between an example and its nearest example. This model presents the following advantages: it only required lineal time for its execution, it is not sensible to noisy affected examples and it works with nominal and continuous variables.
- B&B-BFF [17,51]: It uses the branch and bound method for generating subset of candidate variable and the Bobrowski's method [7] for evaluating the feature subset. It is based on the monotony principle, that is, a variable subset can not be better that other variable subset when the first is contained on the second subset. BFF applies a search process based on the heuristic strategy *best first*. This model obtains the optimum subset when the evaluation function is monotonous. However, normally the functions used do not satisfy this property.
- DTM (Decision Tree Methods) [11,29]: They use decision trees for determining the feature subset, considering that the variables that do not appear in any node of the tree, are not relevant for the learning process. These methods are very quick in computation but they do not work appropriately with a noisy training set.
- FOCUS [5]: It implements *Min-Feature bias* using the consistency hypothesis. We can find two variants of this algorithm: MIFES [48] and the Schlimmer's method [46]. These algorithms do not present a good behavior on noise affected training sets.
- Furthermore, there are other algorithms, such as, LVF [35] that uses a random search process based on the *Las Vegas* algorithm, POE+ACC [41] that obtains the feature subset appending one variable in each step and it takes two measures for determining the goodness of each subset: the probability of including mistakes in the description of the system (POE) and the average of the correlation coefficient between the variables (ACC), and *MDLM (Minimum Description Length Methods)* [45] that tries to split the feature set in two subsets: one of them with the relevant features and the other with the irrelevant features. It use an information measured and an exhaustive search for finding the best partition.

b) **Wrapper model**: In the same idea of the filter model, the feature selection on the wrapper model occurs outside the basic induction method, but use that method as a subroutine, rather than as a postprocessor (figure 2). The typical wrapper algorithm searches the same space of feature subsets as filter methods, but it evaluates alternatives sets by running some induction algorithm on the training data and using the estimated accuracy of the resulting classifier as its metric.

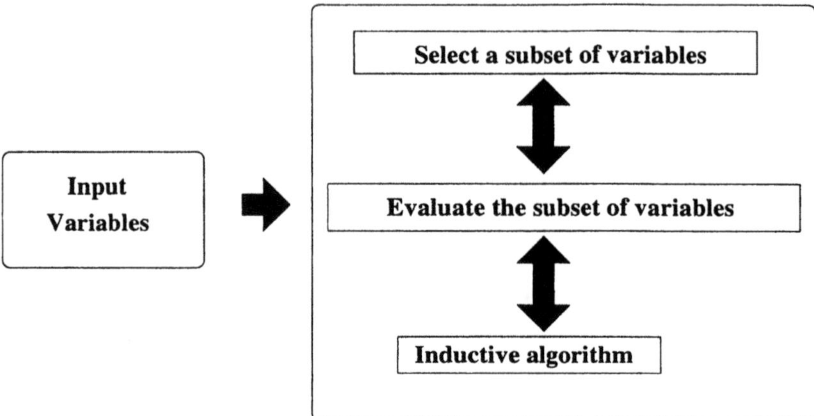

Fig. 2. Wrapper model

The general argument for wrapper approaches is that the induction method that will use the feature subset should provide a better estimate of accuracy than a separate measure that may have an entirely different inductive bias.
The major disadvantage of wrapper methods over filter methods is the associated computational cost, which results from calling the induction algorithm for each feature set considered.
Among the different approaches we describe briefly some of them:

- Kohavi and John wrapper model [28]: In this model, the inductive algorithm is considered a black box that evaluates the goodness of the feature subset. It uses a hill-climbing strategy and two operators: one of them includes new variables and the other eliminates variables in the feature subset. The optimum subset is the feature subset that presents the best accuracy on the examples with the knowledge obtained by the inductive algorithm.
- OBLIVION [33]: It algorithm combines the wrapper idea with the simple nearest-neighbor method, which assigns to new instances the class of the nearest case stored in memory during learning. The feature-selection process effectively alters the distance metric used in these

decisions, taking into account the features judged relevant and ignoring the others. OBLIVION carries out a backward elimination search through the space of feature sets, starting with all features and iteratively removing the one that leads to the greatest improvement in estimated accuracy. The algorithm continues this process until the estimated accuracy actually declines. For estimating accuracy, the process uses leave-one-out cross-validation on each feature set on novel test cases.
   - Aha and Bankert wrapper model [2]: This algorithm describes a technique much like OBLIVION, but it starts with a randomly selected subset of features and includes an option for beam search rather than greedy decisions.
   - Most research on wrapper methods has focused on classification, but some works, such as [40,49], combine this idea with k-nearest neighbor for numeric prediction.

c) **Embedded model**: Methods for inducing logical descriptions provide the clearest example of feature selection methods embedded within a basic induction algorithm. In fact, many algorithms for inducing conjunctions do little more than add or remove feature from the concept description in response to prediction errors on new instances.
   Similar operations for adding and removing features form the core of methods for inducing more understandable description on the concepts, but some methods also involve routines for combining features into richer descriptions.
   The most known algorithms that include, in any sense, embedded feature selection are the algorithms based on decision trees such as CART [9] and C4.5 [44] and some algorithms that represent the knowledge by DNF rules such as the AQ family [39], CN2 [12], SIA [50] and SLAVE [23].

In general, the filter and wrapper approaches can be applied to any inductive learning algorithm. On the other hand, the embedded approach need to be defined during the development stage of the learning algorithm. However, we can find some general feature selection models that can be applied to inductive learning algorithms for permitting to select feature subset during the learning process.

In the next section, we describe a general model for selection features embedded in learning algorithm based on genetic iterative approach.

A general embedded feature selection model for learning algorithms based on genetic iterative approach The main component of the genetic iterative learning algorithms is the module of rule extraction. In this module, a genetic algorithm attempts to find the best combination of value assignment to the antecedent variables of a rule within the particular concept.

For the definition of the general embedded feature selection model, we suppose the existence of a genetic iterative learning algorithm with the following characteristics:

1. It uses DNF rules with binary coding for representing the information about the problem.
2. In the evaluation function there is a component that includes a preference criterion in a similar way to the criterion 2 described in the Section 3, that is, it defines a criterion for selecting the rules with the least number of variables involved in its antecedent.
3. There are an appropriate set of genetic operators for working with the genetic representation of the DNF rules.
4. The genetic algorithm uses a way to create an initial population of solutions taking information from the training data.

The combination of the above first two characteristics gives to the algorithm the capacity of detecting irrelevant variables. However, the task of transforming relevant variables into irrelevant variables in the genetic algorithm depends of the size of the coding of each variable and the number of predictive variables involved in the learning problem. So, with the basic representation of a DNF rule, it is more difficult to eliminate a variable than to add it, and the system might take a long time to find a good solution.

Consequently, if we want to improve the detection of the irrelevant variables, we first need to change the genetic representation. So, our goal consists of obtaining a better representation of the genetic solutions to make a feature selection for each possible rule.

The idea is to include a new value associated to each antecedent variable in order to discover if the variable will be considered as part of the antecedent of the rule or not.

Let X_i be a predictive variable involved in a learning problem which domain is composed by m_i labels. In the new representation, we propose coding the X_i variable using m_i+1 values, where one of these values determines the relevance or irrelevance of the variable, and the others determine the values assigned to the variable if it is considered to be relevant. With this new coding, a simple mutation on the new component changes its relevance status. This coding therefore reduces the computational cost associated with the detection of the irrelevant variables.

By considering this process for all the antecedent variables, we have two structures to represent the complete antecedent of a rule: One codifies the relevance of the variables and the other codify the assignment variable/value. With this decomposition, the genetic representation has a complex chromosome composed of two structures: a variable chromosome and a value chromosome. This division allows us to clearly distinguish between the two different tasks that are simultaneously carried out in the genetic algorithm (search for the appropriate variables and search for the appropriate value assignments) and we can associate the most appropriate set of genetic operators and set these operators on the variable chromosome.

The variable chromosome keeps information that tries to estimate the relevance degree of each variable with respect to the class that must be learned. With the objective of extracting knowledge about the relevance degree of the variables on the training set for defining the initial population, we define the following information measurement:

$$\tau(X,Y) = \frac{I(X,Y)}{H(X,Y)}$$

where

$$I(X,Y) = \sum_x \sum_y p(x,y) log_2 (\frac{p(x,y)}{p(x).p(y)})$$

is the Kullback's information measure [31] and

$$H(X,Y) = \sum_x \sum_y p(x,y) log_2 p(x,y)$$

is the Shanon Entropy over two variables.

The τ measure estimates the dependence between variables X and Y in the following way: Values of $\tau(X,Y)$ close to zero determine a high degree of independence of both variables, whereas values close to one demonstrate a high degree of functional dependency between them.

In the iterative learning algorithm, we select rules fixing a value of the consequent variable. We therefore need to restrict the previous measure for the particular class that is being learned, and we define

$$\tau_C(X) = \frac{I(X,Y=C)}{H(X,Y=C)}$$

where X is a predictive variable, Y is a consequent variable and C is a particular value of the consequent variable.

Like $\tau_C(X)$ measures the dependence or independence degree between the X variables and the C value of the consequent variable, we can interpret this value as the relevance degree of each predictive variable with respect to the C class of the Y variable.

When linguistic variables are used, we must define an adequate way for calculating these values. In [19], we can find a general methodology of belief calculation using fuzzy information.

Thus, we can use $\tau_C(X_i)$ as a measure of the initial relevance of variable X_i for the C class. This relevance measure will be included in the code for the variable chromosome on each individual of the initial population and during the execution, the evolutionary process must define the tendency toward the relevance or irrelevance of the variables.

For an adequate representation of these values a real coding is used and a set of genetic operators for this coding is defined, for example, in [25] the nonuniform mutation and BLX_α crossover operators are used as genetic operator on the variable chromosome.

A possible interpretation is obtained if we include an activation threshold inside the variable chromosome. The inclusion of relevance measures and activation thresholds has previously been used in feature selection processes, such as [27], but in our case, a different activation threshold will be assigned to each chromosome and learned during the evolution process.

Thus, a X_i variable will be considered to be a component of the antecedent of a rule for the C class if $\tau_C(X_i) \leq T_j$ where T_j represents the activation threshold for the different X_i variables of the j individual of the population. Otherwise, the variable will be considered to be irrelevant for the antecedent. Initially, the activation threshold is randomly defined for each element of the population taking a value in the interval $[\min_i \tau_C(X_i), \max_i \tau_C(X_i)]$.

The values T_j are affected by the genetic operators during the evolution of the genetic algorithm and therefore the activation threshold are learned by the algorithm.

In [25] are shown the results obtained by the application of this general model of feature selection on a genetic iterative learning algorithm called SLAVE. The experimentation shows that the modified algorithm improves clearly the feature selection of the original algorithm obtaining a more reduced set of simpler rules and maintaining, and in some cases improving, the accuracy on the databases used[2].

This general feature selection model can be extended to other machine learning algorithms that use a genetic algorithm as search algorithm. In [1] we can find an extension for adapting this model to a learning algorithm based on the Pittsburgh approach.

3 Improving the accuracy keeping the interpretability in LFM

The objective of the LFM algorithms consists of obtaining fuzzy models with a good interpretability and an acceptable accuracy. In the previous section we have described some tools that permit to increase the interpretability of the knowledge learned. In this section, we are interested in describing some

[2] In theory, the selection of a subset of variables reduces the search space, but it does not imply to improve the prediction capacity of the system obtained, since if we can obtain certain degree of accuracy with a subset of variables, with the whole set of variables, at least, we can obtain the same degree of accuracy. However, in practice where the time is limited, it is easier to find a better solution in a reduced search space than in the whole search space of rules. This fact is the reason because in some cases the feature selection helps to the algorithm to improve the accuracy.

elements and processes for improving the knowledge accuracy without loosing its interpretability. In concrete, we show a mechanism for solving a problem that appears in the algorithm based on any iterative strategy about the co-operation/competition relations of the learned rules. On the other hand, we describe and analyze some proposals for modifying the semantic of the linguistic labels that define the domain of the variables involved in the learning problem and we focus our description in the semantic hedges.

3.1 The cooperation/competition problem

The iterative model decomposes the learning problem in a subset of sub-problems that must be learned. The global solution is constructed as the aggregation of the partial solutions obtained. This way for constructing the global solution provides correct descriptions of the concepts involved in the learning problem when there are not interactions among the concepts, that is, each rule learned verifies the consistency condition and each set of learned rules for a concept satisfies the completeness condition.

However, in the real world problems are very difficult to keep previous conditions. When the completeness condition is not verified and fundamentally when the consistency condition is not satisfied, then the global solution obtained by the aggregation of the partial solutions presents an imprecise description of the system, in the following sense: there are rules of different concepts that can be applied to the same zone of the example space.

In these situations may be convenient to keep the global solution near of verifying the consistency and completeness conditions, relaxing the satisfaction of these conditions in the partial solutions, but for doing this, it is necessary that the learning algorithm has a global vision about the concepts that must be learned.

This global vision of the problem is not only useful when the examples are affected by noise or imprecision. Knowing that the training data is not affected by noise or imprecision it can be convenient to lost the satisfaction of the consistency condition for improving the comprehensibility of the descriptions, keeping the accuracy on the example set. Typical examples of this situation are the rules that are exceptions to other rules (exception rules).

In the iterative model, the learned rules only establish cooperation relations among the rule of a same concept with the objective of satisfying the completeness condition. However, among the rules with different concept do not establish any relations and without these relations is not possible to detect exception rules and therefore it is not possible to select simpler rules.

For this reason, it is interesting to include in the learning process cooperation/competition relations among the rules of different concepts, where the cooperation among them permits to relax the consistency condition when each rule is being learned and the competition permit to relax the completeness condition.

In the literature, we can find several works that try to solve this problem. In [23] is proposed the inclusion of these relations by the redefinition of the negative and positive example concepts. The new definitions take into account the inference process used for reasoning with the learned rules. During the learning process is not known the complete rule base for solving the problem, but it is known a partial information about its. The new definitions of negative and positive examples use this partial information for establishing cooperation/competition relations with the new rules that must be learned.

Other alternative can be found in [13]. In this case, the learning process is composed by three stages: rule extraction, rule simplification and tuning. In the stage of the rule extraction is selected a set of rules using a genetic iterative algorithm. This first stage finishes when all the examples are sufficiently covered by the rule set. The rule simplification stage is defined as a competitive process for improving the cooperation between the rules which objective is to reduce the number of rules and to increase the accuracy. The last stage is a tuning process of the linguistic labels for improving the accuracy. A modified version of the previous process can be found in [14].

In a similar sense of the previous idea, some post processing mechanisms such as the refinement theory algorithms can be considered as mechanisms that permit to improve the cooperation/competition relations among the rules. We analyze these algorithm in the section 5.

3.2 Semantic hedges

One of the main problems when we try to use a learning algorithm consists of fixing a correct discretization for the continuous variables. The input of some algorithms [14,23,36,37] is a discretization based on the use of fuzzy domains. In many cases, the information provided by an expert is basic for the correct behaviour of these learning algorithms. A bad discretization of some variable may generate a strong limitation in the learning algorithm, since its predictive capacity is related with this discretization. Thus, in order to improve the learning algorithm behavior, it is basic to include the possibility of either to generate appropriate discretizations or to modify an initially proposed discretization.

There are different learning algorithms that can make up or modify this discretization of the domains:

- Some of them can make up the discretization during the learning algorithm. For example, we can consider decision tree learning algorithms like C4.5 [44] or we can consider rule learning algorithms like CN2 [12]. Usually, they obtain intervals using statistical methods.
- Other learning algorithms use an initial discretization, and once we have obtained a particular knowledge, a later process modify this initial knowledge by tuning the discretization. This idea has been widely used in the fuzzy rule learning literature. An example of these systems is [3,4,13].

A third option consists of defining an initial discretization, for example proposed by an expert or a uniform distribution of the labels, and during the learning process to modify the semantic of this discretization. Thus, the learning algorithm uses an initial data base (membership function of the labels), but it has the capability to adapt the semantic of the different labels to the training examples. An interesting way for modifying the semantic of the linguistic labels is made through the use of linguistic hedges [53]. Therefore, the solution consists of extending the usual basic domain of the linguistic variables (composed by a finite domain) to a richer description. In order to include this modification, researchers have presented several proposals of rule models that include the linguistic hedges.

A semantic modifier is a β function on the set of labels, D, that takes values in a set of fuzzy sets $\mathcal{R}$, that is,

$$\begin{aligned} \beta\colon D &\longrightarrow \mathcal{R} \\ \mathrm{L} &\longrightarrow \beta(L) \end{aligned}$$

where $\mu_{\beta(L)}(x) = \beta(\mu_L(x)) \in [0,1]$ is the modified label of the original label L

Using this general kind of linguistic hedges, the model of fuzzy rule that we consider is

$$\text{IF } X_1 \text{ IS } \beta_1(A_1) \text{ AND } X_2 \text{ IS } \beta_2(A_2) \ldots X_n \text{ IS } \beta_n(A_n) \text{ THEN } Y \text{ IS } B.$$

We can find several works where the semantic modifiers has been used for improving the accuracy without lost interpretability of the knowledge learned. A deep study about the inclusion of the linguistic hedges in a learning algorithm was made in [43]. Next, we describe some of them:

- **Shape modifying linguistic hedge**: Under this category we enclosed those semantic modifiers that modify the possibility distribution of the membership function of the label but maintain the support of the linguistic label.

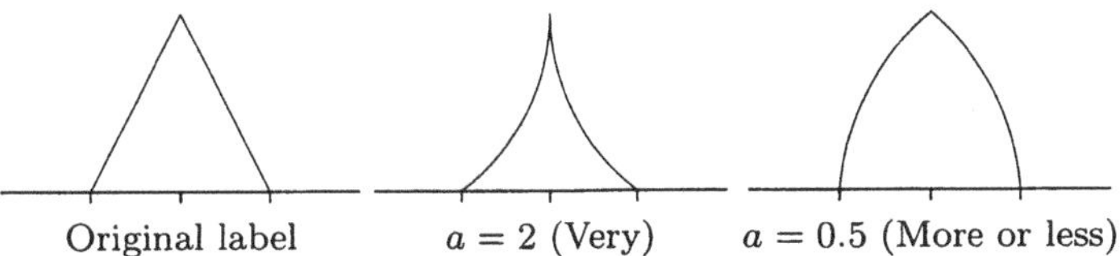

Fig. 3. Powered modifiers

Among them, the semantic modifiers that verify to be of the class $\beta(L) = L^a$ are used in a great number of algorithms and they are called **Powered modifiers**. These linguistic hedges verify the following interesting

property: if $\nu_L(x) \geq \nu_L(y)$ then $\beta(\nu_L(x)) \geq \beta(\nu_L(y))$, that is, they maintain the order of the support of the original label with respect to the membership of each element, and therefore the modified linguistic label has been obtained by a slight alteration on the initial label. This operator has two different interpretations depending of the value of a parameter (see figure 3: for values greater than 1 the modified label is more precise than the original one, on the contrary, for value between 0 and 1, the effect is the opposite, that is, the label increase the imprecision on the original label.
Two particular cases of this operator are when $a = 2$ that represents the linguistic hedge "VERY" and when $a = \frac{1}{2}$ that represents the linguistic hedge "MORE OR LESS" [52]. These two semantic modifiers have been used in works such as [14,24,34]
Another class of modifiers that alters the shape of the label without change its support is the **modifiers of symmetry** (see figure 4). Basically, these operators modify the symmetry of the label moving its core along of its support. They are very useful in modeling problems, however it is difficult to associate them a linguistic interpretation

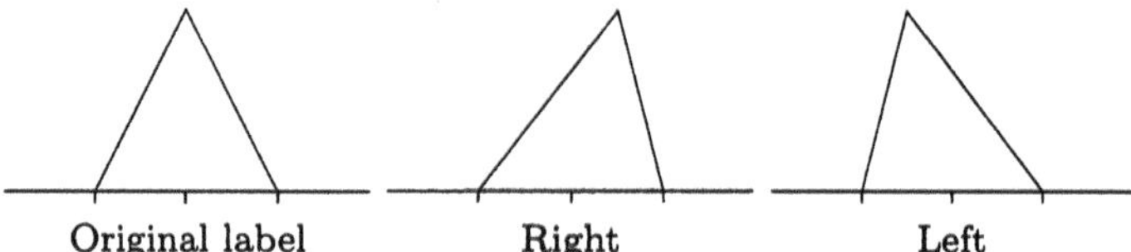

Fig. 4. Modifiers for changing the symmetry

- **Expansive/reduced modifiers** [8]: These modifiers change the support and core sets of the fuzzy sets enlarging or reducing them, but trying to keep a center of gravity similar to the original (see 5).

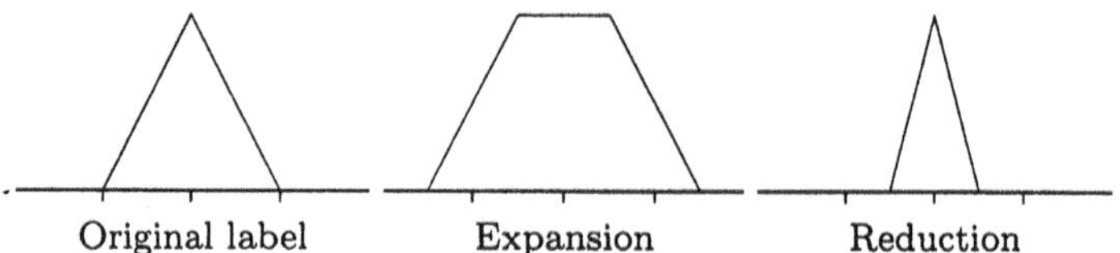

Fig. 5. Expansive/reduced modifiers

The linguistic label obtained by the application of these modifiers can be considered as a modification that allow us to increase or to decrease the imprecision of the label, in an alternative way to the Powered modifiers. Thus, they may be a variation of "VERY" or "MORE OR LESS" linguistic hedges.

- **Shifted modifiers** [8,32]: These modifiers move the linguistic labels along their domains (figure 6). Normally, the application of these modifier is restricted for keeping the relative order in the labels distribution on the domain.

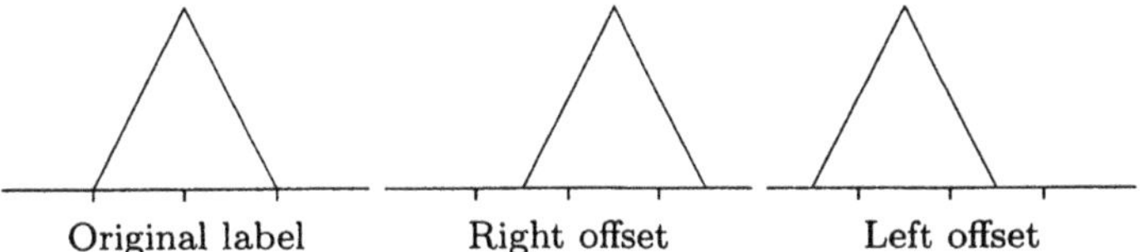

Fig. 6. Shifted modifiers

A linguistic interpretation of these modifiers may be "SOMETHING MORE" or "SOMETHING LESS".

The inclusion of one of the previous modifiers improves in an important way the capacity of knowledge representation of a learning algorithm. However, only one of the above linguistic hedges may be not sufficient for representing correctly some problems. This fact supposes that in complex problems is necessary more than one kind of previous modifiers. In these sense, a DNF rule model that includes parametrized extensions of some linguistic hedges is proposed in [43]. Furthermore, this work proposes a comprehensible combination of the modifiers and an embedded method on the learning algorithm for obtaining the appropriate configuration of these linguistic hedges that allow to improve the accuracy of the knowledge, keeping or reducing the number of rules.

4 Improving the accuracy and the interpretability in LFM

The accuracy and interpretability of a DNF fuzzy rule set can be improved simultaneously using a refinement theory algorithm. The main idea is to start from a theory, that is, a initial set of DNF rules (that can be empty, incorrect or incomplete) obtained by an expert or a learning algorithm, and to try of "improving" this set using an example set and processes of simplification, addition, elimination, or merge of rules. The main difference among the different refinement algorithms is the heuristic used for the previous processes. Thus, the heuristic used in the EITHER [42] algorithm improves the rule set but minimizing the number of changes on the initial knowledge. However, we can find another algorithm such as JoJo [16] where its heuristic is only guided by the accuracy of the rule set. This previous algorithms were designed for working with crisp DNF rules. In [22] a refinement algorithm of DNF fuzzy rules was proposed. This algorithm is useful to improve the accuracy and interpretability of the knowledge obtained by LFM algorithms.

The refinement algorithm proposed in [22] tries to adapt the initial fuzzy rule set to the information contained in the example set and simultaneously to obtain a simple and accurate new theory.

The improvement in the theory is analyzed by two point of view:

- to improve the description of the theory, that is, to make more understandable the theory,
- to improve the accuracy of the theory on the considered set of examples.

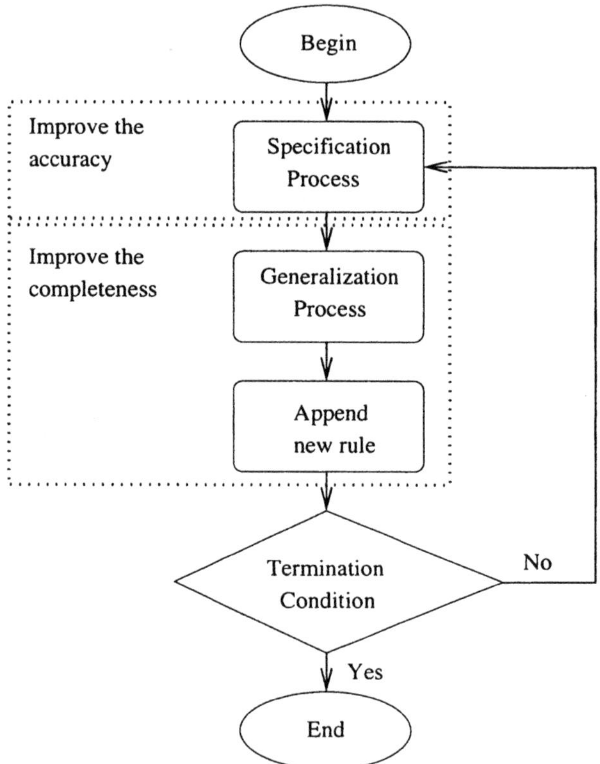

Fig. 7. The refinement theories process

The process consists of a heuristic search that allow to generalize, specify, add and eliminate rules. In each step, the algorithm tries to apply some of these actions in order to improve the interpretability of the theory, defined by a set of DNF rules, and each local change will be accepted if it doesn't make worse the accuracy. The refinement algorithm uses a heuristic and a hill-climbing strategy for selecting the most promising action in each step of the algorithm toward a good solution. The heuristic components selects rules in a particular order. This order is based on the weight of each rule. The weight is defined as a rate between accuracy of the individual rule and the number of covered examples and calculated on the example set. The rule with the highest weight is considered the first one and the most relevant, and so on.

The complete process is shown in figure 7 where from the initial set of rules, the algorithm tries to improve the accuracy by specifying the rules, after it tries to improve the completeness in two steps, first by generalizing rules and later adding new rules.

Now, we will describe the different steps in the refinement algorithm with more detail.

1) Specification process

In this step the refinement algorithm tries to improve the accuracy of the theory. The idea is to specify the rules in order to reduce the number of failures in each rule. Thus, this process begins by specifying the rules in the order in which they appear in the rule set. We shall say that a value for a variable is active if this value appears for that variable in the description of the rule. So, we shall say that a value for a variable is not active if this value does not appear for that variable in the description of the rule. For example, given the following rule R_1:

IF
  X_1 is (a_1 or a_2 or a_3) and
  X_2 is b_2
THEN
  Y is c_2

where $X_1 = \{a_1, a_2, a_3, a_4, a_5\}$, $X_2 = \{b_1, b_2, b_3\}$ and $Y = \{c_1, c_2, c_3\}$. From the above rule can be generated a new rule by the elimination of an active value. For example, removing a_1 of the X_1 variable we obtain the following rule R_2:

IF
  X_1 is (a_2 or a_3) and
  X_2 is b_2
THEN
  Y is c_2

Because $R_2 \subset R_1$ we can say that the R_2 is more specific than R_1 and furthermore R_1 is a generalization of the R_2 rule.

In this way, a specification operation can be considered as changing to non-active a value that previously was active and we will say that this value is deactivated. This process can be described in the following way:

1. The most relevant rule is selected.
2. Select a variable and do the following:
   - 2.1. Select an active value.
   - 2.2. This value is deactivated.
   - 2.3. If the overall accuracy is improved or maintained, the modification is accepted. Otherwise, the value is activated.
   - 2.4. If there are other active values, go to 2.1.
   - 2.5. If all the values of the variable are deactivated, then this rule is removed from the rule set.
   - 2.6. If there are other unselected variables, go to 2.
3. If there are more unselected rules, select the next most relevant rule and go to 2., otherwise the process finishes.

Furthermore, we must note that this step allows us to remove rules from the rule set. Therefore, the elimination of a rule from the rule set is considered as a particular case of specification and it is caused when all the values of an antecedent variable are deactivated. Following the previous example, the elimination of the value b_2 of the X_2 variable of the R_1 rule implies the elimination of this rule of the rule set.

2) Generalization process

In this second step, the refinement algorithm attempts to improve the completeness of the rules by using an ordered process of generalization. In this process, new values are appended to the antecedent variables for each rule. This generalization is done when the overall accuracy is kept or strictly improved, using the rule set that includes the modified rule. There are two reasons for the improvement in accuracy:

a) Unclassified examples can be correctly covered by generalizing one of the rules.

b) Examples incorrectly classified by one rule with a lower order can be correctly covered when a rule with a higher order is generalized.

Consequently, this step of the algorithm improves the completeness and the accuracy of some of the rules of the rule set.

This procedure has a similar structure to the previous algorithm, but the order of rules is reversed and it begins by selecting the least relevant rule:

1. The least relevant rule is selected.
2. Select a variable and do the following:
   2.1. Select a deactivated value.
   2.2. This value is activated.
   2.3. If the overall accuracy is strictly improved or maintained, the modification is accepted. Otherwise, the value is deactivated.
   2.4. If there are other deactivated values, go to 2.1.
   2.5. If there are other unselected variables, go to 2.
3. If there are more unselected rules, select the next most relevant rule and go to 2., otherwise the process finishes.

In this process, the inverse order is used for the following reason: the inference method used favors the rules that appear in the first positions in the rule set when the conflict problems appear. A rule that has the T position in the rule set may include in its description the descriptions of all the rules that have a lower position than T, if it does not reduce the accuracy of the systems. This fact means that the first rules are more specific than the last rules in the rule set. So, the last rules have a higher probability of being generalized than the first rules.

3) Append new rules

The next step in the algorithm, consists of appending new rules to cover the examples that are not covered by any rule and which the previous process cannot cover. This task adds the most specific rule with the best adaptation for each example that is not covered in the training set in the position with less relevance in the rule set. The addition of new rules on the rule set may decrease the interpretability but it is important for improving the accuracy. Furthermore, this lost of interpretability may be temporal, since it is possible that with the inclusion of a new rule, other rules with less accuracy could be removed in the next iteration.

The previous steps are repeated until the rule set is stable. In the algorithm, the rule set is determined as being stable when the number of rules and the accuracy is maintained in two consecutive iterations.

The refinement algorithm obtains a reasonably small set of accuracy DNF fuzzy rules which can be easily understood from a human point of view. The main idea of the process is the use of a particular heuristic to combine the processes of specification, generalization, addition and elimination of rules.

5 Conclusions

The main goal of this chapter has been to gather different proposals about the improving of accuracy and interpretability in LFM. Many of these proposals were applied on the learning algorithm SLAVE along different papers, but the idea is to show how these proposals are really useful in many others fuzzy rule learning algorithms. Thus the description of the proposals has tried to be enough general to be applied on other algorithms. These proposals have been ordered in three groups, those that improve interpretability, those that improve accuracy keeping the interpretability and finally those that improve both characteristics simultaneously. These proposals are mainly based in the use of the linguistic DNF rule model. In the chapter we have are considered important techniques for machine learning such as the selection features, semantic modifiers and refinement theories.

References

1. Aguirre E., González A., Pérez R. (2001) A feature selection model for learning algorithms based on Pittsburgh approach (in Spanish). Accepted in First Spanish Congress of Evolutionary and Bio-inspired Algorithms, Merida, Spain.
2. Aha D.W., Bankert R.L. (1996) A comparative evaluation of sequential feature selection algorithms. Artificial Intelligence and Statistical V (D.Fisher and J-H Lens eds), Springer, New York.
3. Alcalá R., Casillas J., Castro J.L., González A., Herrera F. (2001) A multicriteria genetic tuning for fuzzy logic controllers. Mathware and Soft Computing, **8(2)**, 179–201.
4. Alcalá R., Benítez J.M., Casillas J., Cordón O., Pérez R., Fuzzy Control of HVAC Systems Optimized by Genetic Algorithms. Accepted in Applied Intelligence.
5. Almuallin H., Dietterich T.G. (1992) Learning with many irrelevant features. Proc. 9th National Conference on Artificial Intelligence, MIT Press, Massachusetts, 547–552.
6. Blum A.L., Langley P. (1997) Selection of relevant features and examples in machine learning. Artificial Intelligence, **97**, 245–271.
7. Bobrowski L. (1988) Feature selection based on some homogeneity coefficient. Proc. 9th International Conference on Patter Recognition, 544–546.
8. Bouchon-Meunier B., Jia Y. (1992) Linguistic modifiers and imprecise categories. Internation Journal of Intelligent Systems, **7**, 25–36.
9. Breiman L, Friedman J.H., Olshen R.A., Stone C.J. (1984) Classification and Regression Trees. Wadsworth Belmont, CA.
10. Castillo L., González A., Pérez R. (2001) Including a simplicity criterion in the selection of the best rule in a genetic fuzzy learning algorithm. Fuzzy Sets and Systems, **120(2)**, 309–321.
11. Cardie C. (1993) Using decision trees to improve case-based learning. Proc. 10th International Conference on Machine Learning, 25–32.
12. Clark P., Niblett T. (1986) Learning if-then rules in noisy domains. TIRM 86-019, The turing institute, Glasgow, (1986).

13. Cordón O., Herrera F. (1997) A three-stage evolutionary process for learning descriptive and approximate fuzzy-logic-controller knowledge bases from examples,. International Journal of Approximate Reasoning, **17**, 369–407.
14. Cordón O., del Jesús M.J., Herrera F. (1998) Genetic learning of fuzzy rule-based classification systems cooperating with fuzzy reasoning methods. International Journal of Intelligence Systems, **13**, 1025–1053.
15. De Jong K.A., Spears W.M., Gordon D.F. (1993) Using Genetic Algorithms for Concept Learning. Machine Learning, **13**, 161–188.
16. Fensel D., Wiese M. (1993) Refinement of Rule Sets with JoJo. Lectures Notes in Artificial Intelligence, **677**, 378–383.
17. Foroutan I., Sklansky J. (1987) Feature selection for automatic classification of non-gaussian data. IEEE Transactions on Systems, Man and Cybernetics, **17(2)**, 187–198.
18. González A., Pérez R., Verdegay J.L. (1994) Learning the structure of a fuzzy rule: a genetic approach. Fuzzy Systems and Artificial Intelligence, **3(1)**, 57–70.
19. González A. (1995) A learning methodology in uncertain and imprecise environment. International Journal of Intelligence Systems, **19**, 357–371.
20. González A., Herrera F. (1997) Multi-stage Genetic Fuzzy Systems Based on the Iterative Rule Learning Approach. Mathware and Soft Computing, **4(3)**, 233–249.
21. González A., Pérez R. (1998) Completeness and consistency conditions for learning fuzzy rules. Fuzzy Set and Systems, **96(1)**, 37–51.
22. González A., Pérez R. (1998) A fuzzy theory refinement algorithm. International Journal of Approximate Reasoning, **19**, 193–200.
23. González A., Pérez R. (1999) SLAVE: A genetic learning system base on an iterative approach. IEEE Transaction on Fuzzy Systems, **7(2)**, 176–191.
24. González A., Pérez R. (1999) A study about the inclusion of linguistic hedges in a fuzzy rule learning algorithm. International Journal of Uncertainty, Fuzziness and Knowledge-Based Systems, **7(3)**, 257–266.
25. González A., Pérez R. (2001) Selection of relevant features in fuzzy genetic learning algorithm. IEEE Transaction on Systems, Man and Cybernetics, **31(3)**, 417–425.
26. Janikow C.Z. (1993) A Knowledge-intensive Genetic Algorithm for Supervised Learning. Machine Learning, **13**, 189–228.
27. Kira K., Rendell L.A. (1992) The feature selection problem: Traditional methods and a new algorithm. Proc. 9th National Conference on Artificial Intelligence, **129**.
28. Kohavi R., John G.H. (1997) Wrapper for feature subset selection. Artificial Intelligence, **97**, 273–324.
29. Koller D., Sahami M. (1996) Toward optimal feature selection. Proc. 13th International Conference on Machine Learning.
30. Kononenko I. (1994) Estimating attributes: Analysis and extension of RELIEF. Proc. European Conference on Machine Learning, 171–182.
31. Kullback S. (1968) Information theory and statistics. New York: Dover.
32. Lakoff G. (2001) Hedges: a study in meaning criteria and the logic of fuzzy modeling. Fuzzy Sets and Systems, **123(3)**, 343–358.
33. Langley P., Sage S. (1994) Oblivious decision trees and abstract cases. Working notes of the AAAI-94, 113–117.

34. Liu B.D., Chen C.Y., Tsao J.Y. (2001) Design of adaptive fuzzy logic controller based on linguistic-hedge concepts and genetic algorithms. IEEE Transaction on Systems, Man and Cybernetics Part B, **31(1)**, 32–53.
35. Liu H., Setiono R. (1996) A probabilistic approach to feature selection- a filter solution. Proc. International Conference on Machine Learning, 319–327.
36. Magdalena L. (1997) Adapting the Gain of an FLC with Genetic Algorithms. International Journal of Approximate Reasoning, **17**, 327–349.
37. Magdalena L., Monasterio F. (1997) A Fuzzy Logic Controller with Learning Through the Evolution of its Knowledge Base. International Journal of Approximate Reasoning, **16**, 335–358.
38. Mamdani E.H. (1974) Applications of fuzzy algorithms for control a simple dynamic plant. Proceedings of the IEEE 121, 1585–1588.
39. Michalski R.S. (1981) Theory and Methodology of inductive learning. Machine Learning: An artificial intelligence approach, 33–56.
40. Moore A.W., Lee M.S. (1994) Efficient algorithms for minimizing cross validation error. Proc. 11th International Conference on Machine Learning, 190–198.
41. Mucciardi A.N., Gose E.E. (1971) A comparison of sever techniques for choosing subset of patter recognition. IEEE Transactions on Computers, **C-20**, 1023–1031.
42. Ourston D., Mooney R.J. (1994) Theory refinement combining analytical and empirical methods. Artificial Intelligence, **66**, 273–309.
43. Pérez R. (1997) Learning fuzzy rules using genetic algorithms (in Spanish). PhD dissertation, Dpto. Ciencias de la Computación e Inteligencia Artificial, Universidad de Granada (Spain).
44. Quinlan J.R. (1993) C4.5: Programs for machine learning. Morgan Kaufmann, San Mateo, CA.
45. Rissanen J. (1978) Modeling by shortest data description. Automatica, **14**, 465–471.
46. Schlimmer J.C. (1993) Efficiently inducing determinations: A complete and systematic search algorithm that uses optimal pruning. Proc. 10th International Conference on Machine Learning, (P. B. Brazdil, ed), 284–290.
47. Segen J. (1984) Feature selection and constructive inference. Proc. 7th International Conference on Patter Recognition, 1344–1346.
48. Sheinvald J., Dom B., Niblack W. (1990) A modeling approach to feature selection. Proc. 10th International Conference on Patter Recognition, **1**, 535–539.
49. Townsend-Weber T., Kibler D. (1994) Instance-based prediction of continuous values. Working notes of the AAAI-94, Workshop on Case-Based Reasoning, Seattle, WA, 30–35.
50. Venturini G. (1993) SIA: a Supervised Inductive Algorithm with Genetic Search for Learning Attributes based Concepts. Machine Learning: ECML-93, 280–296.
51. Xu L., Yan P., Chang T. (1988) Best first strategy for feature selection. Proc 9th International Conference on Patter Recognition, 706–708.
52. Zadeh L.A. (1972) A fuzzy-set theoretic interpretation of linguistic hedges. Journal of Cybernetics, **2(2)**, 4–34.
53. Zadeh L.A. (1975) The concept of a linguistic variable and its applications to approximate reasoning. Part I Information Sciences, **8**, 199–249. Part II Information Sciences, **8**, 301–357, Part III Information Sciences, **9**, 43–80.

An Iterative Learning Methodology to Design Hierarchical Systems of Linguistic Rules for Linguistic Modeling

Rafael Alcalá[1], Oscar Cordón[2], Francisco Herrera[2], and Igor Zwir[3]

[1] Department of Computer Science,
University of Jaén, Jaén, Spain
e-mail: alcala@ujaen.es

[2] Department of Computer Science and Artificial Intelligence,
University of Granada, Granada, Spain
e-mails: {casillas,ocordon,herrera}@decsai.ugr.es

[3] Department of Computer Science,
University of Buenos Aires, Buenos Aires, Argentina
e-mail: zwir@dc.uba.ar

Abstract. Among the different Fuzzy logic-based techniques applied to System Modeling, Linguistic Modeling refers to the use of Fuzzy Rule-based Systems with higher description ability. However, these kinds of systems suffer from a weakness which is its lack of accuracy in some complex problems. This feature is a consequence of the inflexible linguistic rule structure -composing the Fuzzy Rule-based Systems Knowledge Bases- and the biases of the typical identification methods designed to learn these linguistic rules from examples.

In this chapter we present a more flexible Knowledge Base structure that allows us to improve the accuracy of linguistic models without losing their interpretability to a high degree, the Hierarchical Knowledge Base. Furthermore, a proper Iterative Hierarchical Systems of Linguistic Rules Learning Methodology is designed for inductively identify the former structure, trying to improve current learning methods and to avoid their lacks as prototype identification algorithms.

The behavior of the final hierarchical linguistic models are analyzed from the accuracy and interpretability points of view on a real-world insurance application. Comparisons with other system modeling techniques are also performed.

1 Introduction

One of the most important applications of Fuzzy Rule-Based Systems (FRBSs) is *System Modeling* [21]. *Linguistic Modeling* [26] is the type of System Modeling with Fuzzy Logic techniques where the main requirement is the interpretability of the final model designed. Mamdani-type FRBSs become the typical example of linguistic models that present the highest description level, i.e., they are composed of fuzzy rules that are globally interpretable [5,14,30].

Although linguistic models are very comprehensible, they also come with a problem which is its lack of accuracy in some complex problems. This fact is due to some problems related to the linguistic rule structure considered

[29], which are a consequence of the inflexibility of the concept of linguistic variable [3]. Besides, those methods that usually learn these models from data are also biased by the former structure and by their nature, which is close to prototype identification algorithms [4,25].

In order to deal with these problems of Linguistic Modeling and trying to improve the accuracy of the stated models, in this chapter we consider an extension of the Knowledge Base (KB) of linguistic FRBSs by means of the concept of "*layers*" [6]. In this extension –which is also a generalization– the KB is composed of a set of layers containing linguistic partitions, with different granularity levels, and linguistic rules, whose linguistic variables take values in these partitions. This KB was called Hierarchical Knowledge Base (HKB) and it is formed by:

- a Hierarchical Data Base (HDB), containing linguistic partitions of the said type.
- a Hierarchical Rule Base (HRB), with the corresponding linguistic rules.

We present an Iterative Hierarchical Systems of Linguistic Rules Learning Methodology (I-HSLR-LM) which is a general purpose hierarchical inductive learning metamethodology. I-HSLR-LM is designed to derive the former HKB from examples and to avoid the typical drawbacks of biased linguistic rules generation methods (LRG-methods). All of these performed in a localized and iterative way.

In this chapter we will also show how can the basic algorithm introduce different policies to perform the best hierarchical fuzzy decomposition and, afterwards, the corresponding integration into a compact HKB:

- *Generation policies*, considering weighted and double-consequent reinforced linguistic rules.
- *Expansion policies*, viewing the hierarchical process as a replacement or a reinforcement of bad performance linguistic rules.
- *Selection policies*, allowing different criteria –accuracy or trade-off accuracy-complexity oriented– to summarize the more compact set of linguistic rules by genetic algorithms (GAs).

This hierarchical environment –hierarchical representation and learning strategy– will allow us to develop very accurate models, as well as solutions constituting intermediate trade-offs between accuracy and interpretability. Its behavior will be evaluated based on the former criteria on a real-world insurance application.

In order to do so, this chapter is set up as follows. In Section 2, the HKB philosophy is introduced and the lacks of LRG-methods are also highlighted. In Section 3, the localized I-HSLR-LM is described in detail, and different policies concerning to the algorithm performance are studied. In Section 4, HSLR models –obtained from the I-HSLR-LM– are applied to solve a real-world insurance application. The achieved results are analyzed from the accuracy and interpretability point of view and compared with other system modeling techniques. In Section 5, we summarize the contain of the chapter. Finally, the appendix contains the basic LRG-method considered in the experiments.

2 Framework

There are many reasons that encourage the use of hierarchical representations. From the theoretical point of view, the theory of fuzzy sets offers an excellent tool for representing the uncertainty associated with the hierarchical decomposition task and for providing smooth transitions between the individual local submodels [2]. Moreover, considering the lack of truth-functionality in many "logics of uncertainty", results of the research on human plausible reasoning conducted by Michalski [15] show that people derive a combined certainty of a conclusion from uncertain premises by taking into consideration structural (or semantic) relations among these premises, based on a hierarchical knowledge representation.

From the practical point of view, it has been seen that the KB structure usually employed in the field of Linguistic Modeling has the drawback of its lack of accuracy when working with very complex systems [3]. Moreover, the typical learning methods usually employed to obtain the KB of FRBSs suffer from their lacks as prototype-identification algorithms [6,31].

The current approach is oriented to produce a more general and well defined structure, the said HKB. This structure should be flexible enough to allow a wide variety of linguistic models, from very accurate to properly interpretable ones. Our purpose is to preserve the descriptive abilities of previous models, increasing their accuracy by the use of different hierarchical levels. All of this performed by simplifying the inference mechanism adopted by other previous hierarchical approaches [13,22,28], activating independently each rule as it is done in the conventional inference mechanism. Besides, we use summarization processes to obtain a compact set of rules that have good cooperation between them.

2.1 Hierarchical Knowledge Base Philosophy

The lack of accuracy of linguistic models is due to some problems related to the linguistic rule structure considered in their KB, which are a consequence of the inflexibility of the concept of linguistic variable [29]. A summary of these problems may be found in [3], being the rigid partitioning of the input and output spaces the most influential factor.

Therefore, we present a more flexible KB structure that allows us to improve the accuracy of linguistic models without losing their interpretability to a high degree: the HKB [6,9]. It is composed of a set of layers, and each layer is defined by its components in the following way:

$$layer(t, n(t)) = DB(t, n(t)) + RB(t, n(t))$$

with:

- $n(t)$ being the number of linguistic terms in the partitions of layer t.
- $DB(t, n(t))$ being the Data Base (DB) which contains the linguistic partitions with granularity level $n(t)$ of layer t.
  Generically, we could say that a DB from a layer $t+1$ is obtained from its predecessor as:

$$DB(t, n(t)) \rightarrow DB(t+1, 2 \cdot n(t) - 1)$$

  which means that a linguistic partition in $DB(t, n(t))$ with $n(t)$ linguistic terms becomes a linguistic partition in $DB(t+1, 2 \cdot n(t) - 1)$ [6] (see Figure 1 and Table 1). In order to satisfy this requirement, each linguistic term $S_k^{n(t)}$ –term of order k from the linguistic partition in $DB(t, n(t))$– is mapped into $S_{2k-1}^{2 \cdot n(t)-1}$, preserving the former modal points, and a set of *n(t)-1* new terms is created, each one between $S_k^{n(t)}$ and $S_{k+1}^{n(t)}$ ($k = 1, ..., n(t) - 1$) (see Table 2). *This approach tries to preserve as much as possible the structure of the fuzzy sets from one layer to the next in the hierarchy, maintaining their "local" interpretability and description* [6].
  In this view, we can generalize this two-level successive layer definition for $n(t)$, for all layers t, in the following way:

$$n(t) = (N-1) \cdot 2^{t-1} + 1$$

  with $N = n(1)$, i.e., the number of linguistic terms in the initial layer partitions.
- $RB(t, n(t))$ being the RB formed by those linguistic rules whose linguistic variables take values in the former partitions.

The main purpose of developing an HRB is to model the problem space in a more accurate way. To do so, those linguistic rules from $RB(t, n(t))$ that model a subspace with bad performance are expanded into a set of linguistic rules, which become their image in $RB(t+1, 2 \cdot n(t) - 1)$. This set of rules models the same subspace that the former one and replaces it.

From now on and for the sake of simplicity, we are going to refer to the components of a $DB(t, n(t))$ and $RB(t, n(t))$ as *t-linguistic partitions* and *t-linguistic rules*, respectively. This set of layers is organized as a hierarchy, where the order is given by the granularity level of the linguistic partition defined in each layer. That is, given two successive layers t and $t+1$, the granularity level of the linguistic partitions of layer $t+1$ is greater than the ones of layer t. This fact causes a refinement of the previous layer linguistic partitions. As a consequence of the previous definitions, we could now define the HKB as the union of every layer t:

$$HKB = \cup_t layer(t, n(t))$$

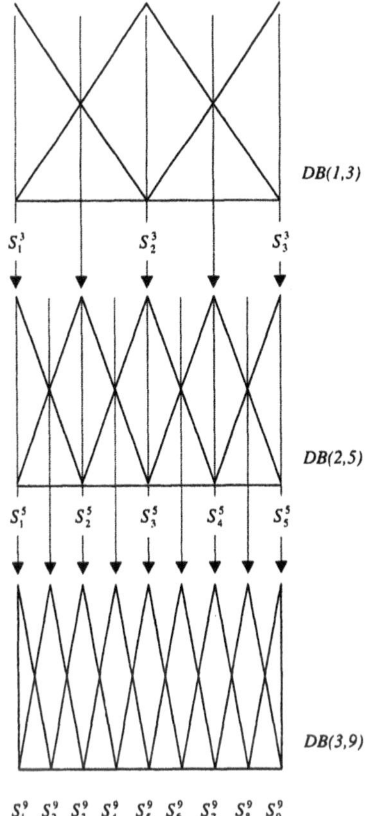

Figure 1: Three layers of linguistic partitions which compose the HDB.

DB(t,n(t))		**DB(t,n(t))**
$DB(1,2)$		$DB(1,4)$
$DB(2,3)$		$DB(2,7)$
$DB(3,5)$		$DB(3,13)$
$DB(4,9)$	*or*	$DB(4,25)$
$\vdots$		$\vdots$
$DB(6,33)$		$DB(6,97)$
$\vdots$		$\vdots$

Table 1: Hierarchy of DBs starting from 2 or 4 initial terms.

DB(t,n(t))		**DB(t+1,2·n(t)-1)**
$S_{k-1}^{n(t)}$	$\longrightarrow$	$S_{2k-3}^{2\cdot n(t)-1}$
		$S_{2k-2}^{2\cdot n(t)-1}$
$S_{k}^{n(t)}$	$\longrightarrow$	$S_{2k-1}^{2\cdot n(t)-1}$
		$S_{2k}^{2\cdot n(t)-1}$
$S_{k+1}^{n(t)}$	$\longrightarrow$	$S_{2k+1}^{2\cdot n(t)-1}$

Table 2: Mapping between terms from successive DBs

2.2 Linguistic Rule Generation Methods

In order to characterize LRG-methods, and regarding [7,31], we can say that basically an LRG-method does its job as a prototype-identification algorithm. It performs the optimization of a functional that measures the extent (matching degree) by which a simple or extended prototype (fuzzy rule structure) fits a subset of the object (data) being described (see Figure 2).

From this perspective, the linguistic rule identification problem is formulated as a clustering problem in the sense that extracted subsets (objects, data) meet, to some extent, the requirements imposed by the model collection (prototypes or rule set) in the same way that elements of a clustering partition satisfy the constraint that their members be as similar as possible [4,25].

Therefore, most of LRG-methods have the same drawbacks that prototype identification methods have and try to solve them based on their own summarization treatment:

- Simple formulation of the prototype-identification (i.e., rule-identification) problem as an optimization of a functional would simply result in a large collection of very specific rules with small extent and high accuracy, but with poor generalization ability. Smaller rather than larger significant prototypes with high generalization power would be preferred.
- The determination of a complete clustering or an exhaustive partition of the dataset into a fixed number of prototypes or rules becomes a problem difficult to be solved. On the one hand –as possibilistic clustering indicates–, *not all of the rules obtained might be interesting*, but a subset could be [4,31]. Otherwise, removing the requirement for a prior knowledge of their number would be desired [11].

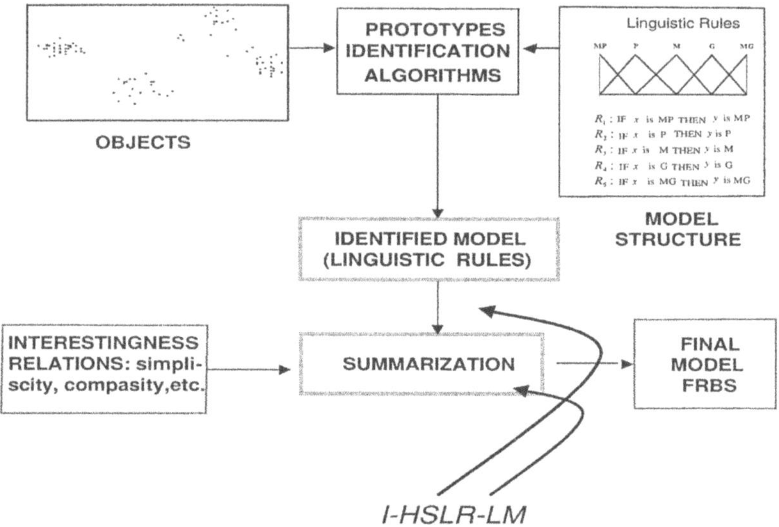

Figure 2: LRG-methods as prototype identification algorithms

All of these motivated the development of a I-HSLR-LM, which can be capable to implement a sort of trade-off between the extensionality and the accuracy of the models generated. That means, to adopt a more general treatment than that of a typical LRG-method, emphasizing the sequential isolation of individual rules without assuming a priori biased number of them [7]. It has been designed as a localized methodology which will modify the initial model identified by an LRG-method in an iterative way, performing gradual refinements on. All of these metamethodogical treatment is adopted to preserve the intrinsic skills of each learning method.

3 An Iterative Localized HSLR Learning Methodology

I-HSLR-LM has been thought as a methodology to perform a local hierarchical treatment of those problem subspaces which are bad modeled by those linguistic models generated from conventional LRG-methods. It is a parameterized methodology which works on the initial solutions of the former methods in an iterative way to obtain a variety of linguistic models, from highly accurate to highly interpretable ones. To do so, a trade-off between the following parameters is managed by the I-HSLR-LM (see Figure 3):

- *The factor of expansion* α is used to control the level of "weakness" that a rule should have to be expanded in more specific ones. That is, an adaptable threshold whose high values produce a less rule expansion [1], which means a smaller number of rules and, thus, a more interpretable model. It is noteworthy that this adaptation is not linear and, as a consequence, *the expansion of more rules does not ensure the decrease of the global error of the modeled system.*
- *The maximum number of iterations of the algorithm* $Kmax$ is used to control the granularity level of those more specific hierarchical rules which replace the ones with bad performance. A high $Kmax$-value could produce more *accurate but overfitted model.* However, I-HSLR-LM provides tools for avoiding this undesirable results [17].
- *The factor of hierarchical neighborhood* δ determines the desired scope where the more specific hierarchical rules should be generated. The narrow the scope, the simpler and more interpretable the model. *It should be noted that a wider or global* [13] *scope does not always guarantee to obtain a more accurate model* [6].

In the following we will first present the I-HSLR-LM algorithm. Afterwards, we propose different alternative policies to its most important components. These policies, which can be combined with the basic localized iterative strategy implemented in the methodology, are: HRB generation policies, HRB expansion policies and HRB selection policies. All of this compose a flexible hierarchical framework to deal with complex problems based on different requirements of accuracy and/or interpretability.

[1] We do not isolate the rule for measuring it performance, but compare it fitness with the whole set behavior. For example, $\alpha = 1.1$ means that a *t-linguistic rule* with an error of a 10 % higher than the error of the entire RB should be expanded.

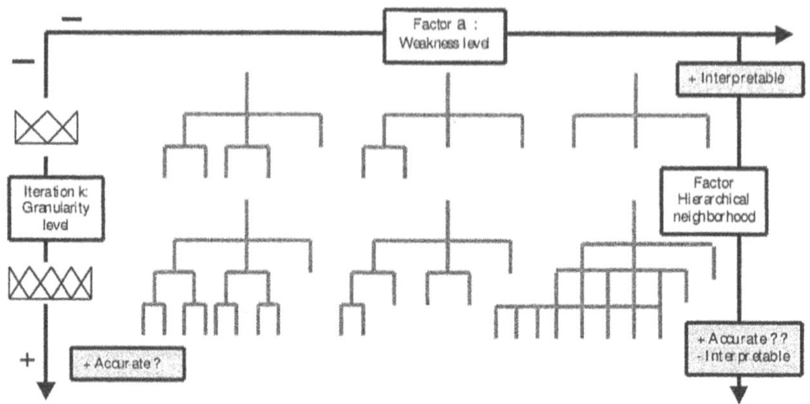

Figure 3: Trade-offs between interpretability and accuracy

3.1 Algorithm

In this Subsection we present our iterative methodology to generate an HKB [6,9]. To do so, we use an LRG-method, which –as an inductive method– is based on the existence of a set of input-output training data E_{TDS} and a previously defined $DB(1, n(1))$. The data set $E_{TDS} = \{e^1, ..., e^l, ..., e^q\}$ is composed of q input-output data pairs $e^l = (ex_1^l, \ldots, ex_m^l, ey^l)$, which represent the behavior of the system being modeled. The measure of error used in the algorithm is the Mean Square Error (MSE) calculated over E_{TDS}. We should note that any other local error measure can be considered with no change in our methodology, such as the one shown in [6]. The algorithm basically consists of the following steps which may be also graphically seen in Figure 4.

Initialization Process

Step 0. **RB(1,n(1)) Generation Process.** *Generate RB(1,n(1)),* where the rules of the initial layer are obtained from the fuzzy partitions in $DB(1, n(1))$ and an LRG-method:

$$HRB^1 = RB(1,n(1)) = LRG\text{-}method(DB(1, n(1)), E_{TDS})$$

with $n(1) = N$ and the initial $DB(1, n(1))$, given by an expert or taken from a normalization process considering a small number of terms; $k = 1$, with k being the iteration counter, and $p = 1$, where p stands for the last layer generated.

Iteration Process *(iteration k)*

Step 1. **HRB Generation Process.** *Generate* HRB^{k+1} *from* $RB(t, n(t))$, $DB(t, n(t))$ and $DB(t + 1, 2 \cdot n(t + 1) - 1)$, with $1 \leq t \leq p \leq k + 1$ (see Figure 5).

(a) *Select the bad performance t-linguistic rules from HRB^k,* which will be expanded in $RB(t+1, 2 \cdot n(t) - 1)$.

i. *Calculate the error of HRB^k as a whole:*

$$MSE\left(E_{TDS}, HRB^k\right) = \frac{\sum_{e^l \in E_{TDS}} (ey^l - s(ex^l))^2}{2 \cdot |E_{TDS}|}$$

with $s(ex^l)$ being the output value obtained from the HRB^k, when the input variable values are $ex^l = (ex_1^l, \ldots, ex_m^l)$, and ey^l is the known desired value.

ii. *Calculate the error of each individual t-linguistic rule:*

$$MSE\left(E_i, R_i^{n(t)}\right) = \frac{\sum_{e^l \in E_i} (ey^l - s_i(ex^l))^2}{2 \cdot |E_i|}$$

with $s_i(ex^l)$ being the output value obtained when inferring with $R_i^{n(t)}$ and E_i being a set of the examples matching the antecedents of the rule i to a specific degree τ $(0, 1]$:

$$E_i = \{e^l \in E_{TDS} \,/\, Min(\mu_{S_{i1}^{n(t)}}(ex_1^l), \ldots, \mu_{S_{im}^{n(t)}}(ex_m^l)) \geq \tau\}$$

iii. *Select the t-linguistic rules with bad performance,* which are going to be expanded, making the difference from the good ones:

$$HRB^k_{bad} = \{R_i^{n(t)} \,/\, MSE(E_i, R_i^{n(t)}) \geq \alpha \cdot MSE(E_{TDS}, HRB^k)\}$$
$$HRB^k_{good} = \{R_i^{n(t)} \,/\, MSE(E_i, R_i^{n(t)}) < \alpha \cdot MSE(E_{TDS}, HRB^k)\}$$

with α being the threshold that represents a percentage of the error of the whole RB which determines the expansion of a rule. For example, $\alpha = 1.1$ means that a *t-linguistic rule* with an MSE a 10 % higher than the MSE of the entire HRB^k should be expanded.

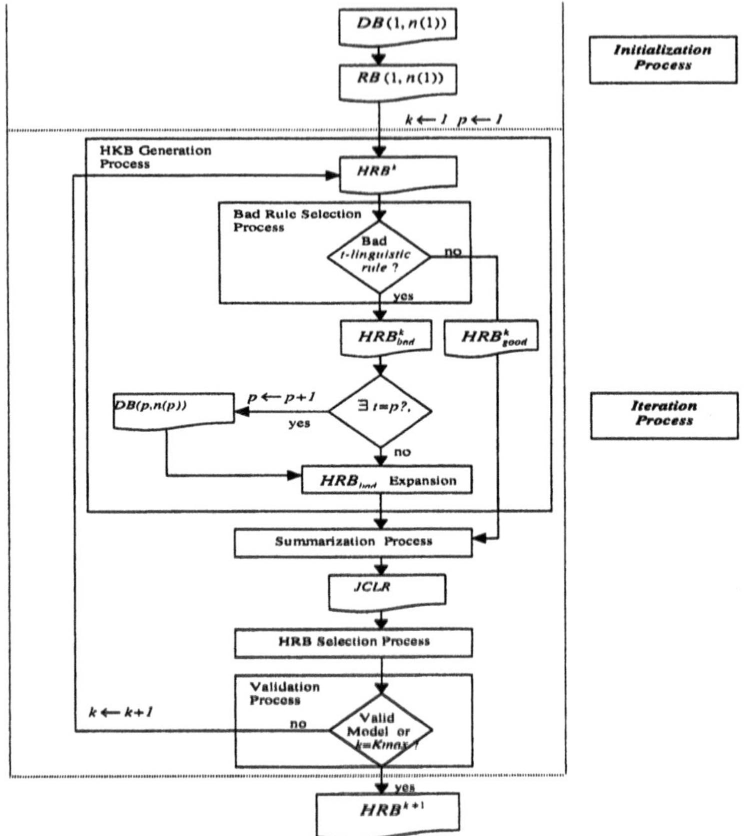

Figure 4: Algorithm of the I-HSLR-LM design process

Now, for each $R_i^{n(t)} \in HRB_{bad}^k$:

(b) *Select DB(t+1,2·n(t) − 1).* If $(t = p)$ then: $p \longleftarrow p + 1$; create $DB_{x_j}(p,n(p))$[2], for all input linguistic variables x_j $(j = 1, ..., m)$, and $DB_y(p,n(p))$, for the output linguistic variable y; and make $HDB^p \longleftarrow HDB^{p-1} \cup DB(p,n(p))$. That is, if the *t-linguistic partitions* corresponding to the bad rule have reached the maximal DB granularity level available in the present HDB, then generate the next layer DB.

(c) *Expand the bad performance t-linguistic rule.*

[2] $DB(t, n(t))$ is referred to as $DB_{x_j}(t, n(t))$ $(j = 1, ..., m)$, meaning that it contains the *t-linguistic partition* where the input linguistic variable x_j takes values, and as $DB_y(t, n(t))$ for the output variable y.

i. *Select the (t+1)-linguistic partition terms from $DB(t+1, 2 \cdot n(t) - 1)$ that δ-intersect the ones of the bad performance t-linguistic rules:*
$I(S_{ij}^{n(t)}) = \{S_h^{2 \cdot n(t)-1} \in DB_{x_j}(t+1, 2 \cdot n(t) - 1) \ /$
$\quad Max_{u \in U_j} Min \left\{ \mu_{S_{ij}^{n(t)}}(u), \mu_{S_h^{2 \cdot n(t)-1}}(u) \right\} \geq \delta\}$
$I(B_i^{n(t)}) = \{B_h^{2 \cdot n(t)-1} \in DB_y(t+1, 2 \cdot n(t) - 1) \ /$
$\quad Max_{v \in V} Min \left\{ \mu_{B_i^{n(t)}}(v), \mu_{B_h^{2 \cdot n(t)-1}}(v) \right\} \geq \delta\}$
where $\delta \in [0, 1]$ represents a cross level of *"significant intersection" or hierarchical neighborhood parameter* (see Figure 5: if $\delta = 0.5$, the problem subspace resulting from the bad *t-linguistic rule* expansion is the one represented by the small white square; however if $\delta = 0.1$, it would be composed of the union of the former small white square and the grey one).

ii. *Combine the previously selected m sets $I(S_{ij}^{n(t)})$ and $I(B_i^{n(t)})$:*

$$I(R_i^{n(t)}) = I(S_{i1}^{n(t)}) \times \ldots \times I(S_{im}^{n(t)}) \times I(B_i^{n(t)})$$

with $I(R_i^{n(t)}) \subset DB(t+1, 2 \cdot n(t) - 1)$. That is, create a fuzzy grid in the input fuzzy subspace defined by a bad performance rule being expanded.

iii. *Extract (t+1)-linguistic rules from the combined selected (t+1)-linguistic partition terms,* producing a set of L (t+1)-linguistic rules, which are the expansion (image) of the bad t-linguistic rule $R_i^{n(t)}$:

$$CLR(R_i^{n(t)}) = LRG - method(I(R_i^{n(t)}), E_i) = \{R_{i_1}^{2 \cdot n(t)-1}, \ldots, R_{i_L}^{2 \cdot n(t)-1}\}$$

with $CLR(R_i^{n(t)})$ being the set of candidates to be in the HRB^{k+1} from rule i.

Step 2. **Summarization process.** *Obtain a joined set of candidate linguistic rules (JCLR),* performing the union of the group of the new generated *(t+1)-linguistic rules* and the former good performance *t-linguistic rules:*

$$JCLR = HRB_{good}^{k} \cup (\cup_i CLR(R_i^{n(t)}))$$

with $R_i^{n(t)} \in HRB_{bad}^{k}$.

Step 3. **HRB Selection Process**. *Simplify the set JCLR by removing the unnecessary rules from it and generating an HRB^{k+1} with good cooperation.* In this chapter we consider a genetic process[3] [10,13] (see Subsection 3.2)

[3] GAs have constituted one of the most successfull tools for designing FRBSs, i.e., GFRBSs [10].

to put this task into effect, but any other optimization technique could be considered:

$$HRB^{k+1} = Selection\ Process(JCLR)$$

In the $JCLR$ –where there are coexisting rules of different hierarchical layers– it may happen that a complete set of *(t+1)-linguistic rules,* which replaces an expanded *t-linguistic rule,* does not produce good results. However, a subset of this set of *(t+1)-linguistic rules* may work properly, with less rules that cooperate better among them, and with the good rules from the previous layer (see Figure 6).

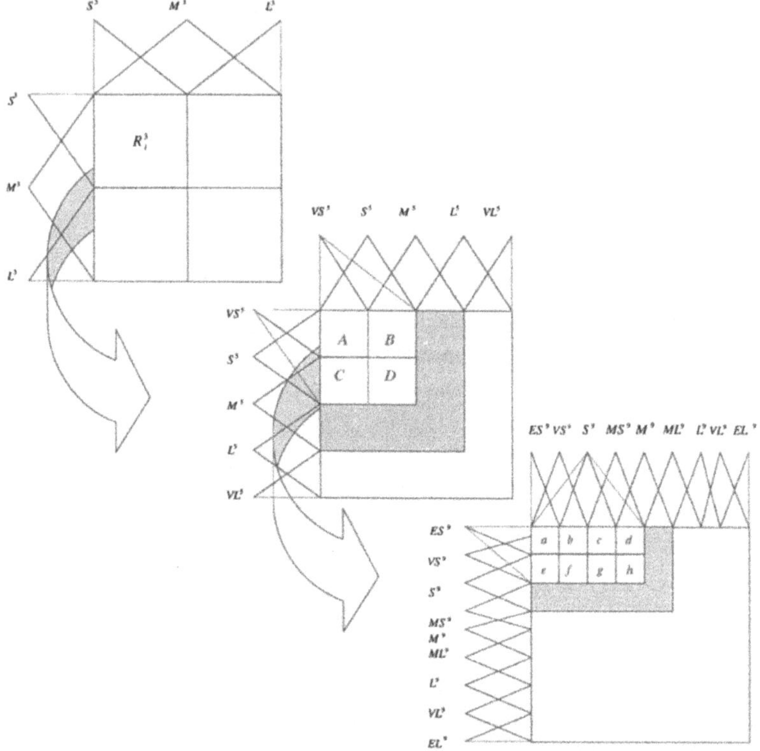

Figure 5: Example of the HRB Generation Process

Step 4. **Model Validation Process.** The current model is either accepted as a proper one for the given purpose, or it is rejected generating another iteration of the hierarchical process. Among the different indices that can be used to measure the quality of linear or nonlinear systems after an identification loop [2], we consider a monotonic MSE measure on the training set, computed as:

$$IF\ (MSE(HRB^{k+1}(E_{TDS})) \leq MSE(HRB^{k}(E_{TDS}))\ and\ (k < Kmax))$$
$$THEN\ k \longleftarrow k+1; Goto\ Step\ 1$$

combined with a previously defined maximum number of iterations $Kmax$, based on a trade-off between the complexity and the accuracy of the model required. Finally, as a consequence of applying this algorithm, the HKB is obtained as:

$$HKB = HDB^{p} + HRB^{k+1}$$

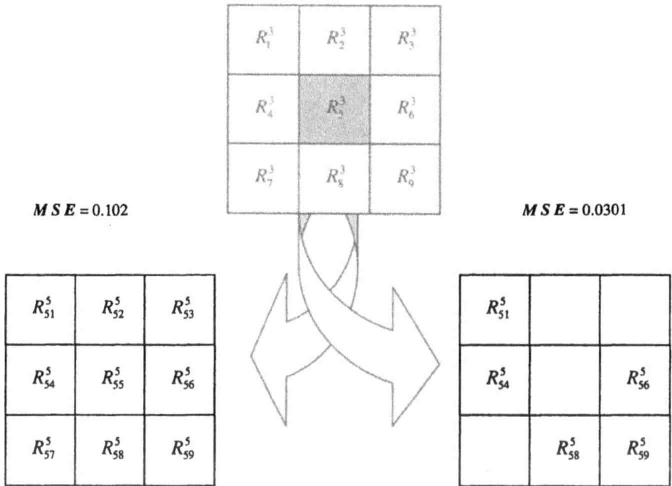

Figure 6: HRB Selection Process

3.2 Design Policies

Different design policies have been found as emergent features of the building process performed by I-HSLR-LM. In the following, we give a brief description of such policies which are basically related with the generation, expansion and selection processes implemented in the former methodology.

Generation Policies. As a consequence of the existence of hard modeling adjacent subspaces, the HRB Generation Process produces two different kinds of special linguistic rules (see Step 1(c)i. of previous algorithm).

Weighted (t+1)-linguistic rules are produced as a consequence of the $DB(t+1, 2 \cdot n(t) - 1)$ generation approach and the stated adjacent places. To illustrate this situation, consider the two dark grey squares in Figure 7 with the *2-linguistic rule:*

$$IF\ x_1\ is\ VL^5\ and\ x_2\ is\ M^5\ THEN\ y\ is\ M^5$$

which is both derived from the expansion of R_2^3 and R_4^3. The overlapping of the expanded rule images is produced by low values of the parameter δ (see Step 1.(c) i. in the algorithm).

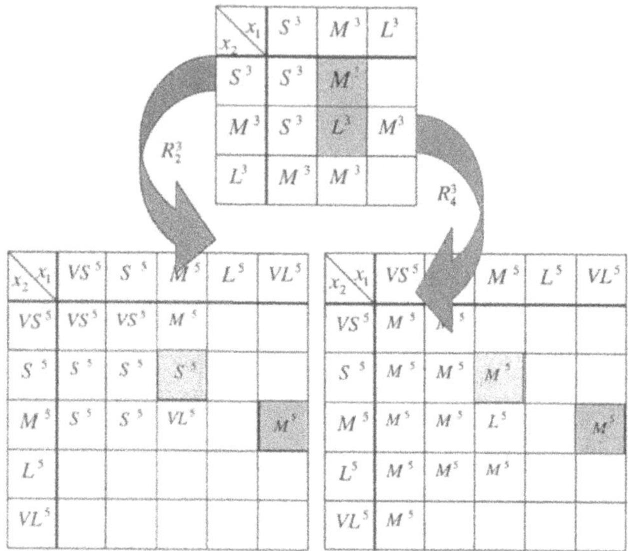

Figure 7: Generation policy: weighted and double-consequent rules.

The use of weighted rules produces a sort of *reinforcement* on the whole subspace of the rule in hard modeling subspaces shared by adjacent rules. These rules are generated twice by the method and sometimes preserved by the selection process. Their influence lies on the use of a defuzzification method that considers the matching degree of the fired rules,e.g., the *Center of Gravity weighted by the matching degree* [5,6] in our case. Take note that these rules do not excessively increase the computational cost of the process, i.e., they are weighted and then processed only once in our implementation of the fuzzy inference.

In spite of their good performance, sometimes these weighted rules can overfit the system modeled in combination with a hierarchical process. So that, we experimentally compared (see Section 4) the effect of their exclusion [6,9] in the input of the selection process of the I-HSLR-LM by modifying the algorithm in the following way:

$$HRB^{k+1} = Selection(Extract\_Repeated(JCLR))$$

Additionally, a specific identification procedure for learning the proper weight of each rule in a hierarchical environment is given in [1].

Double-consequent (t+1)-linguistic rules are an extension of the usual linguistic model structure which allows the KB to present rules where each combination of antecedents may have two or more consequents associated [5,19]. They do not constitute an inconsistency from the interpolative reasoning point of view but only a shift of the main labels making that the final output of the rule lie in an intermediate zone between them both. Let us consider the specific combination of antecedents of Figure 7, "x_1 is M^5 and x_2 is S^5", which has two different consequents associated, M^5 and S^5. From a Linguistic Modeling point of view, the resulting double-consequent rule may be interpreted as follows:

$$IF\ x_1\ is\ M^5\ and\ x_2\ is\ S^5\ THEN\ y\ is\ between\ S^5\ and\ M^5$$

Double-consequent linguistic rules enrich the representational power of linguistic rules postponing the rule selection decisions until the summarization process is performed, i.e., considering the best cooperation between them.

Expansion Policies: hierarchical replacement and hierarchical reinforcement. In the algorithm presented in Section 3, we considered the hierarchical process as a replacement of a *t-linguistic rule* with bad performance. Here we extend the former criteria proposing a different operation mode by preserving both the expanded rule and some of the rules composing its image in the next layer RB. That is, to consider the expansion process as a *hierarchical reinforcement of a bad rule* [6,9,13]. Figure 8 shows both kinds of rule expansion policies and allows us to illustrate how the reinforcement extension (b) modifies our previous replacement approach (a).

The reinforcement policy does not eliminate the concept of "replacement" of the expanded rule, but extends it allowing the Selection Process to eliminate that rule when it cooperates bad with the remainder. Thus, *a bad performance rule is considered as inadequate if both HRB Generation and HRB Selection processes penalize its behavior.* Moreover, it allows I-HSLR-LM to backtrack to reconsider earlier choices, performing a sort of local bidirectional search and solving some of the problems of hierarchical clustering techniques [7,11].

To empirically prove the effect of this kind of refinement (see Section 4), we have designed experiments modifying the Step 3 of the algorithm in the following way:

$$HRB^{k+1} = Selection(HRB^k_{bad} \cup JCLR)$$

Selection Policies: Accuracy-Oriented and Trade-off Accuracy-Complexity-Oriented. The genetic rule selection process corresponding to Step 3 of the I-HSLR-LM algorithm is based on a binary coded GA, in which the selection of the individuals is performed using the stochastic universal sampling procedure, an elitist selection scheme, and the generation of the offspring population is put into effect by using the classical binary multipoint crossover (performed at two points) and uniform mutation operators [16].

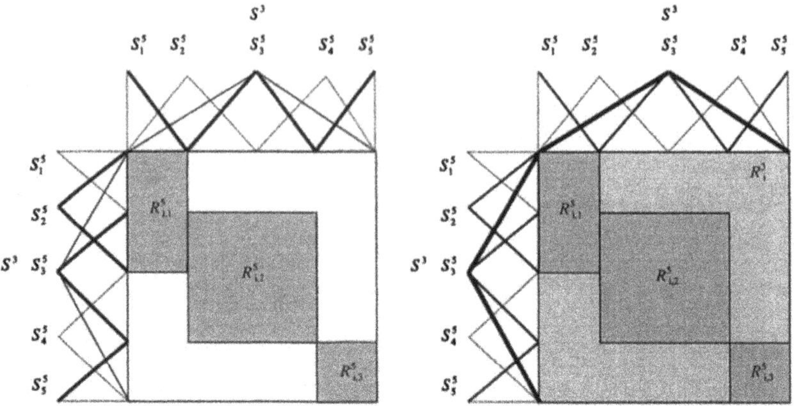
Figure 8: Expansion policies: (a) Replacement (b) Reinforcement

The coding scheme of the GA generates fixed-length chromosomes. Considering the rules contained in $JCLR$ counted from 1 to z, an z-bit string $C = (c_1, ..., c_z)$ represents a subset of rules for the HRB^{k+1}, such that:

$$IF \ c_i = 1 \ THEN \ (R_i \in HRB^{k+1}) \ ELSE \ (R_i \notin HRB^{k+1})$$

The initial population is generated by introducing a chromosome representing the complete previously obtained rule set, i.e., with all $c_i = 1$. The remaining chromosomes are selected at random.

As regards the fitness function $F(C_j)$, we propose two alternative policies. The *accuracy-oriented policy* is based on a global error measure (MSE) over a training data set. It determines the accuracy of the FRBS encoded in the chromosome, depending on the cooperation level of the rules existing in the selected HRB:

$$F(C_j) = MSE(E_{TDS}, C_j) = \frac{\sum_{e^l \in E_{TDS}} (ey^l - s(ex^l))^2}{2 \cdot |E_{TDS}|}$$

with $s(ex^l)$ being the output value obtained from the defuzzification of the RB encoded in the chromosome, when the input variable values are $ex^l = (ex_1^l, \ldots, ex_m^l)$, and ey^l is the known desired value.

The alternative *trade-off accuracy-complexity-oriented policy* is viewed as a kind of post-pruning [17] and rule simplification strategy that considers the following function $F'(C_j)$, penalizing those RBs with a high number of rules:

$$F'(C_j) = w_1 \cdot F(C_j) + w_2 \cdot N_{rules}$$

with $F(C_j)$ being the former fitness function, N_{rules} being the number of rules of that HRB, and with w_1 and w_2 being weights defining the relative importance of each objective. In the present experiments, these constants are initialized in the following way [8,13]:

$$w_1 = 1.0;\ w_2 = 0.1 \cdot \frac{MSE_{initial}}{N_{initial\ rules}}$$

with $MSE_{initial}$ and $N_{initial\ rules}$ being the error and the amount of rules of the candidate HRB to be summarized, respectively.

4 Example of Application: Experiments and Analysis of Results

With the aim of analyzing the behavior of the proposed methodology described in Section 3, we consider an application related on an ongoing research project which is devoted to apply a comprehensive set of techniques to predict some meaningful variables for the insurance company regulatory process [24]. In the following, we first specify the notation and parameters used in the present implementation. Then, we briefly describe the former application and analyze the results obtained by I-HSLR-LM based on accuracy and interpretability criteria. Comparisons with other modeling techniques are finally performed.

4.1 Notation and Parameters

As stated, I-HSLR-LM has been thought as a refinement of simple linguistic models which uses an HKB of some layers. That is, starting from an initial $layer(1, n(1))$, a successive *layer(2,2·n(1)-1)* is created and so on, continuing in an iterative process in order to extract the final system of linguistic rules. For the sake of simplicity and interpretability, we will only consider the generation of HSLRs with up to three hierarchical levels.

In the following application we are going to refer to those experiments produced by the I-HSLR-LM by the following notation:

$$I\text{-}HSLR(LRG - method, n(1), n(p), Kmax)$$

where $n(1)$, and $n(p)$ are the initial and final granularity levels of the HKB, respectively, and *Kmax* is the number of iterations performed by this methodology, e.g., $I-HSLR(WM$-method, 3, 9, 2). We should note that $I\text{-}HSLR(LRG\text{-}method, n(1), n(2), 1)$ represents an interpretability-oriented hierarchical refinement of simple models.

The LRG-method considered for the previous experimentation is the one proposed by Wang and Mendel in [27], noted by WM-method in the following. This method is briefly described in the Appendix. A reference to an application of WM-method is represented by WM*-method*(r), with r being the granularity level of the linguistic partitions used in the method.

In order to distinguish between methods from different policies, we will add a representative suffix of the ones listed in Table 3:

Method	Policies		
	Generation	**Expansion**	**Selection**
I-HSLR	weighted	replacement	accuracy
NW-I-HSLR	non-weighted	replacement	accuracy
HR-I-HSLR	weighted	reinforcement	accuracy
HR-NW-I-HSLR	non-weighted	reinforcement	accuracy
AC-I-HSLR	weighted	replacement	accuracy-complexity
AC-NW-I-HSLR	non-weighted	replacement	accuracy-complexity
AC-HR-I-HSLR	weighted	reinforcement	accuracy-complexity
AC-HR-NW-I-HSLR	non-weighted	reinforcement	accuracy-complexity

Table 3: I-HSLR-LM Methods used in the experiments

The initial DB used for the I-HSLR-LM is constituted by *three linguistic terms* with triangular-shaped fuzzy sets giving meaning to them, i.e., $DB(1,3)$. The values of the parameters used in all of these experiments are listed in Table 4.

Parameter	**Decision**
Generation	
δ : hierarchical neighborhood	0.1
τ: used to calculate E_i	0.5
α : expansion factor	$1.1, 1.9$
Kmax: iteration level	$1-2$
GA Selection	
Number of generations	$500-2000$
Population size	$61-81$
Mutation probability	$0.1-0.2$
Crossover probability	0.6

Table 4: Parameter values

The results obtained in the developed experiments are collected in tables where MSE_{tra} and MSE_{tst} stand for the values obtained in the $\sqrt{MSE}$ measure computed over the training and test data sets, respectively. $\#R$ will stand for the number of simple rules of the corresponding HRB.

Finally, we will try to solve the former application by generating different kinds of models: global linguistic models, classical regression and neural models. To compare the influence of the parameters that make I-HSLR-LM a local methodology, we consider another method which also performs a multigranular treatment of linguistic rules [13]. This global strategy obtains an HSLR by creating several hierarchical linguistic partitions with different granularity levels, generating the complete set of linguistic rules in each one of these partitions, and performing a genetic rule selection on the whole rule set. This global approach will be referred as G-HSLR($LRG-method, n(1), n(2)$).

To apply comparison methods like classical regression, the parameters of the polynomial models were fit by Levenberg-Marquardt, while exponential and linear models were fit by linear least squares. The multilayer perceptron was trained with the Backpropagation algorithm and the number of neurons in the hidden layer was chosen to minimize the test error[4].

4.2 Problem: Modeling of the Mathematical Reserves of the Premiums

With the main purpose of assisting the State Insurance Agency in targeting resources to those insurers in greatest need of regulatory attention, an information system -based on the behavior of the companies- called Insurance Regulatory Information System (IRIS) [18] was created by the National Association of Insurance Commissioners (NAIC) of USA [24]. This regulation is specially needed in company reinsurance or company capitalization process, fusions, etc.

Unfortunately, the application of this regulatory process in the local Argentinian market is limited by the incomplete reported data, the lack of auditing controls which should guarantee the confidence of the information collected from the companies [24].

Nowadays, the Argentinian Insurance Agency is hardly involved in the task of auditing the quality of the insurance controlling process, i.e., trying to avoid as much as possible the current frauds. Therefore, they provided us with historic information about the companies in order to estimate and predict a set of variables of IRIS that describe the status of the companies in a more objective way. In this chapter, we show an illustrative example of the said estimation tasks by modeling the behavior of the *mathematical reserves of the premiums or risk reserves* of the insurance companies in the Argentinian local market. Besides accuracy, interpretability and human-based validation are also hardly required for the models to be accepted by the actuaries.

The variable representing the mathematical reserves of the premiums corresponds to the amount of money –in relation with the premiums emitted– that the insurance companies have in storage, to prevent possible losses or prosecutions. The commissioners referred us that this output variable has not a reliable estimation, that its nature is imprecise and that sometimes it is estimated from a subjective analysis of the company analysts.

To perform the modeling task we should find a relationship between the variables listed in Table 5 for a certain time period. The meaning of the inputs correspond to *the sum of emitted policies in a fixed time period, the loadings from insurance engagement cancellations* and *the reinsured premiums*, respectively [24].

Therefore, and making use of the available data from 479 insurance companies from 1997-1999, we will try to predict the value of the output variable for any other company or time period. The range of the input-output variables is shown in Table 5 with the purpose of understanding their distribution. To compare the performance of the former models, we have randomly divided

[4] See Matlab toolbox.

Symbol	Meaning	Range
x_1	Emitted Premiums	-231.541 to 297.105.728
x_2	Loading of Premiums	-290.439 to 37.761.324
x_3	Reinsurance Cessions	-82.198.928 to 2.113.913
y	Risk Reserves	-69.287.348 to 17.626.323

Table 5: Input-output variables of the risk reserves modeling problem

the sample in two sets of 384 and 95 examples (80% and 20% of the total sample). The measure used in any variable is the Argentinian currency Peso Argentino.

The results obtained with I-HSLR-LM are shown in Tables 6 and 7. For the sake of space restriction, we show the more representative results but a more detailed description with an specific evaluation between the different design options (α,$Kmax$, δ,*LRG-methods,* different initial partitions, etc.) can be found in [6,9].

In view of the results, we can conclude:

- ***From the accuracy point of view,*** that the different models generated from I-HSLR-LM clearly outperform both MSE_{tra} and MSE_{tst} of those models obtained by the WM-method in all iteration and factor of expansion levels. In the former experiments, they also outperform classical regression and neural networks models in the approximation of both data sets, training and test. Global methods are also overcome in generalization results, because of their overfitting lacks.
- ***From the complexity point of view***, I-HSLR-LM has obtained relatively simple models –sometimes even simpler– for the problem with respect to the accuracy improvements achieved over the initial models generated by the WM-method. Moreover, even having a high number of rules, the HKB gives a hierarchical order which can be used in the sense of interpretability. That is, human beings can not understand a hundred of different rules, but can associate a group of them (default rules) with an specific task and deal with more general and subsumed rule sets [15]. All of these, having in mind that there exists more specific exception rules that increase the accuracy of the former comprehensive modeling [23].

The different parameter setting considered in designing hierarchical models produce a variety of trade-off accuracy-interpretability solutions. One boundary condition emphasize the generation of a low number of rules, reducing the complexity and improving the interpretability of the model. Moreover, this reduction prevents the possible model overfitting, i.e., a pre-pruning strategy [20] which also improves the generalization and accuracy of the results. The opposite boundary is oriented to obtain very accurate and well fitted models that use to be more complex. Let us briefly analyze the effect of each individual parameter:

Method	MSE_{tra}	MSE_{tst}	#R.
WM-method(3)	11.982.260	11.662.128	**12**
WM-method(5)	6.365.430	5.876.539	25
WM-method(9)	3.245.889	3.134.790	52
AC-I-HSLR(WM-method,3,5,1)	3.615.872	**2.023.296**	14
I-HSLR(WM-method,3,5,1)	3.585.442	2.140.364	23
AC-I-HSLR-NW(WM-method,3,5,1)	3.691.965	2.334.891	8
I-HSLR-NW(WM-method,3,5,1)	3.593.078	2.163.902	24
AC-I-HSLR-HR(WM-method,3,5,1)	3.615.872	2.334.891	14
I-HSLR-HR(WM-method,3,5,1)	3.585.442	2.140.364	23
AC-I-HSLR-HR-NW(WM-method,3,5,1)	3.649.434	2.438.413	12
I-HSLR-HR-NW(WM-method,3,5,1)	3.593.078	2.163.902	24
AC-I-HSLR(WM-method,3,9,2)	3.529.993	2.148.272	22
I-HSLR(WM-method,3,9,2)	3.375.199	2.487.191	53
AC-I-HSLR-NW(WM-method,3,9,2)	**3.317.158**	2.095.503	10
I-HSLR-NW(WM-method,3,9,2)	3.415.880	2.581.895	52
AC-I-HSLR-HR-NW(WM-method,3,9,2)	3.552.962	2.100.340	22
I-HSLR-HR(WM-method,3,9,2)	3.413.823	2.493.878	51
AC-I-HSLR-HR(WM-method,3,9,2)	3.429.218	2.209.379	29
I-HSLR-HR-NW(WM-method,3,9,2)	3.438.127	2.544.483	63

Table 6: Results obtained with the I-HSLR-LM with α=1.1 (error 10%)

- ***Factor of expansion*** (α). The proposed algorithm seems to be robust for any value of α [6]. As a general rule, when α grows up, the system complexity decreases, i.e., less rules are expanded and thus a simpler –more interpretable– HRB is finally obtained.
- ***Iteration level*** ($Kmax$). We should note that most of those models obtained by the use of more than one iteration perform the best approximation in MSE_{tra}. However, in more complex problems, this configuration can overfit the system modeled also performing a considerable increase in the model complexity.
- ***Factor of hierarchical neighborhood*** (δ)***.*** We have empirically proved that *it is not always true that a set of rules with a higher granularity level performs a better modeling of a problem than another with a lower granularity level.* The localized approach does not only benefit the interpretability by producing simpler models, but also the model accuracy and generalization capabilities [6].

Although the interaction between the former parameters guide the hierarchical learning process, the different policies considered in this chapter also make influence in the interpretability and accuracy of the results. Furthermore, their quality does not only depend on specificity but also on rule

Method	MSE_{tra}	MSE_{tst}	*Complexity*
Lineal	70.250.674	18.331.709	6 nodes, 4 par.
Exponential	72.669.183	19.816.666	11 nodes, 3 par.
Polynomial of 2nd. order	61.972.488	30.201.654	28 nodes, 11 par.
Polynomial of 3rd. order	60.004.603	19.597.421	123 nodes, 16 par.
Perceptron of 3 layers 3-10-1	21.479.077	6.443.723	63 par.
AC-G-HSLR(WM-method,3,5,1)	3.634.025	2.318.720	15 rules
G-HSLR(WM-method,3,9,2)	2.554.492	2.559.914	49 rules
AC-I-HSLR(WM-method,3,5,1)	3.615.872	2.013.296	14 rules
I-HSLR-HR(WM-method,3,9,2)	3.429.123	1.993.020	53 rules

Table 7: Results obtained comparing HSLRs with other system modeling techniques[5]

weights or reinforcements, and moreover, on compasity and cooperation policies between these rules [10]:

- ***Generation policies.*** *Weighted rules* perform the best approximation in MSE_{tra} and do not excessively increase the computational cost, however, sometimes they produce overfitting. As well as these kinds of rules, *double-consequent rules,* also improve the model accuracy reinforcing the scope of the refinement rules. Both types of rules emphasize the *cooperative selection* of "good rules" (good clusters), postponing that decision to the HRB Selection Process.
- ***Expansion policies.*** From the accuracy view and as a general rule, *replacement* can be used in more linear models while *reinforcement* performs better in more non-linear approximations. That is because of the more freedom and independence from different parameters –such as the *granularity of the initial partitions* and the *factor of expansion* α– provided. Considering interpretability features, the reinforcement expansion can be seen as a sort of *default reasoning* [23], where the conjunction of more specific rules and the former default one constitutes the exception.
- ***Selection policies.*** The *accuracy-oriented* approach is obviously related with more accurate and complex models. Otherwise, when complexity is associated with interpretability, the *trade-off accuracy-complexity-oriented* policy is the best decision. However, complexity could be also related with accuracy. Therefore, as it is seen in previous experiments, a model with less number of rules could produce better generalizations, avoiding the undesirable system overfitting.

5 Summary

The HKB and the I-HSLR-LM presented in this chapter offer an excellent tool for representing the vagheness in the model and the uncertainty associated with the decomposition task, respectively. That is, a class of local modeling which attempts to solve a complex modeling problems by decomposing it into a number of simpler hierarchical subproblems and integrating good submodel behaviors with the whole model by the use of a proper rule selection process. All of these, by considering a balance between the interpretability and the accuracy of the system modeled.

Finally, as was said by Goldberg [12], if the future of Computational Intelligence "*lies in the careful integration of the best constituent technologies*", hierarchical and hybrid Fuzzy Systems and GAs require more than simple combinations derived from putting everything together, but a more sophisticated analysis and design of the system components and their features [10].

6 Appendix: The WM Rule Generation Method

The inductive RB generation process proposed by Wang and Mendel in [27] is widely known because of its simplicity and good performance. It is based on working with an input-output training data set, E_{TDS}, representing the behavior of the problem being solved and with a previous definition of the DB composed of the input and output primary linguistic partitions used. The linguistic rule structure considered is the usual Mamdani-type rule with m input variables and one output variable presented in Section 2.

The generation of the linguistic rules of this kind is performed by putting into effect the three following steps:

1. *To generate a preliminary linguistic rule set.* This set will be composed of the linguistic rule best covering each example (input-output data pair) existing in the input-output data set E_{TDS}. The structure of these rules is obtained by taking a specific example, i.e., an $m+1$-dimensional real array (m input and 1 output values), and setting each one of the rule variables to the linguistic label associated to the fuzzy set best covering every array component.
2. *To give a rule matching degree.* Let $R = IF\ x_1\ is\ S_1\ and \ldots x_m\ is\ S_m\ THEN\ y\ is\ B$ be the linguistic rule generated from the example $e_l = (x_1^l, \ldots, x_m^l, y^l)$, $l = 1, \ldots, |E_{TDS}|$. The degree of importance associated to it will be obtained as follows: $G(R) = \mu_{S_1}(x_1^l) \cdot \ldots \cdot \mu_{S_m}(x_m^l) \cdot \mu_B(y^l)$.
3. *To obtain a final RB from the preliminary linguistic rule set.* If all rules presenting the same antecedent values have associated the same consequent in the preliminary set, this linguistic rule is automatically put (only once) into the final RB. On the other hand, if there are conflictive rules, i.e., rules with the same antecedent and different consequent values, the rule considered for the final RB will be the one with higher importance degree.

References

1. R. Alcalá, O. Cordón, F. Herrera, I. Zwir, Hibridazing Hierarchical and Weighted Linguistic Rules, Proceedings of the ACM Symposium on Applied Computing (SAC), Spain, (2002).To appear.
2. R. Babuška, Fuzzy Modeling for Control, Kluwer Academic Publishers (1998).
3. A. Bastian, How to Handle the Flexibility of Linguistic Variables with Applications, International Journal of Uncertainty, Fuzziness and Knowledge-Based Systems 2:4 (1994) 463-484.
4. J. C. Bezdek, Fuzzy Clustering, in E. H. Ruspini, P. P. Bonisone, W. Pedrycz (Ed.), Handbook of Fuzzy Computation, Institute of Physics Press (1998) f6.1:1-f6.6:19.
5. O. Cordón, F. Herrera, A Proposal for Improving the Accuracy of Linguistic Modeling, IEEE Transactions on Fuzzy Systems 8:4 (2000) 335-344.
6. O. Cordón, F. Herrera, I. Zwir, Linguistic Modeling by Hierarchical Systems of Linguistic Rules, IEEE Transactions on Fuzzy Systems, (2001). To appear.
7. O. Cordón, F. Herrera, I. Zwir, Fuzzy Modeling by Hierarchically Built Fuzzy Rule Bases, International Journal of Approximate Reasoning, 27:1 (2001) 61-93.
8. O. Cordón, F. Herrera, P. Villar, A Genetic Learning Process for the Scaling Factors, Granularity and Contexts of the Fuzzy Rule-based System Data Base, Information Science, 136 (2001) 85-107.
9. O. Cordón, F. Herrera, I. Zwir, Analyzing and Extending Hierarchical Systems of Linguistic Rules, Proceedings of the Joint 9th IFSA World Congress and 20th NAFIPS International Conference, Vancouver, Canada (2001) 1121-1126.
10. O. Cordón , F. Herrera, F. Hoffmann, L. Magdalena, Genetic Fuzzy Systems, World Scientific (2001).
11. M. Delgado, A. F. Gómez-Skarmeta, A. Vila, On the Use of Hierarchical Clustering in Fuzzy Modeling, International Journal of Approximate Reasoning, 14 (1996) 237-257.
12. D. E. Goldberg, A Meditation on the Computational Intelligence and Its Future, Illigal Report #2000019, Department of General Engineering, University of Illinois at Urbana-Champaign (2000).
13. H. Ishibuchi, K. Nozaki, N. Yamamoto, H. Tanaka, Selecting Fuzzy If-Then Rules for Classification Problems Using Genetic Algorithms, IEEE Transactions on Fuzzy Systems 3:3 (1995) 260-270.
14. C. T. Leondes (Ed.), Fuzzy Theory Systems, Techniques and Applications, Academic Press (2000).
15. R.S. Michalski, ed., Multistrategy Learning, Kluwer Academic Press (1993).
16. Z. Michalewicz, Genetic Algorithms + Data Structure = Evolutioon Programs, Springer, (1999).
17. T. Mitchell, Machine Learning, McGraw-Hill (1997).
18. NAIC Online Publications, www.naic.org, Journal of Insurance Regulation (1999).
19. K. Nozaki, H. Ishibuchi, H. Tanaka, A Simple but Powerful Heuristic Method for generating Fuzzy Rules from Numerical Data, Fuzzy Sets and Systems 86 (1997) 251-270.
20. T. Oates, D. Jensen, The Effects of Training Set Size on Decision Tree Complexity, en Proceedings of the Fourteenth International Conference on Machine Learning (1997) 254-262.

21. W. Pedrycz (Ed.), Fuzzy Modelling: Paradigms and Practice, Kluwer Academic Press (1996).
22. C.V.S. Raju, J. Zhou, Adaptative Hierarchical Fuzzy Controller, IEEE Transactions on Systems, Man, and Cybernetics 23:4 (1993) 973-980.
23. R. Reiter, A Logic for Default Reasoning. Artificial Intelligence 13(1980) 81-132.
24. M. A. Rodríguez, Indicadores técnicos aplicados a entidades reaseguradoras. Seguros elementales y seguros de vida (in spanish). Foro económico UMSA (1997).
25. E. H. Ruspini, I. Zwir, Automated Generation of Qualitative Representations of Complex Object by Hybrid Soft-computing Methods, in S. K. Pal, A. Pal (Eds.), Lecture Notes in Pattern Recognition, World Scientific Company, Singapore (2002).
26. M. Sugeno, T. Yasukawa, A Fuzzy-logic-based Approach to Qualitative Modeling, IEEE Transactions on Fuzzy Systems 1:1 (1993) 7-31.
27. L.X. Wang, J.M. Mendel, Generating Fuzzy Rules by Learning from Examples, IEEE Transactions on Systems, Man, and Cybernetics 22 (1992) 1414-1427.
28. R.R. Yager, On the Construction of Hierarchical Fuzzy Systems Model, IEEE Transactions on Systems, Man, and Cybernetics 28:1 (1998) 55-66.
29. L.A. Zadeh, The Concept of a Linguistic Variable and its Application to Aproximate Reasoning, Information Science, Part I: 8 (1975) 199-249; Part II: 8 (1975) 301-357 ; Part III: 9 (1975) 43-80.
30. L.A. Zadeh, Toward a Theory of Information Granulation and Its Centrality in Human Reasoning and Fuzzy Logic, Fuzzy Sets and Systems 90 (1997) 111-127.
31. I. Zwir, E. H. Ruspini, Qualitative Object Description: Initial Reports of the Exploration of the Frontier, Proceedings of the EUROFUSE-SIC99, Budapest, Hungary (1999) 485-490.

Learning Default Fuzzy Rules with General and Punctual Exceptions

Pablo Carmona[1], Juan Luis Castro[2], Jose Jesus Castro-Schez[3], and Manuel Laguia[4]

[1] Department of Computer Science,
University of Extremadura, E-06071 Badajoz, Spain
email: pablo@unex.es

[2] Department of Computer Science and Artificial Intelligence,
University of Granada, E-18071 Granada, Spain
email: castro@decsai.ugr.es

[3] Department of Computer Science,
University of Castilla-La Mancha, E-13071 Ciudad Real, Spain
email: JoseJesusCastro@uclm.es

[4] Department of Languages and Computer Science,
University of Cadiz, E-11003 Cadiz, Spain
email: manuel.laguia@uca.es

Abstract. In this paper, a fuzzy rule learning with exceptions is proposed to identify fuzzy models from examples, such a method has been developed trying to achieve a double goal: accuracy and interpretability. In order to do that, we apply a fuzzy repertory table to obtain the linguistic terms of the experts, and then maximal structure fuzzy rules with these terms are obtained. In a second stage, the conflicts generated by the maximal rules are solved, thus increasing the model accuracy and interpretability. Nevertheless, some irregularities (in small regions) can not be covered with the original linguistic terms. In order to solve these conflicts, the way for using these exceptions is learned, and a case-based reasoning system is applied.

1 Introduction

Many of the machine learning algorithms developed to obtain fuzzy rules try mainly to make good numerical approximation [17], while little attention is paid to the interpretability of the resulting model.

In order to get a good accuracy, a lot of rules are obtained and/or non usual membership functions are employed. If we fix a moderate number of linguistic terms for every variable, then the search space is reduced to an small number of different systems. Thus, the accuracy of the best model is often very low. In these cases, it is usual to apply a tuning of the membership functions in order to increment the accuracy level. On the other hand, if a big number of labels are considered, then the number of rules of the learned system grows exponentially in the inverse way that accuracy. On our view, the balance between accuracy and interpretability is very conditioned and limited by the knowledge model. In this way, we think that the best solution for this problem is to expand this model, in such a way that more general, but fully interpretable, rules can be employed.

In this paper we will present a new model of Fuzzy Rule Based System, together with a learning algorithm for this model. We think that an important aspect in order to assure the interpretability are the linguistic labels. On our view they might be the usual terms in the context where the system will be applied. In this sense, we propose to use a fuzzy repertory table in order to obtain the appropriate variables Vs, and their respective linguistic terms $DDVs$.

Furthermore, using these variables and linguistic terms, the model will use general Fuzzy Rules with general and punctual exceptions,. General fuzzy rules will be a set of basic fuzzy rules grouped in a more general fuzzy rule. Concretely, the form of our fuzzy rules will be the following:

R_i: IF V_0 is ZD_{0i} and V_1 is ZD_{1i} and $\ldots V_n$ is ZD_{ni} THEN V is ZD_i

where V_j is a input variable and ZD_{ji} is the definition area of variable V_j in the rule R_i, that is, the values that V_j variable takes in the rule R_i. Each input variable V_i can take values from its domain definition DDV_i, so, each $ZD_{ij} \subseteq DDV_i$.

Moreover, we will consider general exceptions to this rules. Concretely, the form of our fuzzy rules with general exceptions will be the following:

R_i: IF V_0 is ZD_{0i} and V_1 is ZD_{1i} and $\ldots V_n$ is ZD_{ni} THEN V is ZD_i,
excepting IF V_0 is ZD'_{0i} and V_1 is ZD'_{1i} and $\ldots V_n$ is ZD'_{ni}

where ZD'_{li} is a subset of ZD_{li} for every l=1..n.

The underlying idea is that general fuzzy rules with general exceptions will allow to reduce the number of rules (more basic rules can be grouped in a smaller number of general rules with general exceptions).

Finally, punctual exceptions are introduced in order to capture the cases that can not be captured by the rules from the original membership functions. In such a way, on the contrary that tuning methods, the semantic of the rules are completely preserved, and only exceptional (and understandable) particular cases are detailed.

2 Learning Maximal Fuzzy Rules

Many of the machine learning algorithms developed to obtain fuzzy rules try mainly to make good numerical approximation, while little attention is paid to the interpretability of the resulting model. These algorithms use fuzzy sets to partition input variables universes. The most usual techniques from the literature can be classified as grid and tree partitioning methods, fuzzy clustering and nonlinear parameter-optimization approaches (See [4] for a description of each group). With these approaches, the partition process of the domains of the input variables is difficult to understand to user. This could cause that user not accept the results obtained.

When we are interested in using the results obtained for machine learning algorithms in later stages of the knowledge acquisition process, it is very important that the expert understands them. In order to make this easier, we suggest that the results will be expressed as they would by the expert and this involves acquiring the input variables and the values that can take each input variable directly from the expert, that is to say the partition or definition domain of the input variables.

For acquiring this knowledge many manual and automatic methodologies could be applied. Some of them are card or concept sorting, laddered grid, interviewing,... (See [39] for a description of them). In general, any technique that can be used to obtain expert's conceptual structure. Amongst them, repertory grid technique, and more exactly the fuzzy repertory table [13] is a good method for our purpose, since:

- It has a solid foundation in human psychological theory, Kelly's Personal Constructs Theory [24].
- It has demonstrated useful in eliciting and acquiring knowledge from people.
- It obtains the set of variables that the expert uses based in his personal experience in the field.
- It is easy to formalize or implement.
- It makes instantaneous analysis on data just introduced.
- It is intuitive and easy-to-use.
- It interacts directly with the expert to stimulate him to express his knowledge.
- It allows to expert express his knowledge in a freely way.

2.1 Fuzzy Repertory Table

The fuzzy repertory table (FRT) is based on the repertory grid technique (RG). The principal concepts in repertory grids are *elements*, and *constructs*. Elements are people, things, events or experiences which are related to the particular problem or purpose for using the grid (in our problem, they will be the possible outputs). Constructs are dimensions of similarity and differences among elements (in our problem, they will constitute the input variables). The most basic form for a RG is a rectangular matrix with elements as columns and constructs as rows. Each row-column intersection into the grid contains a rating on a rating scale showing how a person apply a given construct to a particular element. This limits its representation capabilities. A variety of extensions have been implemented or proposed to extend the possible ratings [6,7,9,19–21]. However, they have some limitations that are solved by means of the FRT (See [13] for a detail discussion).

FRT allows represent all values (crisp and vague) that an expert uses when he expresses his knowledge. All values of variables domain definition are represented as fuzzy with a trapezoidal membership function, see Figure 1 and Equation 1.

$$\mu_{ij}(x) = \begin{cases} 0 & x < a, \\ \frac{x-a}{b-a} & a \leq x < b, \\ 1 & b \leq x \leq c \\ \frac{d-x}{d-c} & c < x \leq d, \\ 0 & x > d \end{cases} \quad (1)$$

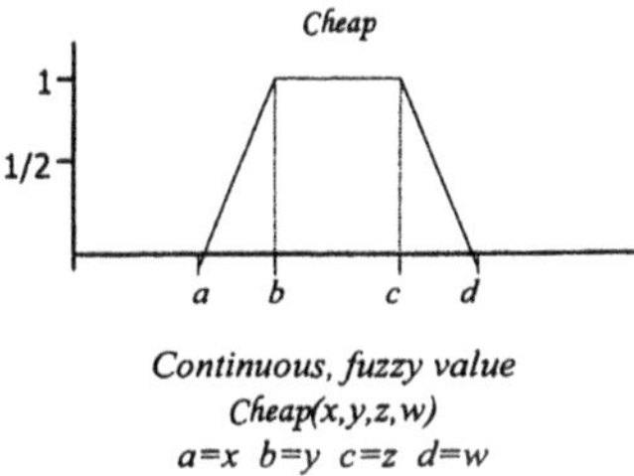

Fig. 1. Example of a trapezoidal function associated to the *cheap* value.

The trapezoidal function has the following merits:

- Adaptable. It can hold any value used by human.
  - Ordered-discrete or ordinal type.
  - Unordered-discrete or nominal type.
  - Boolean type.
  - Raking type.
  - Continuous type.
- Understandable. It is easy for human users to understand.
- Mixed. It can accommodate vague and crisp values.
- Uniform. It can be used by inductive learning algorithms [11,12] and knowledge acquisition tools [13,14].

The FRT performance is shown in Figure 2.

We firstly introduce the set of elements, that in our case will be the possible outputs of the fuzzy rules. Secondly, we identify a set of elements (generally three) that are rated similarly on all constructs. Next, the expert identifies a way (a construct) according to which two elements are similar and the other different (he introduces a input variable). Then, he rates all elements on the new construct (he introduces the values or partitions of variable domain). We repeat this interaction until that the elements are distinguished in a clear manner or the expert finish the interaction.

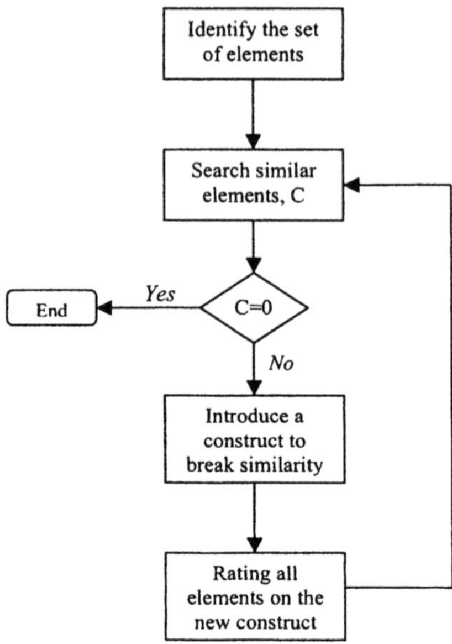

Fig. 2. FRT performance.

The similarity between two elements is assessed by means of a measurement that calculates the area existing between two fuzzy values [13,14]. Two or more elements will be similar if they have similar ratings on many constructs-dimensions.

In this way, we obtain the domain expert's conceptualization. In our problem, it will be the set of input variables $\mathcal{V}$, and the partitions of their universes $\mathcal{DDV}$, that are used by the domain expert. This is a very important fact to subsequent efforts in knowledge acquisition process.

We show a FRT obtained from a buyer agent in an accommodation renting scenario in Figure 3. In that example, the elements of the FRT are the accommodations $a_1, a_2, a_3, a_4, a_5, a_6, a_7$, during the course of developing the FRT, the variables *Rental-rate* and *Distance to work place* were given as a fuzzy concept. In this way, these variables were described by a set of linguistic values. Each value has related a membership function that is obtained from direct interaction with the expert. At the end, from that FRT, we obtain the set of input variables $\mathcal{V}$ and the partitions of their universe $\mathcal{DDV}$ that are shown in Table 1.

2.2 An Inductive Machine Learning Algorithm

Once we have specified the knowledge representation formalism and obtained directly from the expert the input variables ($\mathcal{V}$) and their domain definition

	a_1	a_2	a_3	a_4	a_5	a_7	a_8
Rental rate	Cheap a=100 b=100 c=250 d=300	Expensive a=400 b=450 c=600 d=600	Cheap a=100 b=100 c=250 d=300	Normal a=250 b=300 c=400 d=450	Normal a=250 b=300 c=400 d=450	Normal a=250 b=300 c=400 d=450	Cheap a=100 b=100 c=250 d=300
Distance to work place	Near a=0 b=0 c=15 d=25	Near a=0 b=0 c=15 d=25	Normal a=15 b=25 c=30 d=35	Near a=0 b=0 c=15 d=25	Near a=0 b=0 c=15 d=25	Far a=30 b=35 c=50 d=50	Near a=0 b=0 c=15 d=25
Rental period	x<=6 a=0 b=0 c=6 d=6	x<=6 a=0 b=0 c=6 d=6	x<=6 a=0 b=0 c=6 d=6	6<x<=12 a=6 b=6 c=12 d=12	6<x<=12 a=6 b=6 c=12 d=12	x>=12 a=12 b=12 c=24 d=24	x<=6 a=0 b=0 c=6 d=6
House condition	Good a=b=c=d=4	Good a=b=c=d=4	Normal High a=b=c=d=3	Normal Small a=b=c=d=1	Good a=b=c=d=4	Bad a=b=c=d=0	Suitable a=b=c=d=2

Fig. 3. FRT developed in an accommodation renting scenario, a_i are accommodations and a, b, c and d are the parameters of trapezoidal functions that define each value.

Table 1. Sets $\mathcal{V}$ and $\mathcal{DDV}$ that are acquired using FRT shown in Figure 3

Variable	Variable type	Domain definition
V_1, Rental rate (£)	Continuous (Fuzzy)	DDV_1={Cheap(100, 100, 250, 300), Normal(250, 300, 400, 450), Expensive(400, 450, 600, 600)}
V_2, Distance to work place (Minutes walk)	Continuous (Fuzzy)	DDV_2={Near(0, 0, 15, 25), Normal(15, 25, 30, 35), Far(30, 35, 50, 50)}
V_3, Rental period (months)	Continuous (crisp)	DDV_3={$x \leq 6$(0, 0, 6, 6) $6 < x \leq 12$(6, 6, 12, 12), $x \geq 12$(12, 12, 24, 24) }
V_4, House condition	Ranking (Crisp)	DDV_4={Bad(0, 0, 0, 0), Normal low(1, 1, 1, 1), Suitable(2, 2, 2, 2), Normal high(3, 3, 3, 3), Good(4, 4, 4, 4)}

($\mathcal{DDV}$) that he uses, we must establish the method for acquiring the knowledge inherent to a set of examples defined over $\mathcal{V}$.

We have a set of examples $\mathcal{E} = \{e_1, e_2, \ldots, e_m\}$, each example has the following structure $e_i = ((x_{i1}, x_{i2}, \ldots, x_{in}), o_i)$, where $x_{i1}, x_{i2}, \ldots, x_{in}$ are the values of the input variables belong to $\mathcal{V}$ ($|\mathcal{V}| = n$) and o_i is the value of the output variable. The set of examples will be a subset of $(V_1 \times V_2 \times \ldots \times V_n \times Output)$, where $V_i \in \mathcal{V}$, $\forall i = 1 \ldots n$.

The purpose of the machine learning algorithm will be to approximate the function $\mathcal{V} \longrightarrow O$ that models the system. For doing it, we make use of a set of rules with the structure shown in Section 1.

One of our initial requirements (see section 2) will be that machine learning algorithm must be transparent to expert. In this way, we believe that the expert is more pleasure to accept the results obtained. Since, inductive reasoning is a very familiar method of reasoning, we propose make use of inductive machine learning algorithm for acquiring the knowledge inherent to $\mathcal{E}$. Although a variety of inductive algorithms have been developed, we suggest make use of the inductive machine learning algorithm presented in [11], because it allows us to use the sets $\mathcal{V}$ and $\mathcal{DDV}$ acquired making use of FRT. In addition, it is representative of the class of decision tree learning, but using fuzzy partitions obtained directly from interaction with the expert. Its results will be more understandable that those obtained for decision trees learning algorithms that use crisp partitions (ID3 [33] and C4.5 [34]) and it approximates well the target function.

The algorithm performance is shown in Figure 4.

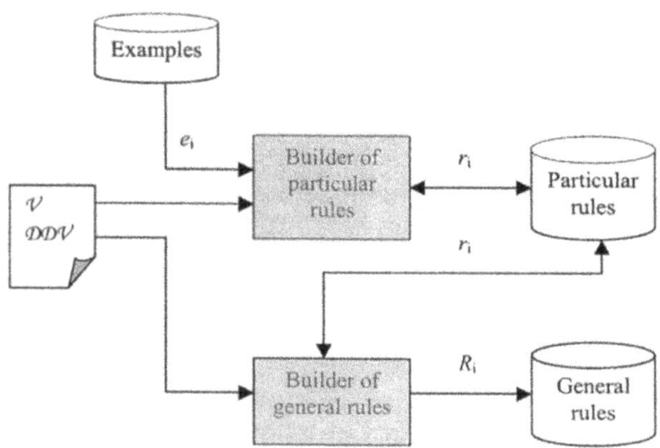

Fig. 4. Suggested scheme for obtaining a set of general rules.

The idea of the suggested algorithm is construct an initial set of rules from training examples (Builder particular rules aim) and then construct a set of maximal rules which are generalizations of those initial rules (Builder general rules aim). Next, we describe briefly both modules.

- *Builder particular rules.* Its mission is take the inputs and determine the degree to which they belong to each of appropriate fuzzy sets via membership functions. Each example from $\mathcal{E}$ is converted in a particular rule, r_i. We remove the particular rules that are repeated.
- *Builder general rules.* It constructs the set of maximal fuzzy rules (in the sense of [11]) that identify the system. To do so, it performs the following:
  1. For each particular rule i, apply a generalization process that consists in the following:
    1.1. Repeat until all input variables are taken into account
      1.1.1. Select the input variable that is less useful for classifying (making use of some entropy measures homogeneity of examples) and has not been taken into account yet, we call it j.
      1.1.2. Add values took from DDV_j to ZD_{ji}, checking whether the new rule constructed maintain a good classification of $\mathcal{E}$. If maintain a good classification then we continue working with the new rule, else we work with the rule before add the value.
    1.2. Introduce the new rule obtained in the set of general rules.
  2. End.

Now, we present an example of the suggested algorithm performance in the accommodation renting scenario mentioned in Section 2.1, (for a more detail description and more complex example, see [11]). We have the sets $\mathcal{V}$ and $\mathcal{DDV}$ that are shown in Table 1 (acquired making use of FRT), and the set of examples that are shown in Table 2.

After, the builder particular rules run over the training set making use of $\mathcal{V}$ and $\mathcal{DDV}$ shown in Table 1, we obtain the set of particular rules that is shown in Table 3.

At the end, the builder general rules obtain the following set of maximal rules:

- R_0 : If distance to work place is *Near* and rental_rate is *Cheap* then rent the accommodation.
- R_1 : If rental_period is $x \leq 6$ and rental_rate is *Cheap* then rent the accommodation.
- R_2 : If rental_rate is *Normal or Expensive* then not rent the accommodation.
- R_3 : If rental_period is $6 < x \leq 24$ and distance to the work place is *Normal or Far* then not rent the accommodation.

These rules are very understandable by the expert and they can be used for facilitating the interviewing domain expert in later phases of knowledge

Table 2. Examples to be used by the inductive machine learning algorithm

House condition	*Rental period*	*Distance to work place*	*Rental rate*	*Decision*
1	$x \leq 6$	7	125	Rent
1	$x \leq 6$	10	100	Rent
1	$x \leq 6$	15	220	Rent
2	$x \leq 6$	10	200	Rent
3	$x \leq 6$	12	210	Rent
4	$x \leq 6$	10	230	Rent
1	$6 < x \leq 12$	5	180	Rent
2	$6 < x \leq 12$	10	175	Rent
3	$6 < x \leq 12$	17	200	Rent
4	$6 < x \leq 12$	2	240	Rent
3	$x \geq 12$	15	200	Rent
3	$x \geq 12$	5	210	Rent
2	$x \leq 6$	10	100	Rent
1	$x \leq 6$	13	150	Rent
4	$x \leq 6$	8	170	Rent
1	$x \leq 6$	45	120	Rent
3	$x \leq 6$	25	200	Rent
2	$x \leq 6$	27	175	Rent
4	$x \leq 6$	11	200	Rent
4	$x \leq 6$	40	180	Rent
1	$x \leq 6$	12	560	Not Rent
3	$6 < x \leq 12$	40	120	Not Rent
1	$6 < x \leq 12$	35	100	Not Rent
2	$6 < x \leq 12$	37	200	Not Rent
4	$6 < x \leq 12$	36	210	Not Rent
1	$6 < x \leq 12$	28	220	Not Rent
1	$x \geq 12$	37	110	Not Rent
4	$x \leq 6$	7	310	Not Rent
3	$x \leq 6$	27	470	Not Rent
3	$6 < x \leq 12$	7	350	Not Rent

Table 3. Particular rules obtained by the Builder particular rules

House condition	*Rental period*	*Distance* to *work place*	*Rental rate*	*Decision*
Nlow	$x \leq 6$	Near	Cheap	Rent
Suitable	$x \leq 6$	Near	Cheap	Rent
Nhigh	$x \leq 6$	Near	Cheap	Rent
Good	$x \leq 6$	Near	Cheap	Rent
Nlow	$6 < x \leq 12$	Near	Cheap	Rent
Suitable	$6 < x \leq 12$	Near	Cheap	Rent
Nhigh	$6 < x \leq 12$	Near	Cheap	Rent
Good	$6 < x \leq 12$	Near	Cheap	Rent
Nhigh	$x \geq 12$	Near	Cheap	Rent
Nlow	$x \leq 6$	Far	Cheap	Rent
Nhigh	$x \leq 6$	Normal	Cheap	Rent
Suitable	$x \leq 6$	Normal	Cheap	Rent
Good	$x \leq 6$	Far	Cheap	Rent
Nlow	$x \leq 6$	Near	Expensive	Not Rent
Nhigh	$6 < x \leq 12$	Far	Cheap	Not Rent
Nlow	$6 < x \leq 12$	Far	Cheap	Not Rent
Suitable	$6 < x \leq 12$	Far	Cheap	Not Rent
Good	$6 < x \leq 12$	Far	Cheap	Not Rent
Nlow	$6 < x \leq 12$	Normal	Cheap	Not Rent
Nlow	$x \geq 12$	Far	Cheap	Not Rent
Good	$x \leq 6$	Near	Normal	Not Rent
Nhigh	$x \leq 6$	Normal	Expensive	Not Rent
Nhigh	$6 < x \leq 12$	Near	Normal	Not Rent

acquisition. Moreover, they contain knowledge that could be used for modelling the system, but some refinement is necessary for reducing the conflicts that generalization process introduces. In next Section, we study how we take advantage of these rules for modelling the system resolving the conflicts in the rules.

3 Conflict resolution in the learning algorithm

In the learning algorithm presented in the previous section, conflicts can arise in several ways. A variety of strategies can be adopted by the expert to solve these conflicts in later stages of the knowledge acquisition process. For example, a strategy that we can apply is the *differentiation strategy* [23], asking to the expert how to resolve the conflicts. However, if the acquired knowledge is to be used for the automatic modelling of the system, we must develop a method to solve these conflicts in an autonomous manner, without the intervention of the human, in order to obtain a more accurate mapping of the system. Besides, since this method will delete the possible contradictions contained in the model, it will allow to increase the interpretability of the final model, taking in care one of the main features of the fuzzy models, as we have remarked previously.

Two types of conflicts can arise during the learning process: conflicting initial rules and conflicting definitive rules. The first one is due to the generalization carried out by the *builder of particular rules* module when translating the examples into the initial rules. The method proposed to solve this type of conflicts consists in selecting one of the conflicting rules attending to some measure of their reliability. The second type of conflicts can appear during the building of general rules. This module tries to obtain rules as general as possible, and obtaining those general rules can provoke that conflicting zones arise where rules with different consequents coexist. In this case, we propose the inclusion of exceptions into the general rules as the way to solve the conflicts, which will also allow to reduce the number of rules in the model and to increase, this way, its interpretability.

The remainder of this section is structured in the following manner: subsection 3.1 describes the method to solve conflicts among initial rules, whereas subsection 3.2 focuses on the method to solve conflicts among the definitive rules generated by the learning algorithm in previous section. Finally, subsection 3.3 presents several strategies to further increase the model interpretability.

3.1 Resolution of conflicting initial rules

The builder of particular rules takes every example and translate each of its values to one of the fuzzy sets associated with the corresponding linguistic variable. In order to do that, it considers the membership degree of these values to the membership functions of the labels and select the label corresponding to the highest membership value.

Example 1. Suppose that after the development of a FRT, the set of input variables $\mathcal{V}$ and the partition of their universes $\mathcal{DDV}$ are the ones shown in Table 4, where the real domains of both variables is [0,1]. Besides, suppose that the output of the model is also a fuzzy value, whose fuzzy partition is shown in Figure 5. Then, the example ((0.09,0.80),0.87) will be translated by the builder of particular rules into the rule $R_i : S, L \rightarrow L$.

Table 4. Sets $\mathcal{V}$ and $\mathcal{DDV}$ in example 1

Variable	Variable type	Domain definition
X_1	Continuous (Fuzzy)	DDV_1={S(0, 0, 0, 0.5), M(0, 0.5, 0.5, 1), L(0.5, 1, 1, 1)}
X_2	Continuous (Fuzzy)	DDV_1={S(0, 0, 0, 0.5), M(0, 0.5, 0.5, 1), L(0.5, 1, 1, 1)}

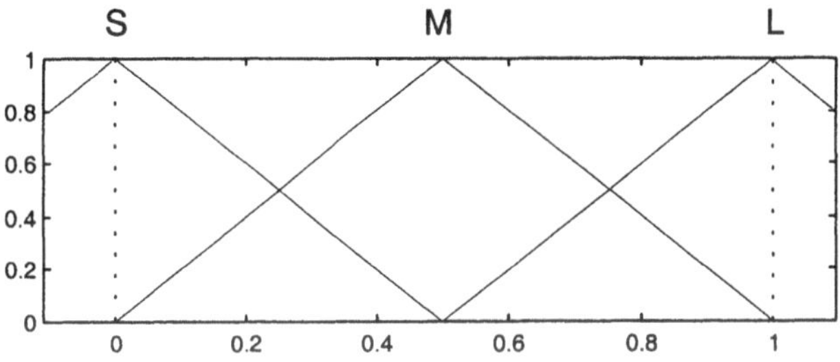

Fig. 5. Fuzzy output in the example 1

This generalization process that transforms the examples into initial rules can provoke the coexistence of conflicting initial rules, that is, rules having the same labels in the antecedent part and a different value in the consequent part. A mechanism will be developed to determine which of those conflicting initial rules must be amplified by the builder of general rules.

The method we propose begins working out the certainty degree of every initial rule from the positive and negative examples that the rule presents in the training set. The certainty measure used for that purpose is an extension of the one proposed by Ishibuchi et al. in [22]. This extension will allow the consequents for taking fuzzy values. Nevertheless, it must be noticed that all the considerations made in this section about that extended measure can be applied also to crisp consequents, since the trapezoidal fuzzy membership functions shown in equation (1) can hold with crisp values, as mentioned in subsection 2.1.

Definition 1. Given a rule $R^{i_1\ldots i_n}_{LY^i} : LX_{1,i_1}, \ldots, LX_{n,i_n} \rightarrow LY^i$ ($LX_{j,\cdot} \in DX_j, LY^i \in DY$) and a set of examples $\mathcal{E} = \{e_1, e_2, \ldots, e_m\}$ where each example takes the form $e_j = ((x^j_1, x^j_2, \ldots, x^j_n), y^j)$, the certainty degree of R^i over $\mathcal{E}$ is defined as

$$\omega(R^{i_1\ldots i_n}_{LY^i}) = \frac{\beta(R^{i_1\ldots i_n}_{LY^i}) - \bar{\beta}(R^{i_1\ldots i_n}_{LY^i})}{\sum\limits_{k=1}^{M} \beta(R^{i_1\ldots i_n}_{LY_k})} \tag{2}$$

where

$$\beta(R^{i_1\ldots i_n}_{LY}) = \sum_{e_j \in \mathcal{E}} \mu_{LX_{1,i_1}}(x^j_1) \times \ldots \times \mu_{LX_{n,i_n}}(x^j_n) \times \mu_{LY}(y^j)$$

$$DY = \{LY_1, LY_2, \ldots, LY_M\}$$

$$\overline{\beta}(R^{i_1\ldots i_n}_{LY^i}) = \sum_{\substack{k=1 \\ LY_k \neq LY^i}}^{M} \frac{\beta(R^{i_1\ldots i_n}_{LY_k})}{M-1}$$

Once the certainty degree of every rule has been obtained, all the initial rules will be processed in descendent order according to their degrees in such a manner that any rule with the same antecedent than another rule previously processed will be rejected. This method, besides solving the conflicts, will introduce a heuristic criterion for processing the initial rules, since it amplifies the rules with highest certainty degree first.

Table 5. Training set in example 2

X_1	X_2	Y
0.0900	0.8000	0.8711
0.1800	0.2000	1.0000
0.1900	0.9100	0.8006
0.2000	0.2300	1.0000
0.2100	0.7600	0.7780
0.2900	0.8300	0.7122
0.4900	0.8000	0.7122
0.7200	0.7800	0.5000
0.8200	0.9500	0.2719
0.9100	1.0000	0.1144

Example 2. Taking into account the assumptions made in the previous example, suppose the training set in Table 5 obtained from the fuzzy relation generated by the model shown in Fig 6.

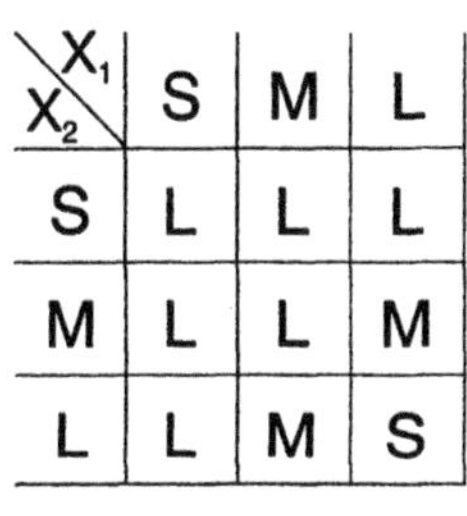

X_2 \ X_1	S	M	L
S	L	L	L
M	L	L	M
L	L	M	S

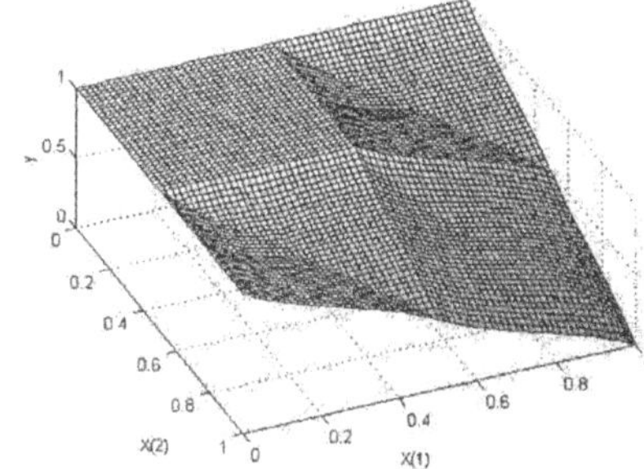

Fig. 6. System to be modeled.

The set of initial rules and their associated certainty degrees once the redundancies are removed and the rest of rules are ordered will be the following:

$$
\begin{array}{l}
R_L^{1,1} : S, S \rightarrow L \ (1.00) \\
R_L^{1,3} : S, L \rightarrow L \ (0.41) \\
R_M^{2,3} : M, L \rightarrow M \ (0.33) \\
R_S^{3,3} : L, L \rightarrow S \ (0.32) \\
R_M^{3,3} : L, L \rightarrow M \ (0.18)
\end{array}
$$

The last rule $R_M^{3,3}$ will not be amplified, since it comes into conflict with the previous initial rule $R_S^{3,3}$.

3.2 Resolution of conflicting definitive rules

The main goal of the learning algorithm presented in section 2 is to obtain maximal rules, i.e., rules as general as possible. This aim can leads to fuzzy model where different consequents coexist in some fuzzy regions of the input space. An example of this behavior can be observed in the fuzzy model obtained in example 2, shown in Table 6.

The basis that will support the approach proposed here to solve the conflicts between definitive rules consists in taking advantage of the information that could be contained in the training set about such regions.

A compound rule of the form presented in section 1 is equivalent to a conjunction of simple rules with one label associated to each input variable. Therefore, the set of simple rules involved in a conflict can be isolated and one of them can be selected based on certain criterion. With that goal, a certainty

Table 6. Conflicts in the identified rule base in example 2

		X_1		
		S	M	L
	S	L	L/M	L/S
X_2	**M**	L	L/M	L/S
	L	L	M	S

degree for each simple rule involved in the conflict will be calculated using the same measure described in equation (2).

Besides, it can occur that none of the conflicting rules in a fuzzy region presents a certainty degree high enough, since due to the aforementioned main goal of the learning algorithm, the coexisting consequents can proceed from initial rules located far away from the subspace under consideration. Hence, a threshold τ will be established on the certainty degree, below of which the search for solving the conflict will be extended to all the possible rules, i.e., all the values in the output domain.

The certainty measure in equation (2) complies with two properties that allow us to enunciate the following proposition (see [10] for more details).

Proposition 1. *Given the example set $\mathcal{E}$, a rule with a certainty degree higher or equal to $(M-2)/[2(M-1)]$ is the rule with the highest certainty degree among the M possible rules with the same antecedent. That is,*

$$\text{if}\, \omega(R_{LY_*}^{i_1\ldots i_n}) \geq \frac{M-2}{2\cdot(M-1)},\ \text{then}\ \omega(R_{LY_*}^{i_1\ldots i_n}) = \max_j \omega(R_{LY_j}^{i_1\ldots i_n})$$

Based in this proposition, an upper bound $\frac{M-2}{2\cdot(M-1)}$ can be established for the value of the threshold τ, since a higher value does not improve the results and would cause an increase in the number of rules to be considered during the conflict resolution, thus increasing the computational cost.

Example 3. Given the assumptions stated in example 2, the threshold will take a maximum value equal to $\frac{3-2}{2\cdot(3-1)} = 0.25$, since no more than one rule can take a certainty degree higher than 0.25.

Once the best rule in the conflicting region has been selected, it will be necessary to modify the rest of compound rules in order to uncover that input region. The method proposed here consists in adding exceptions to the rules. An exception is a n-tuple of labels $(LX_{1,i_1}, \ldots, LX_{n,i_n})$ that defines the fuzzy region of the input space where the compound rule is not applied.

The use of exceptions is interesting for two main reasons:

- Description similar to the human mental model. Several authors from the field of psychology [30,31] have supported that the human mental model for the classification process is based on extracting a set of imperfect rules to which occasional exceptions are added. This fact strengthens the belief that a model describing the system by means of rules with exceptions will be easier to interpret for the human being.

- Reduction in the number of rules. This can be observed in the following example.

Besides, in order to diminish the number of conflicts among the general rules, a new restriction will be imposed to the amplification of a rule: *an amplification will be carried out only if the new input space covered by the rule is not covered yet by a definitive rule with the same consequent.* This strategy tries to avoid the unnecessary overlap among the general rules and, although it is contrary to the maximality goal of the rules, it will lead to a improvement in the obtained fuzzy model once the conflicts are solved.

Example 4. The number of simple rules describing the model in Figure 6 is $3 \times 3 = 9$ rules. A description using the usual technique, which associates an input subspace having the same output (consequent) with the antecedent of each rule gives at least six fuzzy rules. However, the same model can be described with only five rules using exceptions:

$$
\begin{aligned}
R^1 &: \text{if } X_2 \text{ is } \{S, M\} \text{ then } Y \text{ is } L \\
&\quad\ \text{excepting if } X_1 \text{ is } L \text{ and } X_2 \text{ is } M \\
R^2 &: \text{if } X_1 \text{ is } \{S\} \text{ and } X_2 \text{ is } \{L\} \text{ then } Y \text{ is } L \\
R^3 &: \text{if } X_1 \text{ is } \{M\} \text{ and } X_2 \text{ is } \{L\} \text{ then } Y \text{ is } M \\
R^4 &: \text{if } X_1 \text{ is } \{L\} \text{ and } X_2 \text{ is } \{L\} \text{ then } Y \text{ is } S \\
R^5 &: \text{if } X_1 \text{ is } \{L\} \text{ and } X_2 \text{ is } \{M\} \text{ then } Y \text{ is } M
\end{aligned}
$$

equivalent to

$$
\begin{aligned}
R^1 &: \text{if } X_2 \text{ is } \{S, M\} \text{ then } Y \text{ is } L \\
&\quad\ \text{excepting if } X_1 \text{ is } L \text{ and } X_2 \text{ is } M \text{ then } Y \text{ is } M \\
R^2 &: \text{if } X_1 \text{ is } \{S\} \text{ and } X_2 \text{ is } \{L\} \text{ then } Y \text{ is } L \\
R^3 &: \text{if } X_1 \text{ is } \{M\} \text{ and } X_2 \text{ is } \{L\} \text{ then } Y \text{ is } M \\
R^4 &: \text{if } X_1 \text{ is } \{L\} \text{ and } X_2 \text{ is } \{L\} \text{ then } Y \text{ is } S
\end{aligned}
$$

Hence, the following algorithm describes the proposed method to solve conflicts among definitive rules:

1. For each fuzzy region of the input space where two or more different consequents coexist:

  1.1. Work out the certainty degree of the simple rules involved and select the highest (w_1).

  1.2. If w_1 does not reach a threshold τ and there exists a rule with a certainty degree higher than w_1 among the rest of possible rules, select it as the best rule (adding a new compound rule).

1.3. Else, select one of the conflicting rules as follows:

1.3.1. If there are more than one different simple rule with the highest certainty degree (w_1) between the conflicting rules:

1.3.1.1. If all rules appear the same times in the conflicting region, select anyone of them (for example, the first one).

1.3.1.2. Else, select the rule appearing more times.

1.3.2. Else, select the rule with the highest certainty degree (w_1).

1.4. For each deleted simple rule, form the appropriate exception as follows:

1.4.1. If the exception cancels the compound rule, delete the rule.

1.4.2. Else, add the exception to the set of exceptions of the rule.

2. End.

Example 5. Suppose the model obtained in example 2 by the builder of general rules (Table 6):

$$
\begin{aligned}
&R^1 : \{*\}, \{S, M\} \rightarrow L \\
&R^2 : \{S\}, \{*\} \rightarrow L \\
&R^3 : \{M\}, \{*\} \rightarrow M \\
&R^4 : \{L\}, \{*\} \rightarrow S
\end{aligned}
$$

where $\{*\}$ means *"any value"*. Suppose the threshold τ taking the upper bound (0.25).

The conflicts in the fuzzy regions (M, S) and (M, M) are solved adding both exceptions to the rule R^3, since the rules $R_L^{2,1}$ and $R_L^{2,2}$ surpass the threshold in their respective regions ($\omega(R_L^{2,1}) = 1.00$ and $\omega(R_L^{2,2}) = 0.26$). The conflict in (L, S) is solved selecting one of the rules randomly (the first one), since no example covers the region and thus all the rules has a certainty degree equal to 0. Finally, the last conflict in (L, M) provokes the expansion of the search, since none of the conflicting rules reaches the threshold ($\omega(R_L^{3,2}) = -0.50$ and $\omega(R_S^{3,2}) = -0.33$), adding a new compound rule due to the high certainty degree of $R_M^{3,2}$, equal to 0.83.

The final description of the model will be as follows:

$$
\begin{aligned}
&R^1 : \{*\}, \{S, M\} \rightarrow L \\
&\qquad \text{excepting } \{(L, M)\} \\
&R^2 : \{S\}, \{*\} \rightarrow L \\
&R^3 : \{M\}, \{*\} \rightarrow M \\
&\qquad \text{excepting } \{(M, S), (M, M)\} \\
&R^4 : \{L\}, \{*\} \rightarrow S \\
&\qquad \text{excepting } \{(L, S), (L, M)\} \\
&R^5 : \{L\}, \{M\} \rightarrow M
\end{aligned}
$$

coinciding with the system being modeled (Figure 6).

3.3 Interpretability improvement

Although the resolution of conflicts proposed in the previous subsections improve the accuracy of the model and entails a way to improve its interpretability, some strategies can be considers in order to increase the interpretability of the model even more. Next, several of such strategies are described.

Merging definitive rules A first way to improve the interpretability of the model is related with the addition of new rules during the conflict resolution among definitive rules. This can take place when none of the conflicting rules has a certainty degree high enough in the conflicting region and the search is extended. Sometimes, these new added rules could be merged with other existing rules, thus, decreasing the number of final rules.

The possibility (capability) for two rules being merged is defined in the following proposition:

Proposition 2. *A rule $R^i : A_1^i, A_2^i, ..., A_n^i \rightarrow LY^i$ with exceptions $E^i = \{E_1^i, ..., E_{p_i}^i\}$ could be merged with another rule $R^j : A_1^j, A_2^j, ..., A_n^j \rightarrow LY^j$ with exceptions $E^j = \{E_1^j, ..., E_{p_j}^j\}$ if the following is fulfilled:*

1. *$LY^i = LY^j$.*
2. *There exists an r so that $A_r^i \neq A_r^j$.*
3. *$A_s^i = A_s^j$, for all $s \neq r$.*

In order to achieve the aforementioned goal, we propose the following algorithm, which must be called after the addition of every rule during the conflict resolution algorithm:

1. Given the compound rule $R^i : A_1^i, A_2^i, ..., A_n^i \rightarrow LY^i$ with exceptions $E^i = \{E_1^i, ..., E_{p_i}^i\}$ trying to be merged.

2. If there is another rule $R^j : A_1^j, A_2^j, ..., A_n^j \rightarrow LY^j$ with exceptions $E^j = \{E_1^j, ..., E_{p_j}^j\}$ in the set of definitive rules so that it is possible to be merged with R^i:

   2.1. Replace the rules R^i and R^j by the rule $R^* : A_1^i, ..., A_r^i \cup A_r^j, ..., A_n^i \rightarrow LY^i$ with exceptions $E^* = E^i \cup E^j$.

   2.2. Try to merge R^*.

This algorithm is recursive, since the merged rule could satisfy the merging condition with respect to some other definitive rule in the rule base.

Example 6. Suppose the following rule base

$$\begin{aligned} R^1 &: \{S, L\}, \{M, L\}, \{S, L\} \rightarrow L \\ &\quad \text{excepting } \{(L, L, S), (L, M, L)\} \\ R^2 &: \{S, L\}, \{L\}, \{M\} \rightarrow L \\ R^3 &: \{S\}, \{M\}, \{M\} \rightarrow L \end{aligned}$$

and the rule $R^4 : \{L\}, \{M\}, \{M\} \to L$ added during the conflict resolution.

The rules R^4 and R^3 can be merged, since they only differ in the set of labels related to X_1, resulting in $R^{3'} : \{S, L\}, \{M\}, \{M\} \to L$. Next, this rule can be merged with R^2, since they only differ in the set of labels associated with X_2, giving the rule $R^{2'} : \{S, L\}, \{M, L\}, \{M\} \to L$. Lastly, this rule merges with R^1, from which it only differs in the set of labels related to X_3, finally merging the four rules in

$$R^{1'} : \{S, L\}, \{M, L\}, \{S, M, L\} \to L$$
$$\text{excepting } \{(L, L, S), (L, M, L)\}$$

Reducing definitive rules The exceptions in a rule clips the input subspace covered by the antecedent of the rule. This can provoke an input label to be cancelled by a subset of exceptions if the $(n-1)$−dimensional subspace delimited by the label is totally excluded due to subset of exceptions. Next, we propose an algorithm for reducing such rules deleting both the label and the subset of exceptions in these rules:

1. Given the compound rule $R^i : A_1^i, A_2^i, ..., A_n^i \to LY^i$ with exceptions $E^i = \{E_1^i, ..., E_p^i\}$, where $E_p^i = (LX_{1,p_1}, ..., LX_{n,p_n})$ is the new exception added to that rule.
2. For each d from 1 to n:
   - **2.1.** Set up a set of exceptions E^* taking LX_{d,p_d} in the d-th position of every exception and taking the different combinations of the labels from $\{A_1^i, A_2^i, ..., A_{d-1}^i, A_{d+1}^i, ..., A_n^i\}$ in the rest of positions. That is, $E^* \equiv A_1^i \times ... \times A_{d-1}^i \times LX_{d,p_d} \times A_{d+1}^i \times ... \times A_n^i$.
   - **2.2.** If $E^* \subseteq E^i$, then set $E^i = E^i - E^*$, $A_d^i = A_d^i - \{LX_{d,p_d}\}$, and go to step 3.
3. If the reduced rule subsumes in some other compound rule, delete it.
4. Else, try to merge the reduced rule.

This algorithm must be called after the addition of each new exception and compares the subset that each of its labels covers on the input subspace delimited by the antecedent of the rule, deleting both the subset from the exceptions and the label from the antecedent if one of these subsets is included in the set of exceptions. As it can be seen, the reduced rule is tried to be merged, because loosing a label in the antecedent can become true the merging condition.

Example 7. During the conflict resolution in example 5 the rule $R^3 : \{M\}, \{*\} \to M$ received the exception (M, S). The reduction algorithm takes the label M from the new added exception and form the subset $E^* = \{(M, S), (M, M), (M, L)\}$. Since this subset is not included in the set of exceptions of the rule ($\{M, S\}$) the second label S from the new added exception is taken to set up the subset $E^* = \{(M, S)\}$. Now, the subset is included in the set of exceptions of the rule and, thus, the subset $\{(M, S)\}$ is deleted from the set of exceptions and the label S is deleted from the second input variable in the antecedent, generating the reduced rule $R^3 : \{M\}, \{M, L\} \rightarrow M$.

In a similar manner the second exception in R^3 and the both ones in R^4 will be reduced, resulting the reduced model:

$$\begin{aligned} R^1 &: \{*\}, \{S, M\} \rightarrow L \\ &\quad \text{excepting } \{(L, M)\} \\ R^2 &: \{S\}, \{*\} \rightarrow L \\ R^3 &: \{M\}, \{L\} \rightarrow M \\ R^4 &: \{L\}, \{L\} \rightarrow S \\ R^5 &: \{L\}, \{M\} \rightarrow M \end{aligned}$$

In this case, none of the reduced rules has satisfied the merging condition.

Merging exceptions Finally, the interpretability of the model can be further improved using the idea of merging definitive rules when exceptions are added to the rules. In order to do that, it is necessary to change the description of the exceptions as *single exception* with their description as *compound exceptions*. A compound exception will be defined as an n-tuple $E_i = (E_{i,1}, ..., E_{i,n})$, where $E_{i,k} \subseteq DX_k$, i.e., $E_{i,k}$ is a subset of the sets of labels in the corresponding input domain.

The following algorithm describes the method for merging exceptions:

1. Given the set of exceptions $E = \{E_1, ..., E_p\}$ and the exception trying to be merged $E_i = (E_{i,1}, ..., E_{i,n})$.

2. If there exists a $j \neq i$, so that it is possible to merge E_i and E_j:

   2.1. Replace the exceptions E_i and E_j by the exception $E_* = (E_{i,1}, ..., E_{i,r} \cup E_{j,r}, ..., E_{i,n})$.

   2.2. Try to merge E_*.

The possibility (capability) for two exceptions being merged is defined in the following proposition:

Proposition 3. *An exception $E_i = (E_{i,1}, ..., E_{i,n})$ could merge with another one $E_j = (E_{j,1}, ..., E_{j,n})$ if the following is fulfilled:*

1. *There exists an r so that $E_{i,r} \neq E_{j,r}$.*
2. *$E_{i,s} = E_{j,s}$, for all $s \neq r$.*

Unlike the reduction of definitive rules, this algorithm will be called in an *offline* manner (i.e., once the final exceptions of every rule have been obtained) and will be called for each rule while any of its exceptions can be merged. This is due to the use of simple exceptions in the rule reduction procedure.

Example 8. Suppose that a rule with the set of exceptions below results from the conflict resolution:

$$\begin{array}{ll} E_1 = (\{S\},\{S\}) & E_4 = (\{M\},\{M\}) \\ E_2 = (\{S\},\{M\}) & E_5 = (\{M\},\{L\}) \\ E_3 = (\{M\},\{S\}) & \end{array}$$

The first run of the algorithm above tries to merge E_1 producing the set of merged exceptions:

$$\begin{array}{ll} E_1' = (\{S\},\{S,M\}) & E_4 = (\{M\},\{M\}) \\ E_3 = (\{M\},\{S\}) & E_5 = (\{M\},\{L\}) \end{array}$$

In the second run, the exceptions E_3 and E_4 will be merged firstly and, due to the recursive nature of the algorithm, the resulting exception will be merged with E_1', giving the set

$$\begin{array}{l} E_1'' = (\{S,M\},\{S,M\}) \\ E_5 = (\{M\},\{L\}) \end{array}$$

Since the exceptions can not be merged anymore, this set will be the final set of exceptions.

4 Punctual exceptions

The mechanisms of the previous sections provide maximal fuzzy rules with exceptions that achieve accuracy and interpretability. But we can obtain some conflicting rules with equal (or very similar) certainty degrees. The examples can also show irregularities in small regions that cannot be covered with this kind of rules or even with the original linguistic terms.

One plausible solution in these situations is to come back to the set of examples and trying to extract an output from the original source of information. Here, a wide number of techniques can be employed, but we propose a Case–Based Reasoning (CBR) system [26] [25] [1].

Case–based Reasoning is founded on the idea of exploiting the information of previous experiences. A CBR system retains the examples of his past experiences together with its solutions. When it must provide a solution for a new problem, the system retrieves the past experiences more similar to the new problem. Then it reuses the previous solutions of these similar situations and makes adaptation to obtain a solution for the new problem. And finally, it revises this new proposed solution, evaluating its fitting to the problem and, if necessary, repairs the solution. After this process, the new problem and its solution can be added to the Case Base (the set of known experiences) for a possible future use.

In fact we propose a simple CBR system which falls into the category of Instance–based learning because we basically employ a vector of features, we do not revise the solution and the adaptation is little of non–existent.

We can retain the examples of the areas that cannot be properly modeled with maximal fuzzy rules with exceptions. When we need an output for a new instance we try to apply the rules, but if rules cannot provide a clear solution then we use CBR in this area, search the most similar examples and provide an output based on these most similar examples. With this method we can improve the accuracy in this kind of areas and at the same time maintain a good interpretability. Experts usually establish analogies between different situations and obtain solutions using adaptation of previous solutions from other similar situations. Therefore an expert can easily understand "e is very similar to the examples $e_{i_1}, e_{i_2}, \ldots, e_{i_k}$ with outputs $o_{i_1}, o_{i_2}, \ldots, o_{i_k}$ respectively, so the proposed output is o".

The first problem we must face is how to define the similarity between two past experiences. In our problem we have obtained from the expert the input variables ($\mathcal{V}$), and for each example we know the values of these variables and the output: $e_i = ((x_{i1}, x_{i2}, \ldots, x_{in}), o_i)$.

Instead of computing a similarity degree between instances, we can compute their distance [28]. Ideally this function of distance should measure a distance between classes. The distance between two instances should be 0 if they have the same output, and greater than 0 if it is different. But in practice we only know the values of the input variables and we are implicitly employing a principle of locality: if the attributes of two cases are similar, probably their classes will be similar. In general the function of distance can be any function which measures dissimilarity (low values represent cases are similar and high values represent cases are different).

We should clarify that we are employing the term distance in a wide sense. Usually this function of distance is the Euclidean distance, but we only need a function that group points with equal or similar output and separates points with different outputs. Therefore, we do not need a function that verifies the properties of a function of distance from a geometrical point of view. We need a function that group points according with its output. Therefore, ideally, this function should depend on the problem. For example, if in our problem the points are grouped according into bands, it is preferable a function which groups points in this way to the function of the Euclidean distance.

We can define partial measures of distance along each attribute and a method of combination or aggregation of this information into one global measure of distance. We can define the partial distance over a continuous or ordinal domain, for instance, as the absolute value of difference between the values; over a boolean or nominal domain, at worst we can define a matrix with the difference or distance between each pair of values; or even over a fuzzy domain we can compute the distance or difference between two fuzzy sets. At worst we could always define an "ad hoc" partial distance that works properly in that domain. To obtain one global measure of distance we must perform some combination of the values of the partial distances (similar or completely different to the employed by the geometrical distances), like for example the addition (Chebychev distance), the maximum (Manhattan distance), the square root of the addition of the squares of the partial distances (Euclidean distance), or anything else.

The simplest method based on distances is the Nearest Neighbor (1–NN) [18] [15] [16]. To classify a new instance or, in general, assign an output, 1–NN assigns the class or the output of the example which is at a lower distance (according to the function which measures the distance). We could think that this method is too simple and bad, but beyond theoretical considerations, the fact is that experimentally it works.

1–NN methods suffer from some problems. Firstly, if we employ the optimal function of distance for a given problem, we obtain an optimal classifier, but finding this function is equivalent to the original problem. Secondly, once we accept that, in general, we cannot know the optimal distance, we must assume that often the class of the nearest neighbor will not be the right classification, i.e., depending on the base, some points violate the locality principle (1–NN has strong difficulties with outliers and noisy data).

Some variants of 1–NN methods have been studied in the literature, including the IBx series [3] [2] to reduce the storage requirements and increase noise tolerance, Nested Generalized Exemplar (NGE) theory [38] where hyperrectangles are used instead of points, or Value Difference Metric (VDM) [40] that statistically derive the distances for nominal attributes. Other interesting distance is the Asymmetric Anisotropic Similarity Metric (AASM) [35], which defines a local distance that varies along the space and it is asymmetric.

Instead of using only one point, k Nearest Neighbor (k–NN) methods consider the k nearest (or similar) points to the new problem [42] [44]. Here the locality principle can be more flexible in the sense that if the class of the nearest neighbor were not the right one, perhaps it would be the class of some of the k nearest neighbors. Therefore, in general, it is supposed that the measure of distance is less critical in k–NN. But two new questions arise with k–NN: how many cases must we consider? (k), and how to combine the information of these k more similar instances? The easier way to combine the information of the k nearest neighbors is assigning the class of the majority. But better results are obtained with the "Weighted Vote k–NN" [42]: consider the distance of the k points, and inversely average the importance or weight of each one by their distance to the new instance.

The value of k may be fixed or estimated by Cross–Validation [41] [42] [44]. It is unusual to compute a local value of k for each new point instead of using one single value of k in the whole space [42] [43].

k–NN can suffer some problems related with the value of k. If k is too big or the new instance is in a sparsely populated region, k–NN will consider distant points that probably are not very relevant. On the other hand, if k is too little or the new case is in a densely populated region, k–NN will ignore some close points that probably are relevant. Weighted Vote k–NN can reduce the undesirable consequences of a bad choice of k, but it can be still in troubles if for example the optimal value of k is no constant along the space.

Some papers merge k–NN and neural networks [8], or focus on similarity notion instead of employing distances like [32] that uses and interesting entropy–based assessment, but there is a relationship between the notion of similarity and distance [36] [37] [28].

To overcome the difficulties of the k–NN classifiers we propose the opposite point of view. We propose considering the points that are in the neighborhood of the new case e, instead of selecting a fixed number of nearest points. When a new case must be classified we propose establishing a ball around that point and taking into account the points inside that ball. In this way we always guarantee considering only close points to the new case independently of the density of points of the case base: when e is in a sparse area of the space we take into account fewer points than when it is in a dense area.

The rest of this section is organized in the following way. Subsection 4.1 introduces the ε–balls methods. Subsection 4.2 proposes a heuristic that in previous works had achieved good performance. And subsection 4.3 proposes the use of the ε–ball method with the heuristic in the conflicting areas where maximal fuzzy rules with exceptions cannot provide a clear response, or in the areas where there are irregularities that cannot be covered with the original linguistic terms.

4.1 The ε–balls Methods

We choose a fixed real value ε as the radius of the ball, thus we control the size of the ball and we take into account the points that are at a distance lower or equal to ε. We define the *ε–ball* of an instance e and the *ε–ball of the class C_i* of an instance e respectively as:

$$E(e,\varepsilon) = \{e\prime \in \mathcal{E} \text{ such as } d(e\prime, e) \leq \varepsilon\}$$

$$E_i(e,\varepsilon) = \{e\prime \in \mathcal{E} \text{ such as } class(e\prime) = C_i \text{ and } d(e\prime, e) \leq \varepsilon\}$$

where $\mathcal{E}$ is the set of known examples or Case Base and $C = \{C_1, C_2, \ldots, C_M\}$ is the set of classes.

In the measure of the influence of the ε–ball of a class it is important to consider the influence of the number of instances of the ε–ball that belong to that class and their distances to e. We define the *measure of influence of an ε–ball* $E(e,\varepsilon)$ and $E_i(e,\varepsilon)$ respectively as:

$$|E(e,\varepsilon)|_d = \sum_{e\prime \in E(e,\varepsilon)} \exp(-\alpha d(e, e\prime)) \qquad \text{where } \alpha = \frac{4}{\varepsilon^2}$$

$$|E_i(e,\varepsilon)|_d = \sum_{e\prime \in E_i(e,\varepsilon)} \exp(-\alpha d(e, e\prime)) \qquad \text{where } \alpha = \frac{4}{\varepsilon^2}$$

$|E(e,\varepsilon)|_d$ measures the influence of all the points of the ε–ball, and $|E_i(e,\varepsilon)|_d$ the influence of the points of one class C_i. It verifies:

$$|E(e,\varepsilon)|_d = \sum_{i=1}^{M} |E_i(e,\varepsilon)|_d$$

To classify a new case e, we calculate the measure of influence of the ε-ball of each class, and assign to e the class with the greatest influence. We have observed that it is desirable the soft behavior of the exponential function from nearer (and probably significant) points to further (and probably less important) points.

If the value of ε is too little it can occur that there will be no points in the ε–ball, and the new case will be "non–classified". Choosing a big ε is less critical because if we consider new further points their influence is much lower due to the behavior of the exponential function.

One way to partially response these questions consists in choosing the class of the nearest neighbor when the ε–ball is empty, i.e., we classify with the ε–ball method and when it fails we use the 1–NN method, which always provides a nearest neighbor and its class. We name this method ε–ball^{NN} to distinguish it from the previous one.

In [27] we have performed a wide comparison between 1–NN, k–NN, ε–ball and ε–ball^{NN} methods with 67 data sets. We have observed that, in general, ε–ball^{NN} is preferable over the others, and k–NN attains better results than ε–ball^{NN} only with some kinds of problems.

If we have a discrete number of possible outputs we can directly apply the above ε–ball^{NN} method. But if we have a continuous output, we can consider the set of linguistic terms of the output, and to obtain the measure of influence of each fuzzy output $LY_i \in DY = \{LY_1, LY_2, \ldots, LY_M\}$, we should multiply the measure of influence that contributes each example of the ε–ball by its membership value to the fuzzy output LY_i:

$$|E_i(e,\varepsilon)|_d = \sum_{e\prime \in E(e,\varepsilon)} (\exp(-\alpha d(e, e\prime)) \times \mu_{LY_i}(o_{e\prime}))$$

where $\alpha = \frac{4}{\varepsilon^2}$, $o_{e\prime}$ is the output of the example $e\prime$, and μ_{LY_i} is the fuzzy membership function of the fuzzy output LY_i.

4.2 Considering a heuristic

The results obtained in previous works by the distances [27] lead us to think that 1–NN methods can be useful to extract characteristics about the data sets and the distribution of points along the space. There is also a certain relationship between the accuracy of the basic distances and the accuracy of the other methods based in those distances.

For these reasons, and considering that 1–NN methods are relatively fast, if we have some basic distances, we proposed the following heuristic [27]: *train and test the available 1–NN methods (distances) and use the desired family of methods (k–NN, ε–ball, ε–ball^{NN}...) with the distance which offers the best accuracy.*

We named k–NN Heur, ε–ball Heur and ε–ball^{NN} Heur to the classifiers of each family that employed this heuristic to choose their base distance. This heuristic is observed in a great number of data sets.

To use this heuristic to select the distance and shorten the trials with the basic distances (especially when we propose a large number of them) is feasible using the ideas of the Racing Algorithm [29]. This algorithm discards the less promising models and can also be used to find relevant features. Basically, it tests the set of models in parallel, quickly discards those models that are clearly inferior using statistical bounds, and concentrates the computational effort on differentiating among the better models.

In [27] we have confirmed the good behavior of this heuristic. With all the families of classifiers (1–NN, k–NN, ε–ball and ε–ball^{NN}) the better results are obtained with the heuristic, and, in general, the classifier that attains better results is ε–ball^{NN} Heur. Therefore, and with no previous information, we recommend the use of this classifier.

4.3 Maximal Fuzzy Rules with Exceptions and Punctual Exceptions

Maximal fuzzy rules with exceptions is a very powerful method that can achieve at the same time high accuracy and good interpretability. Nevertheless, some irregularities in small regions cannot be covered with the original linguistic terms. Moreover, when there are some conflicting rules with equal or very similar certainty degree, there are not a clear output in that region. In this situation, it looks reasonable using the punctual exceptions (the CBR system) in that region to determine the output for a new instance.

When can we use the CBR system to solve this kind of conflicts?:

1. There are two or more conflicting rules and none of the possible rules in this region reaches τ (step 1.2 of the algorithm of section 3.2).
2. There are two or more conflicting rules with equal (or very similar) certainty degree and one appears more times than the others (step 1.3.1.2 of the algorithm of section 3.2).
3. There are two or more conflicting rules with equal certainty degree (or very similar), and all appear the same times (step 1.3.1.1 of the algorithm of section 3.2).

This description is made in order of difficulty for the maximal fuzzy rules with exceptions, i.e., it makes more sense applying the CBR system in the last situation than in the previous ones. It is more likely that fuzzy rules with exceptions found themselves in problems in the last situation than in the first.

In the first situation, the algorithm of section 3.2 probably selects a good rule and we can employ it without problems, but if we want we would mark this region as conflicting and let the CBR system to determine the output for each new instance.

In the second situation, the algorithm of section 3.2 has more difficulties. There is no clear output in that region. Selecting the simple rule that appears more times has no warranty. We simply have expanded more rules with that output to this region, so it is the more frequent rule. Here, if we do not want taking risks, we can let this region as conflicting and let the CBR system to determine the output (between the conflicting outputs or between all the possible outputs).

In the third situation, the algorithm of section 3.2 randomly selects an output between the conflicting rules. Obviously, it is better employing a system guided by the knowledge of the set of examples, like the CBR system, to determine the output.

When we decide to apply the CBR system, we can let the region with conflicting rules as a "conflicting region". We can even employ special rules with more than one output to indicate these conflicting regions and store the conflicting outputs that caused the conflict. If these regions are frequent, we can even employ the algorithms of the subsection 3.3 to improve the interpretability, and for example merging conflicting regions.

When we need an output for a new instance we try to apply the rules, but if rules cannot provide a clear solution (conflicting region), we use CBR in this area. With this method we can improve the accuracy in these regions and maintain a good interpretability, because an expert can easily understand "e is very similar to the examples $e_{i_1}, e_{i_2}, \ldots, e_{i_k}$ with outputs $o_{i_1}, o_{i_2}, \ldots, o_{i_k}$, so the proposed output is o".

Of course, the CBR system that we propose is the ε–ball^{NN} Heur introduced in the previous subsections 4.1 and 4.2.

The expert must provide one or more functions to measure distance in the domain of the problem, and a real value ε which determines the radius of the ball (we take into account the points at a distance lower or equal to ε).

If we have a discrete number of possible outputs we can directly apply the ε–ball^{NN} Heur method. If we have a continuous output, we can consider the set DY of linguistic terms of the output, and apply the ε–ball^{NN} Heur method taking into account the membership value to each fuzzy output LY_i.

5 Application of the suggested method to a real problem

We use the suggested machine learning algorithm for knowing how a valuer agent working in a property valuation company carries out his work. We must find the set of variables and their values that he uses when fixes the price of a property. The price of the properties is a fuzzy variable that can take the values shown in Figure 7 and it will be the output of our system.

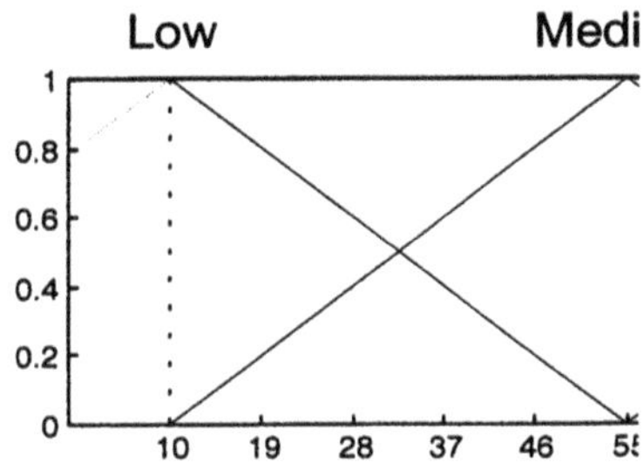

Fig. 7. Fuzzy domain of the output variable *price* (million Spanish pesetas)

For obtaining this set of variables and values, we make use of the FRT. Firstly, the elements of this FRT are a set of properties p_1, p_2, p_3, p_4, p_5 which have been already valued. Secondly, in construction FRT we ask the valuer to describe and analyse the differences and similarities among these properties. To do this, FRT uses the following question:

How are two properties similar and different from the third in the way in which they are valued?

The valuer answers this question by using the attributes of the properties. Thus, through interaction with the valuer, the FRT obtains a set of attributes that identify each one of the properties (see Figure 8). The knowledge acquired with FRT is shown in Table 7. The variables *surface area* and *antique* of the property were given as a fuzzy concept. In this way, these variables were described by a set of linguistic values. The variable *location area* of the property is a ordinal variable, the valuer attaches numbers to each location which show his preferences over each one. Each value has related a membership function that is obtained from direct interaction with the valuer during the FRT performance (see [13]).

Now, we apply the suggested machine learning algorithm explained in Section 2 to the property valuation problem. We have the sets $\mathcal{V}$ and $\mathcal{DDV}$ that are shown in Table 7 (acquired making use of FRT), and the set of examples that are shown in Table 8 valued with regard to the variables acquired by means of the FRT. The elements used in the FRT are also introduced in this set, $p_1 : A, 10, 103$; $p_2 : A, 6, 281$; $p_3 : B, 31, 180$; $p_4 : B, 7, 274$; and $p_5 : C, 16, 166$.

Firstly, the machine learning algorithm begins translating the examples into initial rules by assigning each crisp value to the fuzzy value in which it presents the highest membership degree (*builder particular rules*). This process yields the set of initial rules shown in Table 9, once the redundant rules has been deleted.

Table 7. Sets $\mathcal{V}$ and $\mathcal{DDV}$ that are acquired using FRT shown in Figure 8

Variable	Variable type	Domain definition
V_1, Surface area (m^2)	Continuous (Fuzzy)	DDV_1={Low(50, 50, 85, 190), Medium(85, 190, 190, 330), High(190, 330, 400, 400)}
V_2, Antique of the property (Years)	Continuous (Fuzzy)	DDV_2={Low(0, 0, 4, 20), Medium(4, 20, 20, 40), High(20, 40, 40, 40)}
V_3, Location area	Ordered (Crisp)	DDV_3={A(0, 0, 0, 0), B(1, 1, 1, 1), C(2, 2, 2, 2)}

Table 8. Examples to be used by the inductive machine learning algorithm

Surface	*Antique*	*Location*	*Price*
103	10	A	21
124	32	A	10
124	7	A	28
127	32	A	10
131	11	A	30
145	10	A	35
187	10	A	37
253	11	A	55
281	6	A	74
306	4	A	90
89	20	B	12
106	9	B	51
127	26	B	29
155	36	B	22
180	31	B	31

Surface	*Antique*	*Location*	*Price*
208	10	B	82
264	29	B	57
299	14	B	87
274	7	B	88
99	36	C	31
96	36	C	29
110	39	C	27
183	32	C	70
197	34	C	70
131	10	C	81
152	13	C	82
166	16	C	88
215	23	C	91
180	9	C	94
271	31	C	80

	p_1	p_2	p_3	p_4	p_5
Location area	A $a=b=c=d=0$	A $a=b=c=d=0$	B $a=b=c=d=1$	B $a=b=c=d=1$	C $a=b=c=d=2$
Antique of the property	Low $a=b=0$ $c=4$ $d=20$	Low $a=b=0$ $c=4$ $d=20$	High $a=20$ $b=c=d=40$	Low $a=b=0$ $c=4$ $d=20$	Medium $a=4$ $b=c=20$ $d=40$
Surface area	Low $a=b=50$ $c=85$ $d=190$	High $a=190$ $b=330$ $c=d=400$	Medium $a=85$ $b=c=190$ $d=330$	High $a=190$ $b=330$ $c=d=400$	Medium $a=85$ $b=c=190$ $d=330$

Fig. 8. FRT developed in an property valuation scenario, p_i are properties and a, b, c and d are the parameters of trapezoidal functions that define each value.

Table 9. Initial rules and their certainty degrees

	Surface	*Antique*	*Location*	*Price*	*Certainty*
R_1	Low	Low	A	Low	(0.42)
R_2	Low	Low	B	Medium	(0.87)
R_3	Low	Low	C	High	(0.44)
R_4	Low	Medium	B	Low	(0.58)
R_5	Low	High	A	Low	(1.00)
R_6	Low	High	C	Low	(0.35)
R_7	Medium	Low	A	Medium	(0.34)
R_8	Medium	Low	B	High	(0.36)
R_9	Medium	Low	C	High	(0.60)
R_{10}	Medium	Medium	C	High	(0.43)
R_{11}	Medium	High	B	Low	(0.30)
R_{12}	Medium	High	C	Medium	(0.34)
R_{13}	High	Low	A	Medium	(0.20)
R_{14}	High	Low	A	High	(0.30)
R_{15}	High	Low	B	High	(0.57)
R_{16}	High	Medium	B	Medium	(0.24)
R_{17}	High	Medium	B	High	(0.25)
R_{18}	High	High	C	High	(0.33)

Secondly, the conflict resolution for initial rules proposed in subsection 3.1 removes the rules R_{13} and R_{16} since they have lower certainty degrees than their respective conflicting rules R_{14} and R_{17}.

Thirdly, the *builder general rules* performs the amplification of each initial rule and generates the following conflicting fuzzy model:

$$
\begin{aligned}
&R^1 : \{*\}, \{Medium, High\}, \{A\} \rightarrow Low\\
&R^2 : \{Low, Medium\}, \{Medium, High\}, \{B\} \rightarrow Low\\
&R^3 : \{Low\}, \{Low\}, \{A\} \rightarrow Low\\
&R^4 : \{Low\}, \{Medium, High\}, \{C\} \rightarrow Low\\
&R^5 : \{Medium, High\}, \{High\}, \{B\} \rightarrow Low\\
&R^6 : \{Medium\}, \{High\}, \{A, C\} \rightarrow Medium\\
&R^7 : \{Medium\}, \{Low, Medium\}, \{A\} \rightarrow Medium\\
&R^8 : \{Low\}, \{Low, High\}, \{B\} \rightarrow Medium\\
&R^9 : \{Medium\}, \{Medium\}, \{A, C\} \rightarrow High\\
&R^{10} : \{*\}, \{Low, Medium\}, \{C\} \rightarrow High\\
&R^{11} : \{High\}, \{High\}, \{*\} \rightarrow High\\
&R^{12} : \{High\}, \{Low, Medium\}, \{A\} \rightarrow High\\
&R^{13} : \{Medium, High\}, \{Low, Medium\}, \{B\} \rightarrow High
\end{aligned}
$$

whose spatial representation is shown in Figure 9, and where eight conflicting regions can be observed.

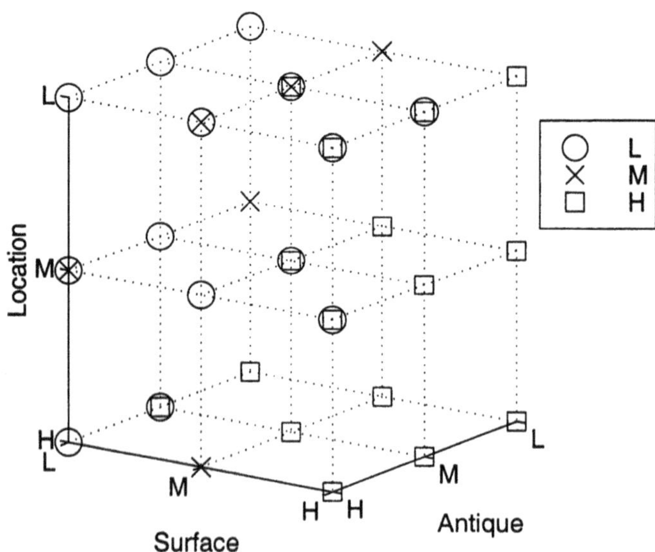

Fig. 9. Conflicting fuzzy model

Finally, the algorithm for solving conflicts among general rules by means of general and punctual exceptions will be launched and its progress is de-

Table 10. Conflict resolution among general rules

Conflicting region	*Conflicting consequents*	*Certainty degrees*	*Extended search*	*Conflict resolution*
(L,M,C)	H,L	0.08,-0.18	M (0.09)	PE (C.1)
(L,H,B)	M,L	0.02,0.48		GE (L)
(M,M,A)	L,H,M	0.25,-0.48,0.23		PE (C.3)
(M,M,B)	L,H	-0.05,-0.19	M (0.23)	PE (C.1)
(M,H,A)	L,M	1.00,-0.50		GE (L)
(H,M,A)	L,H	-0.50,-0.31	M (0.81)	GE (M)
(H,H,A)	L,H	0.00,0.00	M (0.00)	PE (C.1)
(H,H,B)	H,L	-0.43,-0.50	M (0.93)	GE (M)

picted in Table 10 when using a maximum threshold (i.e., $\tau = 0.25$). First column refers to the input fuzzy region where the conflicts occurs, second and third columns show the consequents of the conflicting rules and their certainty degrees, respectively. Fourth column reflects when the search is extended to all possible rules, specifying the consequent of that rule and its certainty degree. The last column shows the method used to resolve the conflict: punctual exceptions (PE) and, in parenthesis, the condition for applying it, or general exceptions (GE) and, in parenthesis, the selected consequent.

Then, the final fuzzy model not considering the fuzzy regions that must be solved by punctual exceptions is described by the rule base,

$$\begin{aligned}
&R^1 : \{Low, Medium\}, \{Medium, High\}, \{A\} \to Low \\
&\quad \text{excepting } \{Medium\}, \{Medium\}, \{A\} \\
&R^2 : \{Low, Medium\}, \{Medium, High\}, \{B\} \to Low \\
&\quad \text{excepting } \{Medium\}, \{Medium\}, \{B\} \\
&R^3 : \{Low\}, \{Low\}, \{A\} \to Low \\
&R^4 : \{Low\}, \{High\}, \{C\} \to Low \\
&R^5 : \{Low\}, \{Low\}, \{B\} \to Medium \\
&R^6 : \{Medium\}, \{High\}, \{C\} \to Medium \\
&R^7 : \{Medium\}, \{Low\}, \{A\} \to Medium \\
&R^8 : \{High\}, \{Medium\}, \{A\} \to Medium \\
&R^9 : \{High\}, \{High\}, \{B\} \to Medium \\
&R^{10} : \{Medium, High\}, \{Low, Medium\}, \{B\} \to High \\
&\quad \text{excepting } \{Medium\}, \{Medium\}, \{B\} \\
&R^{11} : \{High\}, \{High\}, \{C\} \to High \\
&R^{12} : \{*\}, \{Low, Medium\}, \{C\} \to High \\
&\quad \text{excepting } \{Low\}, \{Medium\}, \{C\} \\
&R^{13} : \{High\}, \{Low\}, \{A\} \to High
\end{aligned}$$

If one new property is not valued by the final fuzzy model (i.e., is not covered by the rules), then it is inside one of the conflicting regions that must be solved using punctual exceptions. We recommend employing the ε–ball^{NN} Heur method described in section 4. Here, the expert must take a decision about the function of distance. If the expert has previous knowledge, he can provide the appropriate function of distance, or even an "ad hoc" distance. Otherwise, he can provide some functions of distance and employ the heuristic described in section 4.2. Then some trials must be done to select the distance which offers the better behaviour in this problem.

Let's analyse an example: a new property e of 175 m^2, 25 years and location area B. This property is in the input fuzzy region (M,M,B), one of the conflicting regions that are not covered in the final fuzzy model, and thus, it must be solved employing punctual exceptions.

For simplicity, let's suppose the right distance is the Euclidean distance, but weighting each feature by the absolute value of their correlation with the output.

It does not matter if this distance is directly provided by the expert or as a result of applying the heuristic to a set of functions of distance.

We have employed ε–ball^{NN} with a value of $\varepsilon = 0.4$, and the Euclidean with correlation distance. The correlation of the attributes *Surface area*, *Antique of the property* and *Location area* with the output are 0.6897, -0.4072, and 0.4208 respectively.

Its progress is depicted in table 11. First column refers to the examples in the ε–ball (those which are at distance equal or lower than ε), second column shows the Euclidean with correlation distance of each one of these examples from the new case e. Third, forth and fifth columns show the measure of influence that contributes each example to each fuzzy output. Last row shows the measure of influence of the whole ε–ball for each fuzzy output.

Table 11. Applying punctual exceptions in the conflicting region (M,M,B) to a new instance (175 m^2, 25 years, B area).

Examples in the ε-ball	*distance from the new case*	*Influence of Low*	*Influence of Medium*	*Influence of High*
(89, 20, B, 12)	0.2191	0.0040	0.0002	0.0000
(127, 26, B, 29)	0.1150	0.0326	0.0238	0.0000
(155, 36, B, 22)	0.1818	0.0078	0.0028	0.0000
(180, 31, B, 31)	0.0964	0.0478	0.0419	0.0000
(208, 10, B, 82)	0.2518	0.0000	0.0007	0.0011
(264, 29, B, 57)	0.2206	0.0000	0.0038	0.0002
Total Influence of the outputs		0.0922	0.0732	0.0013

In this example the new property e is assigned an output *Low*, because it has the highest measure of influence (0.0922). The expert can easily understand this output is because of e is very similar to the examples (180, 31, B, 31) and (127,26, B, 29), and quite similar to (155, 36, B, 22), (89, 20, B, 12), (264, 29, B, 57), and (208, 10, B, 82).

Whatever happens, (M,M,B) is a conflicting region and punctual exceptions can cover irregularities and special features inside this kind of regions. With punctual exceptions, different instances inside the same conflicting region can be assigned different outputs. It depends on the attributes of the new instance and the output and distance of the nearest examples. For example, if we consider two new properties (190, 20, B) and (200, 15, B) in this region, they are assigned the output *Medium* and *High* respectively. And this fact makes sense because we are in a conflicting region!

6 Conclusions

A fuzzy rule learning with exceptions has been proposed to identify fuzzy models from examples. Such a method has been developed trying to achieve a double goal: accuracy and interpretability. In order to do that, we apply a fuzzy repertory table to obtain the linguistic terms of the experts, and then maximal structure fuzzy rules with these terms are obtained. In a second stage, the conflicts generated by the maximal rules are solved, thus increasing the model accuracy and interpretability. Nevertheless, some irregularities (in small regions) can not be covered with the original linguistic terms. In order to solve these conflicts, the way for using these exceptions is learned, and a case-based reasoning system is applied. The experiments carried out shown that the algorithm learn with a good accuracy, with comprehensible rules, and with a lower number of rules than classical rule based systems.

References

1. Aamodt, A., Plaza, E. (1994) Case–based reasoning: Foundational issues, methodological variations, and system approaches. *Artificial Intelligence Communications*, 7:39-59.
2. Aha, D.W. (1992) Tolerating noisy, irrelevant and novel attributes in instance–based learning algorithms. *International Journal of Man–Machine Studies*, 36:267-287.
3. Aha, D.W., Kibler, D., Albert, M.K. (1991) Instance–based learning algorithms. *Machine Learning*, 6:37-66.
4. Babuska, R. (1999) Data-driven Fuzzy Modelling: Transparency and Complexity Issues. In: Proceedings European Symposium on Intelligent Techniques ESIT99., AB-01, Crete, Greece.
5. Baldwin, J.F., Lawry, J., Martin, T.P. (2000) Mass Assignment Based Induction on Decision Trees on Words. Uncertainty in intelligent and information systems. Advances in Fuzzy Systems - Applications and Theory. Vol. 20. World Scientific.

6. Bhatia,S.K., Yao, Q. (1993) A New Approach to Knowledge Acquisition by Repertory Grids. In: Bhargava, B. et. al. (Eds.), Proceedings of the Second International Conference on Information and Knowledge Management CIKM 93, ACM Press, Washington, 738-740.
7. Boose, J.H., Bradshaw, J.M. (1987) Expertise transfer and complex problems: using AQUINAS as a knowledge acquisition workbench for knowledge-based systems. International Journal of Man-Machine Studies, 26:3-28.
8. Bottou, L., Vapnik, V. (1992) Local learning algorithms. *Neural Computation*, 4:888-900.
9. Bradshaw, J.M., Ford, K.M., Adams-Webber, J.R., Boose, J.H. (1993) Beyond the repertory grid: new approaches to constructivist knowledge acquisition tool development. International Journal of Intelligent Systems, 8(2):287-333.
10. Carmona, P., Castro, J.L., Zurita, J.M. (2001) FRIwE: Fuzzy rule identification with exceptions. Submitted to IEEE Transactions on Fuzzy Systems.
11. Castro, J.L., Castro-Schez, J.J., Zurita, J.M. (1999) Learning maximal structure rules in fuzzy logic for knowledge acquisition in expert systems. Fuzzy Sets and Systems, 101:345-353.
12. Castro, J.L., Castro-Schez, J.J., Zurita, J.M. (2001) Use of machine learning technique in the knowledge acquisition process. Fuzzy Sets and Systems, 123(3):307-320.
13. Castro, J.L., Castro-Schez, J.J., Zurita, J.M. (2001) Fuzzy repertory table, a method for acquiring knowledge about input variables to machine learning algorithms. Submitted to IEEE Transactions on Fuzzy Systems.
14. Castro-Schez, J.J., Jennings, N.R., Luo, X., Shadbolt, N.R. (2001) Acquiring Domain Knowledge for Negotiating Agents: A Case of Study. Submitted to International Journal of Human and Computer Studies.
15. Cover, T.M., Hart, P.E. (1967) Nearest neighbor pattern classification. *Institute of Electrical and Electronics Engineers Transactions on Information Theory*, 13:21-27.
16. Dasarathy, B.V. (1991). *Nearest Neighbor (NN) Norms: NN Pattern Clasification Techniques.* IEEE Computer Society Press.
17. Dubois, D., Prade, H. (1996) What are fuzzy rules and how to use them. Fuzzy Sets and Systems, 84:169-185.
18. Fix, E., Hodges, J.L.Jr. (1951) Discriminatory analysis, nonparametric discrimination, consistency properties. Technical report, Randolph Field, TX: United States Air Force, School of Aviation Medicine. Technical Report 4.
19. Gaines, B.R., Shaw, M.L.G. (1993) Basing knowledge acquisition tools in personal construct systems. Knowledge Engineering Review, 8(1):49-85.
20. Gaines, B.R., Shaw, M.L.G. (1997) Knowledge acquisition, modeling and inference through the World Wide Web. International Journal of Human-Computer Studies, 46(6):729-759.
21. Hwang, G. (1995) Knowledge Acquisition for Fuzzy Expert Systems. Int. J. Intelligent Systems, 10:541-560.
22. Ishibuchi, H, Nozaki, K., Yamamoto, N., Tanaka, H. (1994) Construction of fuzzy classification systems with rectangular fuzzy rules using genetic algorithms. Fuzzy Sets and Systems, 65:237-253.
23. Kahn, G., Nowlan, S., McDermot, J. (1985) Strategies for knowledge acquisition. IEEE Trans. Pattern Anal. Mach. Intell., 5:511-522.
24. Kelly, G. (1955) The Psychology of Personal Constructs. Norton, New York.

25. Kolodner, J. L. (1992) An introduction to case–based reasoning. *Artificial Intelligence Review*, 6:3-34.
26. Kolodner, J.L. (1993) *Case–Based Reasoning*. Morgan Kaufmann.
27. Laguía, M., Castro, J.L. (2001) Algorithms for classification based on neighborhood: some proposals and a heuristic. Submitted to Machine Learning.
28. Laguía, M., Castro, J.L. (2001) Similarity relations based on distances as fuzzy concepts. In *Conference of the European Society for Fuzzy Logic and Technology (EUSFLAT–2001)*.
29. Maron, O., Moore, A.W. (1997) The racing algorithm: Model selection for lazy learners. *Artificial Intelligence Review*, 11:193-225. Special Issue on "Lazy Learning".
30. Nosofsky, R.M., Palmeri, T.J. (1998) A rule-plus-exception model for classifying objects in continuous-dimension spaces. Psychonomic Bulletin & Review, 5(3):345-369.
31. Nosofsky, R.M., Palmeri, T.J., McKinley, S.C. (1994) Rule-plus-exception model of classification learning. Psychological Review, 101(1):53-79.
32. Plaza, E. López, R., Armengol, E. (1996) On the importance of similitude: An entropy-based assessment. In *Third European Workshop on Case-Based Reasoning (EWCBR–96)*. Springer–Verlag.
33. Quinlan, J.R. (1986) Induction of Decision Trees. Machine Learning, 1, 81-106.
34. Quinlan, J.R. (1993) C4.5: Programs for Machine Learning. Morgan Kaufmann, San Mateo.
35. Ricci, F., Avesani, P. (1995) Learning a local similarity metric for case–based reasoning. In *First International Conference on Case-Based Reasoning (ICCBR–95)*, pages 301-312. Springer–Verlag. Sesimbra, Portugal.
36. Ritcher, M.M. (1992) Classification and learning of similarity measures. In *16. Jahrestagung der Gesellschaft für Klassifikation (GFKL–92)*. Springer–Verlag.
37. Ritcher, M.M. (1995) On the notion of similarity in case-based reasoning. In G. del Viertl, editor, *Mathematical and Statistical Methods in Artificial Intelligence*, pages 171-184. Springer–Verlag.
38. Salzberg, S. (1991) A nearest hyperrectangle learning method. *Machine Learning*, 6:251-276.
39. Schreiber, G., Akkermans, H., Anjewierden, A., Hoog, R., Shadbolt, N., Van de Valde, W., Wielinga, B. (2000) The CommonKADS Methodology. Massachusetts Institute of Technology Press, Cambridge, Massachusets.
40. Stanfill, C., Waltz, D. (1986) Towards memory–based reasoning. *Communications of the ACM*, 29:1213-1228.
41. Weiss, S.M., Kulikowski, C.A. (1991) *Computer Systems that Learn*. Morgan Kaufmann.
42. Wettschereck, D. (1994) *A Study of Distance-Based Machine Learning Algorithms*. PhD thesis, Oregon State University.
43. Wettschereck, D., Dietterich. T.G. (1994) Locally adaptive nearest neighbor algorithms. In Jack D. Cowan, Gerald Tesauro, and Joshua Alspector, editors, *Advances in Neural Information Processing Systems*, volume 6, pages 184-191. Morgan Kaufmann Publishers, Inc.
44. Wettschereck, D., Dietterich, T.G. (1995) An experimental comparison of the nearest-neighbor and nearest-hyperrectangle algorithms. *Machine Learning*, 19:5-28.
45. Zadeh, L. (1983) The role of fuzzy logic in the management of uncertainty in expert systems. Fuzzy Sets and Systems, 11:197-227.

Integration of Fuzzy Knowledge

Tzung-Pei Hong[1], Ching-Hung Wang[2] and Shian-Shyong Tseng[3]

[1] Department of Electrical Engineering, National University of Kaohsiung, Kaohsiung, 811, Taiwan

[2] Chunghwa Telecommunication Laboratories, Chung-Li, 326, Taiwan

[3] Department of Computer and Information Science, National Chiao-Tung University, Hsin-Chu, 300, Taiwan

Summary:
In this chapter, we introduce the problems of knowledge integration and present our research results on developing genetic-algorithm-based (GA-based) integration approaches to shorten the time needed in constructing knowledge bases from multiple knowledge sources. Traditional knowledge-integration systems and GA-based classifier systems are briefly reviewed. Two genetic crisp knowledge integration approaches are then introduced. These approaches can also be applied to fuzzy knowledge integration for fixed membership functions. Two genetic fuzzy knowledge integration approaches are then presented, which generate a fuzzy knowledge base, associated respectively with several sets of local membership functions and with a set of global membership functions.

1 Introduction

Fuzzy knowledge-based systems have been successfully applied to many fields and have shown excellent performance. The acquisition of fuzzy knowledge remains one of the major costs in building systems even though many tools have been developed to help with the acquisition process. The cost of the effort is high and will become prohibitive as we attempt to build larger and larger systems. Reusing and integrating fuzzy knowledge available from a variety of sources thus plays a crucial role in reducing the development cost.

In this chapter, we introduce the problems of knowledge integration and present our research results on developing genetic-algorithm-based integration approaches. We first describe basic concepts of knowledge-based systems, knowledge acquisition, knowledge integration and vague knowledge in Section 2. We then classify the knowledge integration systems, define the problems in integration, and give a brief review of related systems in Section 3. Genetic algorithms are then briefly introduced in Section 4. Two genetic classifier systems, respectively by the Michigan approach and by the Pittsnurgh approach, are also described in that section. GA-based knowledge integration environments are then presented in Section 5. Knowledge encoding and two genetic crisp knowledge integration approaches are also introduced. One uses domain specific operators and the other uses the Michigan approach to refine the rules. These approaches can also be applied to fuzzy knowledge integration for fixed membership functions. Two genetic fuzzy knowledge integration approaches are then presented in Section 6, which generate a fuzzy knowledge base, associated respectively with several sets of local membership functions and with a set of global membership functions. An experimental comparision of these two fuzzy approaches are also made. Finally, conclusions are given in Section 7.

2 Preliminaries

2.1 Knowledge-based Systems and Interpretability

Decision-making is one of the most important activities in the real world. With specific domain knowledge as a background, the task of decision-making is to get an optimal or a nearly optimal solution from input information using an inference procedure. Generally, there are three ways to make a decision in a complex environment:

1. by building a mathematical model;
2. by seeking human expert's advice;
3. by building an expert system or controller.

Among them, building an accurate mathematical model to describe the complex environment is a good way. However, accurate mathematical models do not always exist nor can they be derived for all complex environments because the domain may not be thoroughly understood. The first method is then limited and when it fails, an alternative for making a good decision is to seek human expert's help. However the cost of querying an expert may be high, and there may be no human experts available when the decision must be made.

Recently, expert systems or knowledge-based systems have been widely used in domains for which the first two methods are not suitable [16]. The knowledge base in an expert system can grow incrementally and can be updated dynamically, so that the performance of an expert system will become better and better as it develops. Also, the expert system approach can integrate expertise from many fields, reduce the cost of query, lower the probability of danger occurring, and provide fast response.

The explanation capability is one of the main features in knowledge-based systems. Since the knowledge stored in a knowledge-based system is usually declarative, it can easily be used to interpret the causes to a consequence. The knowledge can easily be verified by designers as well.

2.2 Knowledge Acquisition and Knowledge Integration

The development of a classical knowledge-based system is illustrated in Figure 1 [16].

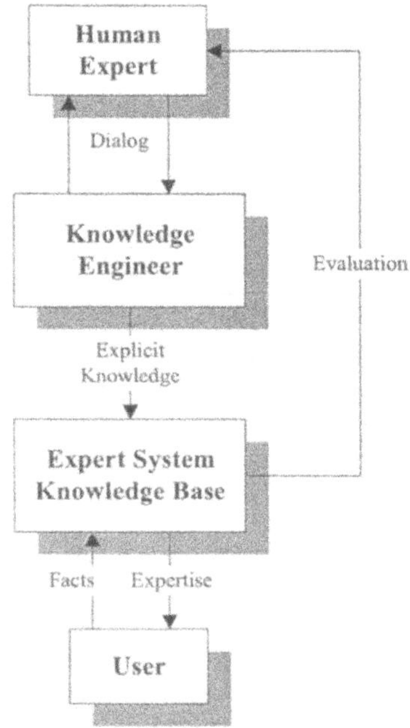

Figure 1: Development of a knowledge-based system

A knowledge engineer first establishes a dialog with a human expert in order to elicit the expert's knowledge. The knowledge engineer then encodes the knowledge for entry into the knowledge base. The expert then evaluates the knowledge-based system and gives a critique to the knowledge engineer. This process continues until the system's performance is judged to be satisfactory by the expert. The user can then supply facts or other information to the knowled-based system and receive advice in response.

Although a wide variety of knowledge-based systems have been built, a development bottleneck occurs in knowledge acquisition. Building a large-scale knowledge-based system involves creating and extending a large knowledge base over the course of many months or years. For instance, the knowledge base of the XCON (R1) expert system has grown over 10 years from 300 component descriptions and 750 rules to 31,000 component descriptions and 10,000 rules [16]. Shortening the development time is then the most important factor for the success of a knowledge-based system.

If a knowledge base is constructed from scratch, the cost of the effort is high and will become prohibitive as we attempt to build larger and larger systems. Reusing and integrating knowledge available from a variety of sources such as domain experts, historical documentary evidence, current records, or existing knowledge bases, thus plays a crucial role in building effective knowledge-based systems [2, 18, 22, 31]. Recently, a great deal of research [2, 29, 31] has been devoted to the study of the reuse and integration of various existing knowledge sources to reduce this bottleneck. Especially for complex application problems, related domain knowledge is usually distributed among multiple sites, and no single site may have complete domain knowledge. The use of knowledge integrated from multiple knowledge sources is thus especially important to ensure comprehensive coverage. Some benefits of integrating multiple knowledge sources in developing a knowledge-based system are described below.

1. Existing knowledge can be reused.
2. Knowledge acquired from different sources has good validity (compared to that from only one source).
3. The resulting knowledge base can have good comprehensive coverage.
4. Integrated knowledge can deal with more complex problems.
5. Knowledge integration may improve the performance of the resulting knowledge base.
6. Integration facilitates fast, inexpensive building of knowledge-based systems.

Finding appropriate ways of reusing and integrating various knowledge sources is thus very important to the design of knowledge-based systems.

2.3 The Management of Vague Knowledge in the Integration Process

Many knowledge integration methods have been developed to integrate knowledge from multiple sources [1, 10, 23, 37, 45]. Most knowledge sources or actual instances in real-world applications usually contain linguistic or ambiguous information. Especially in domains such as medical or control domains, expressions of domain knowledge in linguistic terms are commonly observed. In a knowledge integration process, having the capacity to manage uncertainty and vague information is quite important. Generally, vagueness and ambiguity result from attributes insufficient to appropriately describe knowledge, or when experts, teachers, or users can not clearly describe knowledge. Thus, each knowledge input is expressed as a linguistic description. Each attribute that is used to describe a linguistic concept may be defined as a fuzzy set. As an example, the concept *dangerous dogs* may be expressed as "A dog has a large body and long hairs, and it is dangerous with 0.8 degree of certainty". In the sentence, "large", "long" and "dangerous" are all linguistic terms. Since attributes and classifications used to express domain knowledge represent human perceptions and desires, they are vague by nature.

Several theories such as fuzzy set theory, probability theory, Dempster-Shafer theory [39], and approaches based on certainty factors, have been developed to manage uncertainty and vague information. Fuzzy set theory is more and more frequently used in knowledge-based systems because of its simplicity and similarity to human reasoning. The theory has been applied to many fields such as manufacturing, engineering, diagnosis, economics, and others.

Most fuzzy knowledge-based systems can be seen as special rule-based systems that use fuzzy logic. They contain fuzzy rules in their knowledge bases and derive conclusions from user inputs by fuzzy reasoning processes. They must usually predefine membership functions and fuzzy inference rules to map numeric data into linguistic variable terms and to make fuzzy reasoning work.

3 Knowledge Integration

3.1 Classification of Knowledge-Integration Systems

The construction of a reliable knowledge-based system usually requires the integration of several knowledge sources. A knowledge base integrated from several knowledge sources is usually more solid and more objective than that derived from only a single source. Therefore, the problem of integrating knowledge from different sources has become an interesting and challenging issue. Boose proposed some types of knowledge integration as follows [4]:

1. Integrating knowledge from multiple sources in the same domain,

2. Integrating knowledge from multiple sources in different domains,
3. Integrating knowledge with different knowledge representations (such as rules, decision trees, and semantic nets).

In this chapter, we are concerned with the first type of problems, which is the most common and important to building a knowledge-based system. The other two types are much more complex than the first one.

3.2 Redundancy, Subsumption and Conflict in Knowledge Integration

When knowledge is represented in the form of rules and is integrated from several sources, the problems involving redundancy, subsumption and conflict may arise in the integration process. They are defined as follows.

1. *Redundancy*: two rules have the same situations and have the same conclusions. For example, the following two rules are redundant:

   Rule 1: **IF** A_1 and A_2 **Then** B,
   Rule 2: **IF** A_2 and A_1 **Then** B.

2. *Subsumption*: two rules have the same conclusions, but one contains additional constraints to the other. For example, consider the two rules that follow:

   Rule 1: **IF** A_1 and A_2 **Then** B,
   Rule 2: **IF** A_1 **Then** B.

   Since Rule 1 has an additional constraint A_2, we would say that Rule 1 is *subsumed* by Rule 2. Whenever Rule 1 succeeds, Rule 2 also succeeds.

3. *Conflict*: two rules have the same situations but with conflicting conclusions. That is, the IF parts of two rules are equivalent, but the conclusions are different. For example, the following two rules are conflict:

   Rule 1: **IF** A_1 and A_2 **Then** B,
   Rule 2: **IF** A_2 and A_1 **Then** C.

Redundancy, subsumption and conflict should thus be removed in the knowledge integration process.

3.3 Review of Knowledge-Integration Systems

Recently, several researches have been devoted to the study of reusing and integrating multiple knowledge sources in distributed-knowledge environments [1, 2, 10, 23, 29, 31, 37, 43, 45]. There are two common approaches for integrating multiple knowledge sources. One approach is to reuse large amounts of related knowledge at multiple sources, and then combine them into a large-scale centralized knowledge base [2, 29, 31, 43]. Since knowledge reuse and knowledge integration can increase reliability of the resulting knowledge-based system and can reduce the difficulties in developing a knowledge-based system, many researchers have thought of them as the solutions to the problems of knowledge-acquisition bottleneck [19, 35]. The other approach is to coordinate various knowledge sources from different sites to solve complex tasks in distributed environments [1, 10, 23, 37, 45]. The knowledge in different sites is not integrated. Its goal is to build a framework for knowledge sharing and knowledge reusing based on a multi-agent architecture. "Knowledgeable community" [32] and "knowledge sharing" [30] are the key concepts for such an approach. Examples include KSE [30] and PACT [9].

In this chapter, we focus on the research of reusing and integrating multiple knowledge sources into a centralized knowledge-based system. For this kind of integration, many knowledge acquisition and integration systems [13, 14] based on the Personal Constructs Psychology (PCP) model [25] have been developed. Examples include ETS [5], AQUINAS [3], PLANET [34], KSSO [15], KITTEN [40], and MERGE [22]. The PCP model derives the acquired knowledge based on repertory grids. It is primarily used to find and discuss differences among the knowledge repertory grids acquired from different experts, and to reduce potential opinion conflicts during knowledge integration. Other knowledge-integration systems based on Decision Tables or Integrity Constraints [2, 31] were also proposed. They first combine multiple decision tables acquired from different experts, then remove conflicts by checking the integrity constraints in each table. Some systems related to this research are briefly reviewed as follows.

3.3.1 Maximal Combination of Multiple Knowledge Bases

Baral et al. [2] proposed a method to combine the knowledge present in multiple knowledge-based systems into a single knowledge base. Each knowledge base consists of a set of normal clauses and a set of integrity constraints. The set of integrity constraints is assumed the same for all knowledge bases, but the sets of normal clauses may differ. While combining multiple knowledge bases, this approach has to ensure that the combination is consistent with the integrity constraints.

An application for this method arises in a large company with branches overseas. Each branch manages its own knowledge base. A consistent combination of the knowledge bases is required while a policy maker tries to make a decision about the overall company.

3.3.2 Integration of Multiple Rule-Sets Based on Decision Tables

Ngwenyama and Bryson proposed a formal method [31] to analyze and integrate multiple rule sets. Each rule set is first transformed into a decision table for further analysis. After transformation, a decision table is divided into two parts: one is for condition sets and the other is for action sets. The integration method maintains two decision tables *DE* and *V*, and two auxiliary matrices, *X* and *Y*, for effectively combining multiple rule sets. *DE* contains the different condition sets and the different action sets induced from the given rule sets. *V* is an intermediate decision table for checking whether conflict rules exist or not. *X* is a relationship matrix for describing the relationship of condition sets, and *Y* is a relationship matrix for describing the relationship of conflicting sets. If conflict rules exist, experts must intervene to resolve them until no conflict rules exist.

3.3.3 Integration of Multiple Weighted Knowledge Bases

Knowledge from different sources is often contradictory. For example, in a distributed medical knowledge-based system, different experts may have different opinions on the diagnosis of patient's diseases. Each knowledge source is then considered with a weight to represent the relative degrees of importance to the sources. Lin thus proposed a formal semantic method [29] to integrate multiple knowledge bases, each with its weight. The method adopts the weighted majority principle in case of conflicts. For example, suppose three knowledge bases $T_1=\{a\}$, $T_2=\{a\}$, and $T_3=\{\sim a\}$ with the same weight. The method will return {a} as the results after integrating them.

3.4 Problems of Knowledge-Integration Systems

The following problems may arise when different knowledge sources are intergrated:

1. Domain experts must intervene in the integration process to deal with conflicts and contradictions.
2. Integration is time consuming.
3. The more knowledge sources exist, the more difficult and complex integration becomes.
4. These methods are usually inapplicable to application domains with vagueness and ambiguity.

In this chapter, we will describe knowledge-integration approaches based on genetic algorithms to overcome the above problems.

4 GA-based Classifier Systems

4.1 Review of Genetic Algorithms

The genetic algorithm (GA) is an adaptive search technique based on the mechanics of natural selection and natural genetics. It was proposed by Holland [20] and inspired by Darwin's theory. The genetic algorithm maintains a population of individuals and defines a set of genetic operators to do the evolution process. Each individual is a feasible solution to the problem and is represented by a bit string called a chromosome. In order to evaluate each chromosome's fitness value, an evaluation function is defined. The genetic algorithm searches for an optimal or a near-optimal solution through four stages: population generation, evaluation, selection and manipulation [11, 38, 42]. At the first stage, an initial population of individuals is randomly generated. Each individual in the population is then encoded as a bit string and evaluated by an evaluation function (stage 2). The selection stage obeying the survival of the fittest principle chooses good individuals according to their fitness values and mates them in the manipulation stage.

The manipulation stage uses genetic operators to produce a new population of individuals (offsprings) by manipulating the "genetic information", called genes, possessed by "parents" (members of the current population) [38, 42]. Two basic operators, crossover and mutation, are done probabilistically. The crossover operator takes two parent chromosomes and swaps parts of their genetic information to produce offspring chromosomes. Crossover does not produce new genes but does produce new combinations of genetic material. As Figure 2 shows, the parent strings *P1* and *P2* are swapped to produce two new offspring strings *O1* and *O2* according to a certain crossover point.

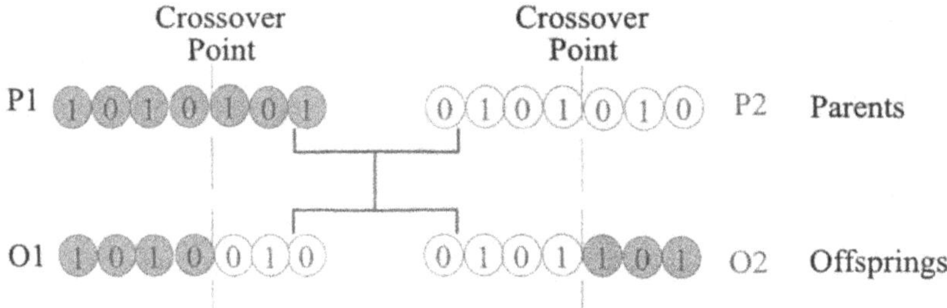

Figure 2: An example of crossover

On the other hand, the mutation operator produces new genes by randomly changing some genes in selected individuals to help the genetic algorithm escape from local-optimum "traps" [38, 42]. For example, a bit "0" in an individual's string could be altered to a bit "1" or vice versa. Figure 3 shows an example of the mutation, which changes the fifth element of a parent string to "1".

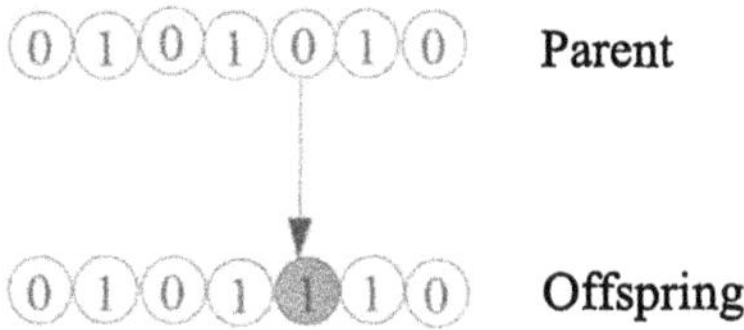

Figure 3: An example of mutation

Offspring chromosomes produced by genetic manipulation form the next population to be evaluated. The four stages form a genetic cycle. Each cycle then produces a new generation of possible solutions as the input to the next cycle. The cycle repeats until a satisfactory solution to the problem is found or termination criteria are met. A simple genetic algorithm [12, 38] is stated below.

The Simple Genetic Algorithm:

Step 1: Define a suitable representation of the problem to be solved.

Step 2: Create an initial population of N individuals for evolution.

Step 3: Define a suitable fitness function for the individuals.

Step 4: Perform genetic operations (crossover and mutation) to generate possible offspring chromosomes.

Step 5: Evaluate the fitness value of each offspring chromosome.

Step 6: Select superior N individuals to form the next generation according to their fitness values.

Step 7: If the termination criteria are not satisfied, go to Step 4; otherwise, stop the algorithm.

Below, we review two famous genetic-based classifier systems, the Michigan approach and the Pittsburgh approach.

4.2 The Michigan Genetic Classifier Systems

Cognitive System One (CS-1) was the first Michigan genetic classifier system. It was devised by Holland and Reitman in 1983 [21]. CS-1 maintains a population in which each individual is a rule, and is encoded as a fixed-length string. A fitness function is defined to evaluate the goodness (called the strength) of each rule in the population. The genetic algorithm then operates on the level of individual rules and selects good parent rules for mating according to their strength values.

The major problem with the Michigan approach is the simultaneous cooperation and competition of the individual rules within the population. During evolution, rules within the population compete with each other, and the ones with high strength values are selected for mating to generate new offspring rules. At the same time, rules within the population must cooperate to solve the given problem. Maintenance and evolution of such a set of co-adapted rules by considering both factors mentioned above is thus crucial to the success of the Michigan approach.

4.3 The Pittsburgh Genetic Classifier Systems

LS-1 was the first Pittsburgh genetic classifier system. It was proposed by Smith in 1980 [17, 41]. LS-1 maintains a population in which each individual is a rule set, and is encoded as a variable-length string. A fitness function is defined to evaluate the goodness of each rule set in the population. The genetic algorithm then operates on the level of individual rule sets and selects good parent rule sets for mating according to their fitness values.

The major problem with the Pittsburgh approach is the maintenance and evaluation of a population of rule sets. It often leads to a much greater computational burden in terms of both memory and processing time. Also, since credit assignment occurs on the level of rule sets by a predefined evaluation function, only the fitness values of rule sets are obtained. It can not help us promote the performance of individual rules.

The Pittsburgh approach is apparently quite different from the Michigan approach. In the next section, two genetic knowledge integration approaches to crisp knowledge integration are presented. They can also be applied to fuzzy knowledge integration for fixed membership functions.

5 GA-based Knowledge Integration Strategies

5.1 Knowledge Integration Environments

The genetic knowledge-integration approaches work in a distributed-knowledge environment shown in Figure 4 [46]. Four types of knowledge and data, including knowledge sets, knowledge dictionaries, data sets and data dictionaries, may be obtained from various sources. Knowledge from each source may be directly obtained by a group of human experts using a knowledge-acquisition tool, or derived from a machine-learning method. All knowledge sources are represented by rules since almost all knowledge derived by knowledge-acquisition (K.A.) tools or induced by machine-learning (M.L.) methods may easily be translated into or represented by rules. Each knowledge set (rule set) is associated with one knowledge dictionary. The knowledge dictionary defines the vocabulary set used in the knowledge set. Each data set is associated with one data dictionary. The data dictionary defines the vocabulary set used in the data set.

Each rule set is first translated into an intermediary representation, and further each intermediary representation is encoded as a string. After encoding, the combined strings form an initial knowledge population, which is then ready for integrating. Genetic knowledge integration approaches are then used to generate an optimal or a nearly optimal rule set from the initial knowledge population.

In this section, two genetic knowledge integration approaches based on the Pittsburgh approach are described. The first one uses domain-specific operators while generating offspring chromosomes. The second one combines the Michigan approach to adjust each individual rule in a rule set. Both of them must encode knowledge before the integration process can start.

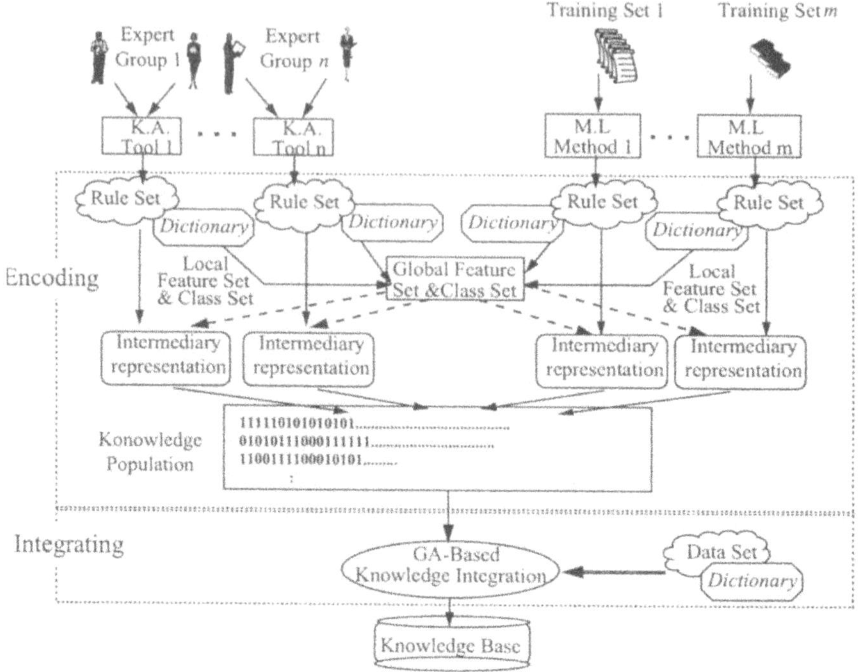

Figure 4: GA-based knowledge-integration framework

5.2 Knowledge Encoding

Since rule sets derived from different knowledge sources generally differ in syntax and size, they must be translated into a uniform syntactical representation before being encoded. The steps for translation of rule sets are described below [46].

1. Collect the features and possible values occurring in the condition parts of rule sets. All features gathered together comprise the global feature set.
2. Collect classes occurring in the conclusion parts of rule sets. All classes gathered together comprise the global class set.
3. Translate each rule into an intermediary representation that retains its essential syntax and semantics. If some features in the feature set are not used by the rule, *dummy* tests are inserted into the condition part of the rule. Each intermediary representation is then composed of *N* *feature tests* and one *class pattern*, where *N* is the number of global features collected. If the *feature tests* or *class patterns* are numerical, they are first discretized into a number of possible regions.

4. Concatenate all intermediary rules to form an intermediary rule set.

After translation, each intermediary rule set is then ready to be encoded as a bit string. Using the intermediary form, we first encode each feature test into a fixed-length binary string, with length equal to the number of possible feature test values. Thus, each bit represents a possible value. Similarly, the class pattern is encoded into a fixed-length binary string with each bit representing a possible class. Each rule in the intermediary representation is then encoded as a fixed-length bit string by:

$$Length\ (rule) = (\sum_{i=1}^{N}\ number\ of\ possible\ value\ of\ feature(i))$$

- number of classes,

where N is the number of features. Since different rule sets may have different numbers of rules, the lengths of the intermediary rule sets may be different. Thus:

$$Length\,(\,rule\ set\,) = \sum_{i=1}^{z} Length\,(rule_i)\,,$$

where z is the number of rules in an intermediary rule set.

5.3 Genetic Knowledge Integration with Domain Specific Operators

The genetic knowledge integration approach with domain-specific operators is first presented. It consists of two parts: *encoding* and *integration*, as shown in Figure 5.

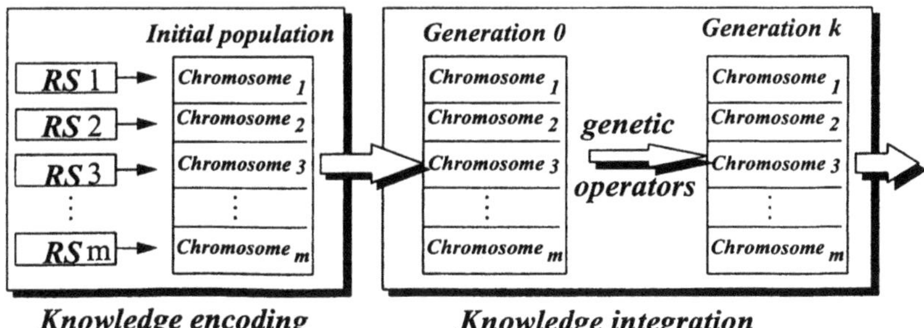

Figure 5: The genetic knowledge integration approach with domain-specific operators

The encoding process transforms each rule set into a bit-string structure. Each rule set represents one individual in the initial population. If the initial number of knowledge sources is small, some *dummy* initial rule sets are randomly generated and inserted into the population to increase the population size. The integration process chooses bit-string rule sets for "mating", gradually creating good offspring rule sets. The offspring rule sets then undergo recursive "evolution" until an optimal or a nearly optimal rule set is found.

An evaluation function and a set of data including documentary evidence, instances or historical records, are then used to qualify the derived rule set. Two important factors, the accuracy and the complexity of the resulting knowledge structure, are used in evaluating derived rule sets.

Four opeators are used in this approach, two fundamental genetic operators (*crossover*, *mutation*) and two domain-specific operators (*fusion*, *fission*). The crossover operator used here selects crossover points differently from the way used in the simple genetic algorithm. The crossover operator in the simple genetic algorithm chooses the same points for both parent chromosomes, but, the crossover operator here need not use the same point positions for both parent chromosomes. Crossover points may be chosen within rule strings or at rule boundaries. The only requirement for crossover points is that they must "match up semantically". That is, if a crossover point is chosen at some rule boundary, then the crossover point of the other parent must also be chosen at the rule boundary. If a crossover point is chosen within a rule string and p bits to left of a rule boundary, the crossover point for the other parent must also be chosen within the rule string and p bits to left of some rule boundary.

The mutation operator is the same as the standard one in the simple genetic algorithm. It randomly changes some elements in a selected rule set and leads to additional genetic diversity to help the integration process escape from local-optimum "traps".

Problems such as *redundancy*, *subsumption*, *contradiction*, and *misclassification* [16] do arise when integrating rule sets. Two domain-specific operators, *fusion* and *fission*, are used to solve these problems. *Fusion* emulates natural evolution's genetic folding phenomenon on chromosomes to eliminate redundancy and subsumption. It uses the logical *"OR"* operation to find subsumptive relationships among rules and then eliminates them by folding the subsumptive rules together. *Fusion* operator can thus reduce rule set complexity by eliminating redundancy and subsumption.

The *fission* operator is used to emulate natural evolution's *schizogenetic* phenomenon on chromosomes to eliminate misclassifications and contradictions. A misclassification occurs when rules classify test objects incorrectly. The *fission* operator selects the "closest" *near-miss* rule [50, 51] to specialize to exclude the wrongly classified object. It also splits a contradictive rule into more specific ones, each concluding to only one class. These new rules will still contradict each other, but can further be removed by the crossover and mutation operators. *Fission* operator thus acts as a specialization operation that resolves misclassification and contradiction problems, and promotes rule accuracy.

In some cases, redundant and contradictory rules can be intended as a local reinforcement of space zones presenting higher complexity [8, 33]. By spliting contradictive rules into more specific ones, the local search activities will increase. The fusion operation for redundant rules will, however, decrease the local search activities and increase the possibility of evolution of other rules.

By designing new genetic operators, the genetic process can take domain-specific characteristics into consideration, thus making the results closer to those desired. This however takes more execution time than using only the original operators.

5.4 Genetic Knowledge Integration Combined with the Michigan Approach

As mentioned above, the role of individuals in a Pittsburgh-approach population is quite different from that in a Michigan-approach population. Both strategies have their advantages and disadvantages. In this section, we introduce a hybrid genetic knowledge integration approach to integrate knowledge from multiple sources into a centralized knowledge base [47]. The integration process is shown in Figure 6.

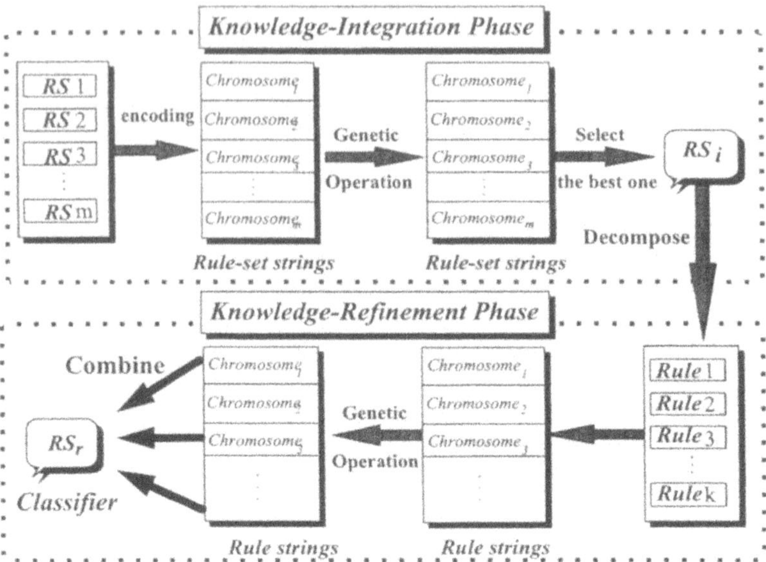

Figure 6: A hybrid knowledge integration process

As Figure 6 shows, RS_1, RS_2, ..., RS_m are the rule sets obtained from different sources, RS_i is the derived rule set after knowledge-integration, $Rule_1$, $Rule_2$, ..., $Rule_k$ are the individual rules decomposed from RS_i, and RS_r is a resulting rule set refined from RS_i. The genetic processes operate at two phases: knowledge integration and knowledge refinement. In the knowledge-integration phase, the genetic algorithm is applied on the rule-set level as in the Pittsburgh approach. However, only crossover and mutation operators are used. Problems such as *redundancy*, *subsumption* and *contradiction* [6] are checked in the refinement phase.

In the *knowledge-refinement phase*, the genetic algorithm is applied on the rule level, as in the Michigan approach. A rule set is decomposed into individual rules, and thought of as a population. Good rules are then selected for genetic operations to produce better offspring rules. After generations, all the rules in a population are then re-combined to form a new rule set. An evaluation function and a set of test objects including documentary evidence, instances or historical records, are then used to qualify the rules. Three important factors including accuracy, necessity and coverage of resulting rules are considered in the evaluation.

6 GA-based Fuzzy Knowledge Integration Strategies

In some application domains, the boundaries of a piece of information may not be clearly defined. Expressing domain knowledge by fuzzy descriptions is thus frequently seen. Appropriately extending the genetic knowledge integration approaches to fuzzy environments is necessary.

Recently, several researchers have investigated automatic generation of fuzzy classification rules and fuzzy membership functions using evolutionary algorithms [6, 26, 28, 44]. These methods can be categorized into the following four types:

1. learning fuzzy membership functions with fixed fuzzy rules [24];
2. learning fuzzy rules with fixed fuzzy membership functions [44];
3. learning fuzzy rules and membership functions in stages [26] (*i.e.*, first evolving good fuzzy rule sets using fixed membership functions, then tuning membership functions using the derived fuzzy rule sets);
4. learning fuzzy rules and membership functions simultaneously [6, 28].

In this section, two genetic fuzzy knowledge integration approaches based on the Pittsburgh approach are described, each of which can integrate multiple fuzzy rule sets and their associated membership functions simultaneously. The first one uses a global set of membership functions for the rules in the resulting fuzzy knowledge base [48]. The second one permits rules to own their local membership functions, as opposed to a global set of membership functions for all rules [49].

6.1 Fuzzy Knowledge Integration Environments

The two approaches work in a fuzzy-knowledge integration environment as shown in Figure 7 [48]. We assume that all fuzzy knowledge sources are represented by fuzzy rules. Each fuzzy knowledge set is associated with one knowledge dictionary and one membership function set. The fuzzy knowledge dictionary defines the features and the classes occurring in the condition and conclusion parts of rule sets, and defines the fuzzy sets of linguistic terms used in the fuzzy rules. Each fuzzy data set is associated with one data dictionary and one membership function set. The fuzzy data dictionary defines the vocabulary and the fuzzy sets of linguistic terms used in the data.
Each fuzzy rule set with its associated membership functions is first translated into an intermediary representation, and is further encoded into a string. After encoding, all the strings form an initial fuzzy knowledge population, which is then ready for integrating. Genetic fuzzy knowledge integration approaches are then used to generate a final set of fuzzy rules and membership functions from the initial fuzzy knowledge population.

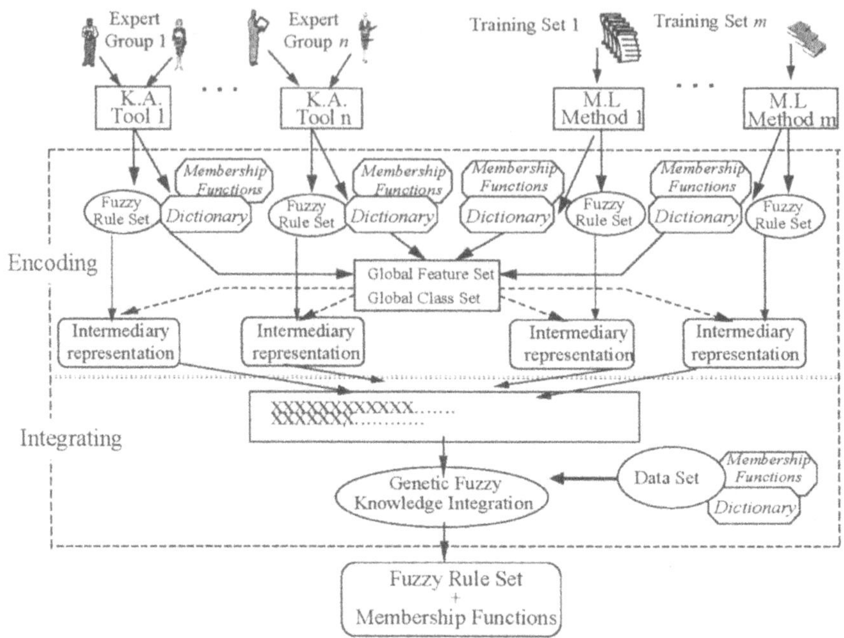

Figure 7: A fuzzy knowledge integration environment

6.2 Genetic Fuzzy Knowledge Integration Approach with Global Membership Functions

In this section, we describe the genetic fuzzy knowledge integration approach with global membership functions (GFKIGM) to manage fuzzy knowledge [49]. GFKIGM consists of two parts: *fuzzy knowledge encoding* and *fuzzy knowledge integ*

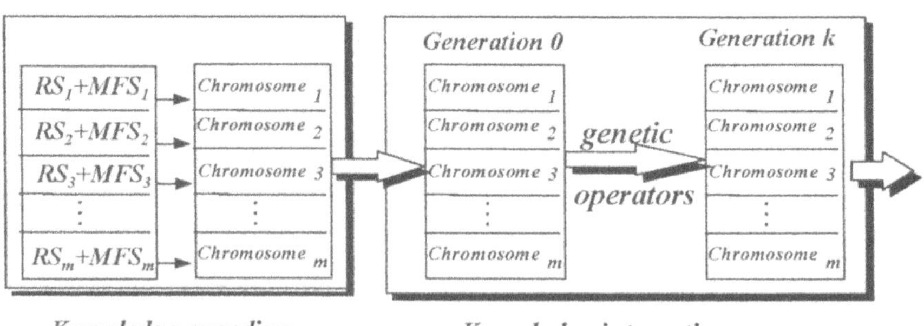

Figure 8: The integration process in GFKIGM

In Figure 8, RS_1+MFS_1, RS_2+MFS_2, ..., RS_m+MFS_m are the fuzzy rule sets with their associated membership function sets, as obtained from different sources for integration. Each rule set has its global membership functions.

We represent each membership function with two parameters as Parodi and Bonelli [36] did. Membership functions applied to a fuzzy rule set are thus assumed to be isosceles-triangle functions, as shown in Figure 9, where c_{ij} is the abscissa with the highest membership value of the j-th linguistic value (a_{ij}) of feature A_i, and w_{ij} represents half the spread of the membership function. A linguistic value a_{ij} of feature A_i is then represented as a pair (c_{ij}, w_{ij}).

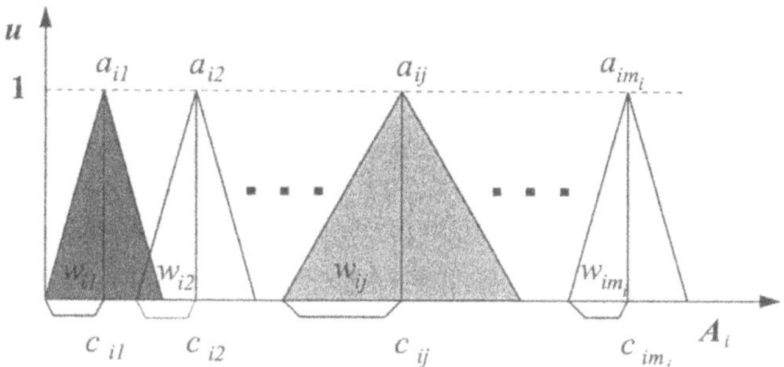

Figure 9: Membership functions of feature A_i

In GFKIGM, each fuzzy rule set and its associated membership function set are encoded as shown in Figure 10. A chromosome for the q-th rule set is divided into two parts: RS_q and MFS_q. RS_q denotes the set of fuzzy rules and MFS_q denotes the set of membership functions associated with RS_q.

In Figure 10, r_{qi} represents the i-th rule in RS_q, T_j represents the j-th test in a rule and is encoded into a fixed-length binary substring $s_{j1}s_{j2}\ldots s_{jm_j}$, where m_j is the

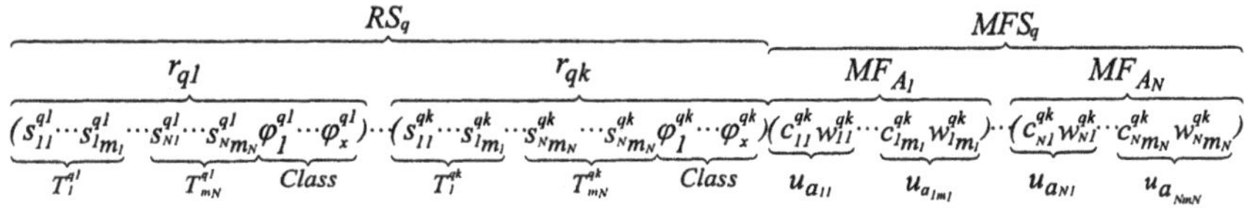

$$\overbrace{\overbrace{(\underbrace{s_{11}^{q1}\cdots s_{1m_1}^{q1}}_{T_1^{q1}}\cdots\underbrace{s_{N1}^{q1}\cdots s_{Nm_N}^{q1}}_{T_{mN}^{q1}}\underbrace{\varphi_1^{q1}\cdots\varphi_x^{q1}}_{Class})}^{r_{q1}}\cdots\overbrace{(\underbrace{s_{11}^{qk}\cdots s_{1m_1}^{qk}}_{T_1^{qk}}\cdots\underbrace{s_{Nm_N}^{qk}\cdots s_{Nm_N}^{qk}}_{T_{mN}^{qk}}\underbrace{\varphi_1^{qk}\cdots\varphi_x^{qk}}_{Class})}^{r_{qk}}}^{RS_q}\overbrace{\overbrace{(\underbrace{c_{11}^{qk}w_{11}^{qk}}_{u_{a_{11}}}\cdots\underbrace{c_{1m_1}^{qk}w_{1m_1}^{qk}}_{u_{a_{1m_1}}})}^{MF_{A_1}}\cdots\overbrace{(\underbrace{c_{N1}^{qk}w_{N1}^{qk}}_{u_{a_{N1}}}\cdots\underbrace{c_{Nm_N}^{qk}w_{Nm_N}^{qk}}_{u_{a_{NmN}}})}^{MF_{A_N}}}^{MFS_q}$$

Figure 10: Representations of a fuzzy rule-set RS_q ***with its associated*** membership functions

number of possible linguistic values of feature A_j. Also, the class pattern in each rule is encoded as a bit substring $\varphi_1\varphi_2\ldots\varphi_x$, where x is the number of possible classes. When the rule concludes to class j, then φ_j is set as 1 and the others are set as 0. N feature tests and one class pattern are then encoded and concatenated as a fixed-length rule substring. In the MFS part, a_{j1}, a_{j2}, ..., a_{jm_j} denote the possible linguistic values of feature A_j, and $u_{a_{ik}}$ denotes the membership function of a_{ik}. Each membership function is represented as a pair (c, w). Thus, all pairs of (c, w)'s for a certain feature are concatenated to represent its membership functions. Since c_{ij} and w_{ij} are both numeric values, MFS_q is thus encoded as a fixed-length real number string rather than a bit string.

Two genetic operators, *two-substring crossover* and *two-point mutation*, and one domain-specific operator, *fusion*, are used in the GFKIGM approach. The *two-substring crossover* operator used here exchanges two substrings of the parents to generate the offspring chromosomes. Three crossover points, cp_{rs}, cp_d and cp_{mf}, are chosen to accomplish this operation. cp_{rs} is located in the rule-set part, and may be chosen within a rule string or at a rule boundary. The cp_{rs} crossover points need not be located at the same point positions for both parent chromosomes. The only requirement for cp_{rs} crossover points is that they must "match up semantically". cp_d is always chosen at the boundary between the rule-set and the membership function set. cp_{mf} is located in the membership function part and may be selected at centers or spreads of membership functions. The cp_{mf} crossover points for both parent chromosomes must however be the same distances to left of the ends of both parent membership-function sets. After offspring fuzzy rule-sets and their membership functions have been generated by the two-substring crossover operation, the order of a newly generated fuzzy membership function and its neighbors may be destroyed. These fuzzy membership functions thus need rearrangement according to their center values.

The two-point mutation operator creates a new fuzzy rule and a new fuzzy membership function from one chromosome. Two mutation points, mp_{rs} and *mpmf*, are selected at random within the rule-set part and the membership function part. The mutation operator applied to the rule-set part is the same as that used in the simple genetic algorithm. It randomly changes bit values in a selected rule set, leading additional genetic diversity to help the integration process escape from local-optimum "traps". The mutation operator applied to the membership function part creates a new fuzzy membership function by adding a random value ε to the center or to the spread of an existing fuzzy membership function. The domain-specific operator, *fusion*, is used to remove *redundancy* and *subsumption* in a way similar to that in Section 4.

The number of offsprings depends on the adopted crossover and mutation rates. A fitness function is defined based on the accuracy and the rule complexity. The superior N individuals among parents and offsprings are then selected according to their fitness values, where N is the population size.

6.3 Genetic Fuzzy Knowledge Integration Approach with Local Membership Functions

In this section, we describe the genetic fuzzy knowledge integration approach with local membership functions (GFKILM) to manage fuzzy knowledge [49]. GFKILM permits rules to own their local membership functions, as opposed to a global set of membership functions for all rules. Assume that the kth rule r_{qk} in the qth rule set is composed of N feature tests and one class pattern. Each feature test in a fuzzy rule is encoded as m_i pairs of (c, w)'s, where m_i is the number of possible linguistic values of feature A_i and a pair (c, w) represents the membership function of one possible linguistic value. For example, if a fuzzy test value "$A_i = a_{ij}$" exists in a rule, the test is then encoded as $(c_{i1}, -w_{i1})(c_{i2}, -w_{i2})\ldots(c_{ij}, w_{ij})\ldots(c_{im_i}, -w_{im_i})$. Only w_{ij} is positive, and the other w values have minus signs added to them. Similarly, if the fuzzy test is ("$A_i = a_{ij}$" or " $A_i = a_{ik}$"), then only w_{ij} and w_{ik} are positive and the other w values are negative. From the sign of w, the encoded string can then correctly represent the test condition of a fuzzy rule. The condition part of each fuzzy rule is then encoded as $(m_1+m_2+\ldots+m_N)$ pairs of (c, w)'s using the proposed encoding method.

Next, the conclusion part of each fuzzy rule is encoded as a bit substring $\varphi_1\varphi_2\ldots\varphi_x$, where x is the number of possible classes. When a rule concludes to class j, φ_j is set as 1 and the others are set as 0. The rule r_{qk} is then encoded as shown in Figure 11.

$$r_{qk}:\ \overbrace{(\underbrace{c_{11}^{qk} w_{11}^{qk}}_{a_{11}}\cdots\underbrace{c_{1m_1}^{qk} w_{1m_1}^{qk}}_{a_{1m_1}})}^{A_1}\cdots\overbrace{(\underbrace{c_{i1}^{qk} w_{i1}^{qk}}_{a_{i1}}\cdots\underbrace{c_{ij}^{qk} w_{ij}^{qk}}_{a_{ij}}\cdots\underbrace{c_{im_i}^{qk} w_{im_i}^{ik}}_{a_{im_i}})}^{A_i}\cdots\overbrace{(\underbrace{c_{N1}^{qk} w_{N1}^{qk}}_{a_{N1}}\cdots\underbrace{c_{Nm_N}^{qk} w_{Nm_N}^{qk}}_{a_{Nm_N}})}^{A_N}\overbrace{(\varphi_1^{qk},\cdots\varphi_x^{qk})}^{Class}$$

***Figure 11: String representations of* $\mathbf{r_{qk}}$**

Since c_{ij} and w_{ij} are both numeric values, each fuzzy rule with its membership functions is then encoded as a fixed-length real number string rather than a bit string. Each fuzzy rule set RS_q that contains k fuzzy rules is then encoded by concatenating strings of k fuzzy rules as shown in Figure 12.

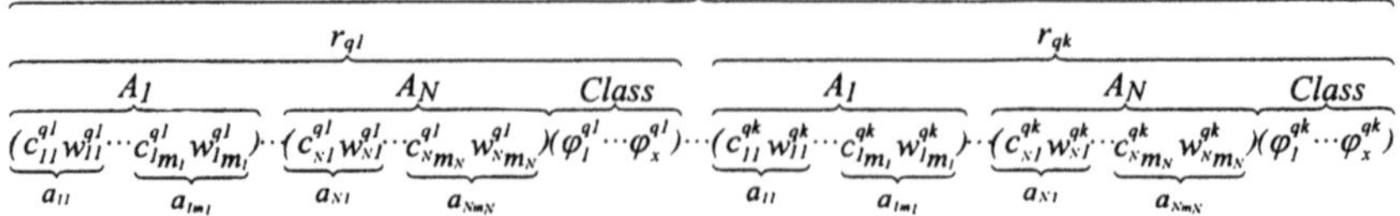

Figure 12: String representations of a fuzzy rule set $\mathbf{RS_q}$

This representation allows genetic operators to easily integrate multiple fuzzy rule-sets and their fuzzy membership function sets at the same time. Furthermore, since fuzzy membership functions are encoded together with each rule (as opposed to using a global collection of membership functions for all rules), rules are permitted to evolve to different degrees of vagueness.

GKFILM uses two fundamental genetic operators (*crossover*, *mutation*) and one domain-specific operators (*fusion*) as GFKIKM does. The crossover points may occur within rule strings or at rule boundaries. The only requirement for crossover points is that they must "match up semantically". The mutation operator is used to create a new fuzzy membership function by adding a random value ε to the center or the spread of an existing fuzzy membership function. The domain-specific operator, *fusion*, is used to remove *redundancy* and *subsumption* in a way similar to that in Section 4. The selection of superior individuals is the same as that in GKFIGM. The advantage of this representation is the expressive power for the derived rules to possess their own specificity in terms of the fuzzy sets they relate to. Especially for multi-dimensional domains that cannot generally use a global collection of membership functions for all rules, the use of local fuzzy sets to perform a linguistic interpretation of individual rules seems valid to overcome "curse of dimensionality". However, this advantage is at the cost of an increase of the search space.

6.4 Comparison of GFKIGM and GFKILM

In this section, we have described how to apply genetic approaches to fuzzy knowledge integration. Two genetic fuzzy knowledge integration approaches, GFKIGM and GFKILM, are described. Both of them can simultaneously integrate multiple fuzzy rule sets with their associated membership functions into one centralized fuzzy knowledge base. The main difference between these two approaches is that GFKILM generates a set of fuzzy rules, each of which is associated with a set of local membership functions, but the GFKIGM approach generates a set of fuzzy rules associated with a global collection of memberships functions for all fuzzy rules. The difference arises due to their different knowledge encoding methods to represent their knowledge sources with their associated membership functions. Two approaches have their advantages and disadvantages.

Table 1 shows a comparison of the two approaches as to accuracy and integration time, over 4000 generations, for diagnosis of the hepatitis. Two possible classes, *Die* or *Live,* from a set of 155 instances in term of 19 features, are to be distinguished. In each run, 70% of the hepatitis cases were selected at random for training, and the remaining 30% of the cases were used for testing. Ten initial rule sets were derived via machine-learning with an average accuracy of 76.88%. The operation frequency for *crossover* and *mutation* was set at 0.9 and 0.04 respectively [48].

Table 1: A comparison of GFKILM and GFKIGM for diagnosis of the hepatitis

GFKILM approach			GFKIGM approach		
Generation	CPU Time	Accuracy	Generation	CPU Time	Accuracy
0	00:00:01	76.88	0	00:00:01	76.88
13	00:00:04	78.44	4	00:00:02	78.67
69	00:00:19	81.32	160	00:00:45	84.50
124	00:00:35	83.28	473	00:02:14	85.42
181	00:00:52	84.32	1495	00:06:46	86.33
261	00:01:15	85.25	1791	00:08:03	86.56
414	00:01:55	86.88	2251	00:10:02	87.21
1401	00:06:07	88.76	2580	00:11:27	88.37
2110	00:09:20	89.49	2710	00:12:02	88.95
3110	00:13:42	89.65	3062	00:13:13	89.10
3550	00:15:35	89.77	3756	00:16:44	90.56
3817	00:16:45	91.83	3847	00:17:09	90.68
4000	00:17:36	92.90	4000	00:17:51	91.61

The GFKILM approach obtained an accuracy rate of 92.90% after 4000 generations. Also, the size of the resulting knowledge base and the CPU time used were respectively 4 and 1056 seconds. On the other hand, the GFKIGM approach produces an accuracy rate of 91.61% after 4000 generations. The size of the resulting knowledge base and the CPU time used were respectively 4 and 1071 seconds. Experimental results show that GFKIGM takes more time than GFKILM since the length of each encoded string in GFKIGM is longer than that of GFKILM. Also, experimental results show that GFKILM has a higher accuracy than GFKIGM. Although GFKILM performs better than GFKIGM in this case, the resulting fuzzy rule set is however associated with a great set of local membership functions that transgress the common sense of fuzzy rule sets. This also causes the inference difficult.

7 Conclusions

In this chapter, we have discussed related issues in fuzzy knowledge integration and described the knowledge integration frameworks and several approaches to automatically integrate multiple knowledge sources into one centralized knowledge base. Two common problems - conflict knowledge management and vague knowledge management, are also discussed and partially solved by the approaches presented. The described approaches differ from some previous remarkable knowledge integration approaches [3, 5, 15] mainly in that they need no human experts' intervention in the integration process. The time spent thus depends on computer execution speed, not on human experts. Much time can thus be saved since experts may be geographically dispersed and experts' discussion is usually very time-consuming [19, 27]. The approaches are also scalable in that they can be applied when the number of rule sets to be integrated increases. Integrating a large number of rule sets may increase the validity of the resulting knowledge base. They are also objective since human experts are not involved in the integration process. In the future, we shall discuss the integration of knowledge sources with heterogeneous knowledge representations.

References

[1] H. Aiba and T. Terano, "A computational Model for distributed knowledge systems with learning mechanisms," Expert Systems With Applications, Vol. 10, No. ¾, pp. 417-427, 1996.

[2] C. Baral, S. Kraus, and J. Minker, "Combining multiple knowledge bases," IEEE Transactions on Knowledge and Data Engineering, Vol. 3, No. 2, pp. 208-220, 1991.

[3] J. H. Boose and J. M. Bardshaw, "Expertise transfer and complex problems: using AQUINAS as a knowledge-acquisition workbench for knowledge-based systems," International Journal of Man-Machine Studies, Vol. 26, pp. 3-28, 1987.

[4] J. H. Boose, "Rapid acquisition and combination of knowledge from multiple experts in the same domain," Future Computing Systems, Vol. 1, pp. 191-216, 1986.

[5] J. H. Boose, "A knowledge acquisition program for expert systems based on personal construct psychology," International Journal of Man-Machine Studies, Vol. 23, pp. 495-525, 1985.

[6] B. Carse, T. C. Fogarty, and A. Munro, "Evolving fuzzy rule based controllers using genetic algorithms," Fuzzy Sets and Systems, Vol. 80, pp. 273-293, 1996.

[7] G. Cestnik, I. Konenenko, I. Bratko, "Assistant-86: a knowledge-elocotation tool for sophisticated users," in Bratko & Lavrac (Eds.) Progress in Machine Learning, Sigma Press, pp. 31-45, 1987.

[8] O. Cordon and F. Herrera, "A proposal for improving the accuracy of linguistic modeling," IEEE Transactions on Fuzzy Systems, Vol. 8, No. 4, pp. 335-344, 2000.
[9] M. R. Cutkosky, R. S. Engelmore, R. E. Fikes, M. R. Genesereth, T. R. Gruber, and W. S. Mark, "PACT: an experiment in integrating concurrent software engineering," IEEE Computer, Vol. 26, No. 1, pp. 28-37, 1993.
[10] S. M. Deen, "An architectural framework for CKBS applications," IEEE Transactions on Knowledge and Data Engineering, Vol. 8, No. 4, pp. 663-671, 1996.
[11] K. A. DeJong, "Learning with genetic algorithm: an overview," Machine Learning, Vol. 3, pp. 121-138, 1988.
[12] K. A. DeJong, W. M. Spears, and D. F. Gordon, "Using genetic algorithms for concept learning," Machine Learning, Vol. 13, pp. 161-188, 1993.
[13] B. R. Gaines, "Integration issues in knowledge supports systems," International Journal of Man-Machine Studies, Vol. 26, pp. 495-515, 1989.
[14] B. R. Gaines and M. L. Shaw, "New directions in the analysis and interactive elicitation of personal construct systems," Internal Journal of Man-Machine Studies, Vol. 13, pp. 81-116, 1980.
[15] B. R. Gaines and M. L. Shaw, "Eliciting knowledge and transferring it effectively to a knowledge-based system," IEEE Transaction on Knowledge and Data Engineering, Vol. 5, No. 1, pp. 4-14, 1993.
[16] J. Giarratano and G. Riley, Expert system principles and programming, PWS Publishing Company, Boston, MA 02116-4324, 1993.
[17] D. E. Golberg and R. Lingle, "Alleles, loci, and the traveling salesman problem," Proceeding of International Conference on Genetic Algorithms and their applications, Lawrence Erlbaum, Hillsdale, N. J., pp. 154-159, 1985.
[18] B. J. Gragun and H. J. Studel, "A decision-table based processor for checking completeness and consistency in rule-base expert-systems," Internal Journal of Man-Machine Studies, Vol. 26, pp. 633-648, 1987.
[19] D. M. Hamilton, "Knowledge acquisition for multiple site, related domain expert systems: Delphi process and application," Expert Systems With Applications, Vol. 11, No. 3, pp. 377-389, 1996.
[20] J. H. Holland, Adaptation in Natural and Artificial Systems, Ann Arbor, MI: University of Michigan Press, 1975.
[21] J. H. Holland and J. S. Reitman, "Cognitive systems based on adaptive algorithms," Machine Learning: An Artificial Intelligence Approach, Los Altos, CA, Morgan Kaufmann Publishers, 1983.
[22] G. J. Hwang, "Knowledge elicitation and integration from multiple experts," Journal of Information Science and Engineering, Vol. 10, No. 1, pp. 99-109, March 1994.
[23] D. Karagiannis and L. Marinos, "Integrating engineering applications via loosely coupled techniques: A knowledge-based approach," Expert Systems With Applications, Vol. 8, No. 2, pp. 303-319, 1995.

[24] C. Karr, "Design of an adaptive fuzzy logic controller using a genetic algorithm," Proceedings of the Fourth International Conference on Genetic Algorithms, pp. 450-457, 1991.

[25] G. A. Kelly, The psychology of personal constructs, Norton, New York, 1955.

[26] J. Kinzel, F. Klawonn, and R. Kruse, "Modifications of genetic algorithms for designing and optimising fuzzy controllers," Proceedings of the First IEEE International Conference on Evolutionary Computation, IEEE Piscataway, NJ, pp. 28-33, 1994.

[27] A. J. LaSalle and L. R. Medsker, "Computerized Conferencing for knowledge acquisition from multiple experts," Expert Systems With Applications, Vol. 3, pp. 517-522, 1991.

[28] M. Lee and H. Takagi, "Integrating design stages of fuzzy systems using genetic algorithms," Proceedings of the Second IEEE International Conference on Fuzzy Systems, IEEE, San Francisco, pp. 612-617, 1993.

[29] J. Lin, "Integration of weighted knowledge bases," Artificial Intelligence, Vol. 83, pp. 363-378, 1996.

[30] R. Neches, R. Fikes, T. Finin, T. Gruber, R. Patil, T. Senator and W. Swartout, "Enabling technology for knowledge sharing," AI Magazine, Vol. 12, No. 3, pp. 36-56, 1991.

[31] O. K. Ngwenyama and N. Bryson, "A formal method for analyzing and integrating the rule-sets of multiple experts," Information Systems, Vol. 17, No. 1, pp. 1-16, 1992.

[32] T. Nishida and H. Takeda, "Towards the knowledgeable community," Proceedings of International Conference on Building and Sharing of Very Large-Scale Knowledge Bases: KB&KS'93, Ohmsha Ltd., Tokyo, pp. 157-166, 1993.

[33] K. Nozaki, H. Ishibuchi and H. Tanaka, "A simple but powerful heuristic method for generating fuzzy rules from numerical data," Fuzzy Sets and Systems, Vol. 86, pp. 251-270, 1997.

[34] L. Mildred and G. Shaw, "PLANET: some experience in creating an integrated system for repertory grid applications on a microcomputer," International Journal of Man-Machine Studies, Vol. 17, pp. 345-360, 1982.

[35] R. Mizoguchi, "Expert systems research in Japan," IEEE Expert, pp. 14-23, 1995.

[36] A. Parodi and P. Bonelli, "A new approach of fuzzy classifier systems," Proceedings of Fifth International Conference on Genetic Algorithms, Morgan Kaufmann, Los Altos, CA, pp. 223-230, 1993.

[37] F. Polat, S. Shekhar, and H. A. Guvenir, "Distributed conflict resolution among cooperating expert systems," Expert Systems, Vol. 10, No. 4, pp. 227-236, 1993.

[38] J. L. Ribeiro Filho and P. C. Treleaven, "Genetic-algorithm programming environment," Computer, pp. 28-43, June, 1994.

[39] G. Shafer and R. Logan, "Implementing Dempster's rules for hierarchical evidence," Artificial Intelligence, Vol. 33, No. 3, pp. 271-298, 1987.
[40] M. L. Shaw and B. R. Gaines, "KITTEN: Knowledge initiation and transfer tools for experts and novices," International Journal of Man-Machine Studies, Vol. 27, pp. 251-280, 1987.
[41] S. F. Smith, A learning system based on genetic adaptive algorithms, Ph.D. Thesis, University of Pittsburgh, 1980.
[42] M. Srinivas, "Genetic-algorithms: A survey," Computer, pp. 17-26, June 1994.
[43] D. M. Steier, R. L. Lewis, and J. F. Lehman, "Combining Multiple Knowledge Sources in an integrated intelligent system," IEEE Expert, pp. 35-43, June 1993.
[44] P. Thrift, "Fuzzy logic synthesis with genetic algorithms," Proceedings of Fourth International Conference on Genetic Algorithms, Morgan Kaufmann, Los Altos, CA, pp. 509-513, 1991.
[45] L. Z. Varga, N. R. Jennings, and D. Cockburn, "Integrating intelligent Systems into a cooperation community for electricity distribution management," Expert Systems With Applications, Vol. 7, No. 4, pp. 563-579, 1994.
[46] C. H. Wang, T. P. Hong, S. S. Tseng and C. M. Liao, "Automatically integrating multiple rule sets in a distributed-knowledge environment," IEEE Transactions on Systems, Man, and Cybernetics, Vol. 28, No. 3, pp. 471-476, 1998.
[47] C. H. Wang, T. P. Hong, and S. S. Tseng, "A hybrid genetic knowledge integration strategy" The IEEE International Conference on Evolutionary Computation, pp. 587-591, 1998.
[48] C. H. Wang, T. P. Hong, and S. S. Tseng, "Integrating fuzzy knowledge by genetic algorithms," IEEE Transactions on Evolutionary Computation, Vol. 2, No. 4, pp. 138-149, 1998.
[49] C. H. Wang, T. P. Hong, and S. S. Tseng, "Integration membership functions and fuzzy rule-sets from multiple knowledge sources," Fuzzy Sets and Systems, Vol. 112, pp. 141-154, 2000.
[50] P. H. Winston, Artificial Intelligence, Third Edition, Addison-Wesley Publishing, 1992.
[51] P. H. Winston, "Learning by augmenting rules and accumulating censors," Machine Learning: An Artificial Intelligence Approach, Vol. 2, pp. 45-61, 1986.

Tuning fuzzy partitions or assigning weights to fuzzy rules: which is better?

Luciano Sánchez[1] and José Otero[2]

Departamento de Informática, Universidad de Oviedo
Sedes Departamentales, Edificio 1. Campus de Viesques
33204 Gijón, Asturias, Spain

Abstract. The accuracy of linguistic classifiers can be improved with several techniques, but they all compromise the interpretability of the rule base up to a certain degree. Assigning weights to fuzzy rules and tuning the memberships associated to linguistic variables are two of the most common methods. In this work we study whether tuning the membership functions in a linguistic classifier is better or not than adjusting rule weights, in terms of the interpretability of the rule base and the complexity of the output.

To make this analysis independent of the learning method, we will use statistical techniques of estimation over a random sets based linguistic classifier (introduced in previous works) that produces a rule base with weighted rules. The loss of precision produced when weights are restricted to binary values will be evaluated by comparing the output of the former method with an integer programming extension of it which is proposed in this chapter. The relative performances of classifiers with tuned membership and binary rules will be compared to those of uniform partition-based weighted rules, and the statistical relevance of these differences measured.

1 Introduction

It is commonly assumed that some precision must be sacrificed to achieve interpretability. Fuzzy classifiers in which memberships are tuned, or rules are weighted, are expected to be more precise than those in which there are not degrees of confidence in the consequents, and also more precise than those in which the semantic values of the linguistic terms (the fuzzy membership functions) are not modified by the learning algorithm. Nevertheless, a comparative study of the gain of performance obtained when memberships are tuned, or rules are weighted, is not immediate [12][9][10]. Neither adhoc fuzzy learning methods like [17][8][13][1], nor modern genetic methods [2] produce the optimum classifier; therefore it is difficult to separate that gain of performance which is inherent to the extra representative power in adjustable classifiers from the gain that is due to the specific properties of the learning algorithm. A method that is not based in heuristics neither stochastic techniques, and that is valid to induce both weighted and not weighted fuzzy classifiers, is needed.

In [15] a random sets based model, numerically identical to a weighted fuzzy rule based classifier, was proposed. All parameters in this last model have statistical sense, thus they can be estimated with conventional procedures: in particular, maximum likelihood can be used to estimate all of them. The numerical algorithm that originates here involves finding the solution of a constrained optimization problem, that can be transformed into an unconstrained one with the help of Lagrange multipliers, and then solved by nonlinear programming (projected gradient method.) We propose now to extend the statistical procedures used in the former work to obtain solutions in which weights are 1 or 0, by means of nonlinear integer programming techniques, so that the difference of performance between fuzzy classifiers with and without degrees of confidence in their consequents can be studied without dependence of the learning algorithm used to induce them from data. It will also be decided whether the loss of linguistic interpretability that is introduced when memberships are tuned is significant.

This chapter is organized as follows: first, we define the concept of random set based, linguistically understandable classifiers, and propose one method for estimating these classifiers' parameters from samples. Then we insert this algorithm into a branch and bound process that produces the best non weighted classifier from that data, and study the statistical significance of the difference between real and integer solutions, corresponding to weighted and not weighted classifiers, in synthetic and real world problems. Finally, we study the effect of tuning the antecedents of the real solution before calculating the integer solution, and decide whether the gain of quality worths the loss of interpretability.

2 Random set based classifiers

2.1 Linguistically understandable classifiers

Let us introduce the notation we will use in this chapter. Suppose we have a set Ω that contains objects ω, each one of them belonging to a class c_i, $i = 1 \ldots C$, and we perform the set of measurements $X(\omega) = (X_1(\omega), \ldots, X_M(\omega))$ over every object. Let us also assume that the mapping X fulfills all necessary conditions to be a random variable. We will say that a classification system is a decision rule that maps every element of $X(\Omega)$ to a class c_i, whose main objective is to produce a low number of errors [6].

Sometimes it is needed to obtain a linguistically understandable classifier. This means we need to setup a decision rule that can be codified in a language that allows it to be linguistically communicated. Fuzzy logic based classifiers have this property [19]. The semantic of a fuzzy classifier depends on the equivalence between linguistic values of attributes and certain fuzzy sets defined over the range of every feature [18], and on a fuzzy inference procedure. For example, the sentence

`if x is` $\widetilde{A}$ `then class = (`c_1 `with conf` p_1, ..., c_C `with conf` p_C`)`

means

$$\text{truth}(\widetilde{A} \to c_1) = p_1, \ \ldots, \ \text{truth}(\widetilde{A} \to c_C) = p_C.$$

where the concept "$\widetilde{A}$" is linked to a fuzzy subset of the feature space.

The same sentence can be given a probabilistic meaning [16]: if A is a *crisp* subset of $X(\Omega)$, then the sentence means

$$P(c_1|A) = p_1, \ldots P(c_C|A) = p_C,$$

with $\sum_{i=1}^{C} p_i = 1$. The probabilistic logic-based view has some advantages. It can be used to give a statistical meaning to the learning of a rule-based classifier from examples, as we will show below. But it is not immediate to compare it to fuzzy logic based rules, in which "$\widetilde{A}$" is a fuzzy set. On the contrary, we will show that fuzzy classifiers can be compared to certain random set-based classifiers, which in turn are defined as the expectation of the probabilistic logic-based ones for a given sample distribution.

2.2 Fuzzy and Random set based classifiers

Suppose we have a machine learning procedure to estimate a crisp partition $\{A_j\}_{j=1,\ldots,S}$ of the feature space, $A_j \cap A_k = \phi$ for $j \neq k$, plus the values $P(c_i|\mathbf{x} \in A_j) = p_{ij}$ that define a probabilistic logic-based classifier. The machine learning task takes as input a random sample $\mathbf{X}$ of classified examples and produces a partition $\{A_1^{\mathbf{X}}, A_2^{\mathbf{X}}, \ldots\}$ and the values $P(c_i|\mathbf{x} \in A_j^{\mathbf{X}}) = p_{ij}^{\mathbf{X}}$. In turn, for an input value $\mathbf{x}$, the classifier that was learned from the sample $\mathbf{X}$ outputs the probabilities of all classes according to the formula

$$P(c_i|\mathbf{x}, \mathbf{X}) = \sum_j p_{ij}^{\mathbf{X}} A_j^{\mathbf{X}}(\mathbf{x}) \tag{1}$$

where $A_j^{\mathbf{X}}(\mathbf{x})$ is 1 if $\mathbf{x} \in A_j^{\mathbf{X}}$ and 0 else.

It is well known that the expected error of this classifier (we will use the notation $\langle\rangle_{\mathbf{X}}$ to denote expectation with respect to the sample distribution of $\mathbf{X}$) is the sum of two positive terms, *bias* plus *variance* [5], where the bias is the error of the average classifier

$$P(c_i|\mathbf{x}) = \left\langle \sum_j p_{ij}^{\mathbf{X}} A_j^{\mathbf{X}}(\mathbf{x}) \right\rangle_{\mathbf{X}}. \tag{2}$$

Since the variance term is positive, the error of this average classifier is lower that the mean error of the individual classifiers. We can reorder the terms in (2):

$$P(c_i|\mathbf{x}) = \sum_j \langle p_{ij}^{\mathbf{X}} A_j^{\mathbf{X}}(\mathbf{x}) \rangle_{\mathbf{X}} \tag{3}$$

and, if the random variables $p_{ij}^{\mathbf{X}}$ and $A_j^{\mathbf{X}}(\mathbf{x})$ (both defined with respect to $\mathbf{X}$) were independent,

$$P(c_i|\mathbf{x}) = \sum_j \langle p_{ij}^{\mathbf{X}} \rangle_{\mathbf{X}} \langle A_j^{\mathbf{X}}(\mathbf{x}) \rangle_{\mathbf{X}} = \sum_j \langle p_{ij}^{\mathbf{X}} \rangle_{\mathbf{X}} \Phi_j(\mathbf{x}) \tag{4}$$

where $\Phi_j(\mathbf{x})$ is the one-point coverage function of the random set $A_j^{\mathbf{X}}$. Observe that $\sum_j A_j^{\mathbf{X}}(\mathbf{x}) = 1$ for all $\mathbf{x}$, so $\left\langle \sum_j A_j^{\mathbf{X}}(\mathbf{x}) \right\rangle_{\mathbf{X}} = 1$ and then $\sum_j \Phi_j = 1$ for all values of $\mathbf{x}$ (in words, the sum of the memberships is always equal to 1).

Expression (4) is very similar to the fuzzy inference formula applied to a fuzzy classifier defined by the rules "truth$(\widetilde{A}_j \to c_i) = t_{ij}$", $j = 1, \ldots, S$:

$$\text{truth}(c_i) = \bigvee_j \text{truth}(\widetilde{A}_j \to c_i) \wedge \widetilde{A}_j(\mathbf{x}) = \bigvee_j t_{ij} \wedge \widetilde{A}_j(\mathbf{x}). \tag{5}$$

The truth value t_{ij} is the counterpart of the value $\langle p_{ij}^{\mathbf{X}} \rangle_{\mathbf{X}}$, and the one-point coverage function $\Phi_j(\mathbf{x})$ is the counterpart of the membership function $\widetilde{A}_j(\mathbf{x})$; the t-norm $\wedge$ is restricted to the product, and the t-conorm $\vee$ is replaced by the sum.

The value $\langle p_{ij}^{\mathbf{X}} \rangle_{\mathbf{X}}$ is the expected probability of class i given the set $\hat{A}_j^{\mathbf{X}}$, this is the mean value of $P(c_i|A_j^{\mathbf{X}})$ for all probabilistic logic-based classifiers with respect to the sample distribution. It can be regarded as the degree of truth of the assertion "All elements in A_j belong to class i". The function $\Phi_j(\mathbf{x})$ is the probability of $\mathbf{x}$ being covered by the random set $A_j^{\mathbf{X}}$, and thus can be associated to the truth of the assertion "$\mathbf{x}$ belongs to A_j". It is possible to assign linguistic labels to the random sets $A_i^{\mathbf{X}}$ and draw their coverage functions in a form that closely resemble a Ruspini fuzzy partition (see Figure 1). In Figure 2 the three types of classifiers that have been discussed (probabilistic, random sets based and fuzzy sets based) are represented along with their inference procedures.

2.3 Approximate and descriptive rules

C rules of the form truth$(\widetilde{A}_j \to c_i)$=$p_{ij}$ can be combined into the assert

```
                if x is Ã_j then
class = (c_1 with conf p_1j, ..., c_C with conf p_Cj).      (6)
```

Unless the concept $\widetilde{A}_j$ can be expressed as the conjunction of independent properties defined over every feature, it is difficult to understand the meaning of this last assert; it is easier to use expressions like

```
      if x_1 is Ã_j1 and ... and x_M is Ã_jM then
class = (c_1 with conf p_1j, ... ,c_C with conf p_Cj)      (7)
```

where all fuzzy sets $\widetilde{A}_{jf}$, $j = 1, \ldots, S$, belong to a given fuzzy partition of the feature f. This is a typical rule structure in the field of fuzzy classifiers.

We will call *linguistic* or *descriptive* to classifiers based on expression 7, and *approximate* fuzzy classifiers to those based on expression 6, following the nomenclature in this book.

Not all approximate fuzzy classifiers can be expressed with linguistic classification rules. The conditions they must fulfill are immediate in fuzzy logic:

$$\widetilde{A}_j(\mathbf{x}) = \widetilde{A}_{j1}(x_1) \wedge \ldots \wedge \widetilde{A}_{jM}(x_M)$$

and exactly the same in probabilistic logic-based rules,

$$\begin{array}{c}\texttt{if } x_1 \texttt{ is } A_{j1} \texttt{ and } \ldots \texttt{ and } x_M \texttt{ is } A_{jM} \texttt{ then} \\ \texttt{class = } (c_1 \texttt{ with conf } p_{1j}, \ \ldots \ , c_C \texttt{ with conf } p_{Cj})\end{array} \tag{8}$$

where the antecedents A_j must be hypercubes in the feature space. If $A_j(\mathbf{x}) = 1$ for all $\mathbf{x} \in A$ and 0 else,

$$A_j(\mathbf{x}) = A_{j1}(x_1) \wedge \ldots \wedge A_{jM}(x_M)$$

and all intervals A_{jk} must be elements of the same crisp partition of the feature k.

Recalling equation (1), a probabilistic rule based classifier obtained from the sample $\mathbf{X}$ outputs the probabilities of all classes according to the formula

$$P(c_i|\mathbf{x}, \mathbf{X}) = \sum_j p_{ij}^{\mathbf{X}} \prod_{k=1}^{M} A_{jk}^{\mathbf{X}}(x_k) \tag{9}$$

where $\prod_{k=1}^{M} A_{jk}^{\mathbf{X}}(x_k)$ is 1 or 0. We obtain that

$$P(c_i|\mathbf{x}) = \sum_j \langle p_{ij}^{\mathbf{X}} \rangle_{\mathbf{X}} \left\langle \prod_{k=1}^{M} A_{jk}^{\mathbf{X}}(x_k) \right\rangle_{\mathbf{X}} . \tag{10}$$

The condition we need to obtain a descriptive random set-based classifier is

$$\Phi_j(\mathbf{x}) = \prod_{k=1}^{M} \Phi_{jk}(x_k) = \left\langle \prod_{k=1}^{M} A_{jk}^{\mathbf{X}}(x_k) \right\rangle_{\mathbf{X}} \tag{11}$$

which is fulfilled when random variables $P(x_k \in A_{jk}^{\mathbf{X}})$ are independent. In this last case,

$$\Phi_{jk}(x_k) = \langle A_{jk}^{\mathbf{X}}(x_k) \rangle_{\mathbf{X}} \tag{12}$$

and the descriptive classifier can then be expressed as a set of rules of the form

$$\begin{array}{c}\texttt{if } x_1 \texttt{ is } \Phi_{j1} \texttt{ and } \ldots \texttt{ and } x_M \texttt{ is } \Phi_{jM} \texttt{ then} \\ \texttt{class = } (c_1 \texttt{ with conf } \langle p_{1j}^{\mathbf{X}} \rangle_{\mathbf{X}}, \ \ldots, \ c_C \texttt{ with conf } \langle p_{Cj}^{\mathbf{X}} \rangle_{\mathbf{X}})\end{array}$$

Selecting a parametric family of coverage functions Φ_{jk} is equivalent to define a family of sample distributions over $\mathbf{X}$. It is immediate that all functions Φ_{jk} can be interpreted as elements of a Ruspini's fuzzy partition of feature k. We will use triangular coverage functions in all numerical examples, as shown in Figure 1.

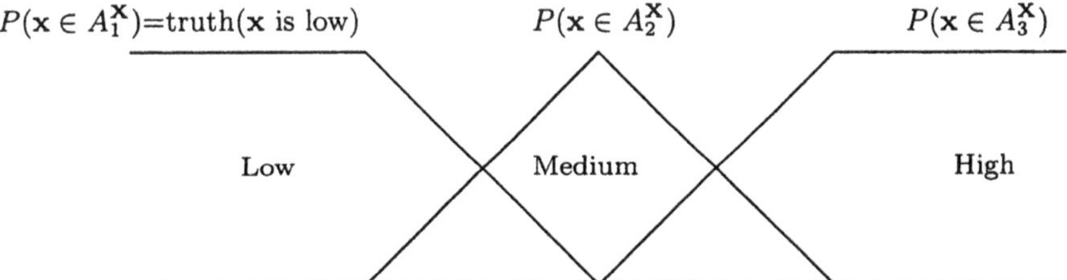

Fig. 1. Graph of the degrees of truth of a numeric value being compatible with the properties "low", "medium" and "high". These labels are associated to random sets $A_1^{\mathbf{X}}, A_2^{\mathbf{X}}, A_3^{\mathbf{X}}$, respectively, and the degree of truth of the assertion "**x** is label" is the probability of **x** being covered by the corresponding random set. For example: truth("80 is low")= $P(80 \in A_1^{\mathbf{X}})$.

3 Inducting weighted rule based classifiers

Let us suppose we know the antecedents of the rules (the functions "truth(A_j is **x**)" or Φ_j, $j = 1, \ldots, S$) and we wish to estimate their consequents (the values "truth($A_j \rightarrow c_i$)" or $\langle p_{ij}^{\mathbf{X}} \rangle_{\mathbf{X}}$.) For simplicity in the notation, let $\theta_{ij} = \langle p_{ij}^{\mathbf{X}} \rangle_{\mathbf{X}}$ and Θ be the parameter vector of all unknown parameters,

$$\Theta = \left(\langle p_{11}^{\mathbf{X}} \rangle_{\mathbf{X}}, \ldots, \langle p_{1S}^{\mathbf{X}} \rangle_{\mathbf{X}}, \ldots, \langle p_{CS}^{\mathbf{X}} \rangle_{\mathbf{X}} \right) = (\theta_{11}, \ldots, \theta_{1S}, \ldots, \theta_{CS}) \,. \tag{13}$$

The likelihood function V_Θ is defined as follows:

$$V_\theta = \prod_{\mathbf{x} \in \mathbf{X}} \sum_{j=1}^{S} \theta_{class(\mathbf{x}),j} \Phi_j(\mathbf{x}) \tag{14}$$

and its maximum, restricted to the conditions $\sum_{i=1}^{C} \theta_{ij} = 1$ for $j = 1, \ldots, S$ and $\theta_{ij} \geq 0$, is a good estimation of the unknowns.

It is easier to work with the logarithm of this function:

$$L(\Theta) = \log(V_\theta) = \sum_{\mathbf{x} \in \mathbf{X}} \log \sum_{j=1}^{S} \theta_{\text{class}(\mathbf{x}),j} \Phi_j(\mathbf{x}). \tag{15}$$

Using Lagrange multipliers to cope with the restrictions, we have to minimize

$$L_1(\Theta, \lambda_1, \ldots, \lambda_S) = \sum_{\mathbf{x} \in \mathbf{X}} \log \sum_{j=1}^{S} \theta_{\text{class}(\mathbf{x}),j} \Phi_j(\mathbf{x}) + \sum_{j=1}^{S} \lambda_j (1 - \sum_{i=1}^{C} \theta_{ij}) \tag{16}$$

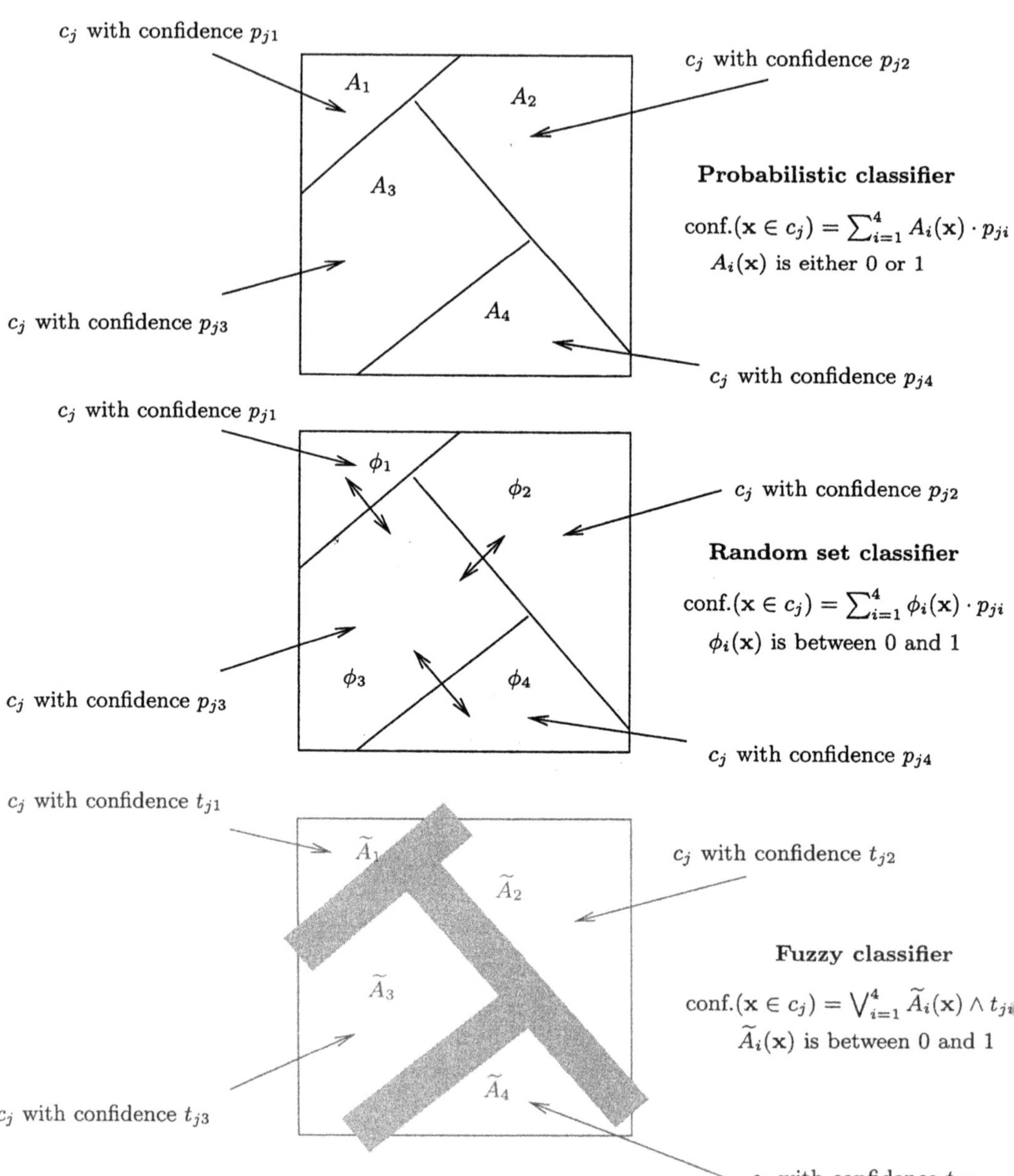

Fig. 2. Three types of classifiers are used in this chapter: Probabilistic classifiers output degrees of confidence depending on a crisp partition. The output of random set classifiers is the expectation of probabilistic classifiers' and confidences are added after multiplying them by the coverage functions. Fuzzy sets use t-norms and t-conorms instead of products and sums. We will compare random set classifiers with fuzzy classifiers in experiments for which fuzzy memberships and coverage functions are numerically identical.

Taking derivatives with respect to θ_{ij} and λ_j, we obtain the following conditions that are true in the minimum:

$$\sum_{\substack{\mathbf{x}\in\mathbf{X}\\ \text{class}(\mathbf{x})=i}} \frac{\Phi_j(\mathbf{x})}{\sum_{j=1}^{S}\theta_{ij}\Phi_j(\mathbf{x})} = \sum_{\substack{\mathbf{x}\in\mathbf{X}\\ \text{class}(\mathbf{x})=k}} \frac{\Phi_j(\mathbf{x})}{\sum_{j=1}^{S}\theta_{kj}\Phi_j(\mathbf{x})} \tag{17}$$

$$\sum_{i=1}^{C}\theta_{ij} = 1, \qquad \text{for } j=1,\ldots,S;\ i,k=1,\ldots,C. \tag{18}$$

The first set of equalities (equation 17) produces $C-1$ equations and the second one (equation 18) leads to S, thus the search of the parameters consists in numerically solving a system of $C \cdot S$ nonlinear equations.

3.1 Incomplete rule bases

Whilst fuzzy rule banks can be incomplete, probabilistic ones can not. But all consequents can be initialized to the value $1/C$ to express the initial absence of knowledge. If the learning algorithm finishes and there are still rules like these, they can be skipped when doing an inference (eq. 4) without affecting the output of the classifier. We can define an "equivalent number of rules" to compare the complexity of a complete probabilistic rule bank to that of an incomplete fuzzy rule bank as the number of rules for which the degree of confidence in any of their consequent parts is different from $1/C$.

Determining which set of $N_r' < S$ rules produces the best classifier has practical relevance. When the number of features is high, the number of parameters grows above all practical linguistic interpretability, and then it is useful to decide whether an approximate solution in which most of the parameters are $1/C$ (thus they can be ignored) is precise enough. Moreover, many fuzzy rule learning algorithms produce incomplete rule banks and it is reasonable to compare them to random set-based classifiers with the same structure, but also with the same "equivalent number of rules".

As far as we know, finding the best set of N_r' rules in polynomial time is an open problem. An heuristic method to obtain banks with a reduced number of rules is shown in Figure 3. This algorithm does not guarantee that there is not a different set of rules with higher likelihood, but has good properties in practical problems. It just selects rules containing the parameters for which the partial derivatives of eq. 16 are higher in the initial point of the minimization ($\theta_{ij} = 1/C$).

3.2 Adjusting the membership functions

For a fuzzy classifier to be linguistically understandable, all fuzzy memberships have to be related to linguistic values of variables, so that a human intuitively can relate a concept with an imprecise range of values. If some

$\mathbf{G}, \mathbf{D} \in \mathbf{R}^{C \times S}, \alpha \in \mathbf{R},\ i \in 1 \ldots C, j \in 1 \ldots S$

function learn-with-weights($\theta \in \mathbf{R}^{C \times S}, S' \subset 1 \ldots S$) **returns** ($\theta' \in \mathbf{R}^{C \times S}$)
  $\theta' =$ **learn-V**(θ,**calculate-V**(S'))
end of learn-with-weights

function calculate-V($S' \in \mathbf{N}$) **returns** ($\mathbf{V} \subset 1 \ldots S$)
  $\theta_{ij} = 1/C$
  $\mathbf{G}_{ij} = C \sum_{\mathbf{x} \in \mathbf{X}, \mathrm{class}(\mathbf{x})=i} \Phi_j(\mathbf{x}) / \sum_{j=1}^{S} \Phi_j(\mathbf{x})$
  $\mathbf{D}_{ij} = \mathbf{G}_{ij} - (S \cdot C)^{-1} \sum_{i=1}^{C} \sum_{j=1}^{S} \mathbf{G}_{ij}$
  $\mathtt{importance}(j) = \sum_{\substack{i=1 \ldots C \\ \mathbf{D}_{ij}>0}} \mathbf{D}_{ij}$
  $\mathbf{V} =$ `set of values` j `of the` S' `rules with higher importance`
end of calculate-V

function learn-V($\theta \in \mathbf{R}^{C \times S}$, $\mathbf{V} \subset 1 \ldots S$) **returns** ($\theta' \in \mathbf{R}^{C \times S}$)
  `repeat`
    $\mathbf{G}_{ij} = \sum_{\mathbf{x} \in \mathbf{X}, \mathrm{class}(\mathbf{x})=i} \Phi_j(\mathbf{x}) / \sum_{j=1}^{S} \theta_{ij} \Phi_j(\mathbf{x})$
    $\mathbf{D}_{ij} = \mathbf{G}_{ij} - (S \cdot C)^{-1} \sum_{i=1}^{C} \sum_{j=1}^{S} \mathbf{G}_{ij}$
    `if` $j \notin \mathbf{V}$ `then` $\mathbf{D}_{ij} = 0$
    `search` α `that minimizes` $L(\mathtt{normalize}(\theta + \alpha \cdot \mathbf{D}))$
    $\theta = \mathtt{normalize}(\theta + \alpha \cdot \mathbf{D})$
  `until` $\alpha ||D|| < \epsilon$
  $\theta' = \theta$
end of learn-V

function normalize($\mathbf{X} \in \mathbf{R}^{C \times S}$) **returns** ($\mathbf{X}' \in \mathbf{R}^{C \times S}$)
  `if` $(\mathbf{X}_{ij} < 0)\ \mathbf{X}_{ij} = 0$
  $\mathbf{X}_{ij} = \mathbf{X}_{ij} / \sum_{i=1}^{C} \mathbf{X}_{ij}$
  $\mathbf{X}' = \mathbf{X}$
end of normalize

Fig. 3. Pseudo code of the numerical algorithm used to approximately solve the set of equations (18), producing at most S' rules. The linear search (determination of the value of α) was implemented with Brent's method. All points examined fulfill eq. (18) because of the function `normalize`, and the algorithm stops when the conditions (17) are approximately true.

changes are allowed to these ranges, this last association is less clear, but the precision of the classifier is increased.

The same relationship between random sets and fuzzy sets that has been used up to now, can be extended to estimate parameters defining the semantic values of the linguistic terms as well. Therefore, antecedents can be induced from data along with the consequent parts of the rules. There are some numerical difficulties with this adjust, arising from the fact of functions 15 and 16 being not differentiable with respect to these parameters: $\phi(\cdot)$ is piecewise linear with respect to them and thus gradient based optimization algorithms are not applicable.

We propose to combine the process shown in Figure 3 with a robust deterministic evolutionary algorithm, Nelder and Mead's simplex [11]. We can setup an objective function for the simplex that depends only on the parameters defining the linguistic partitions. This function calls a full gradient descent algorithm to obtain the consequents that fit the partition being evaluated, and eq. 15 is applied to them to get the return value. This way, the evolutionary search selects changes in the membership function and the right consequents for them are estimated with gradient descent.

4 Inducting rules with binary weights

4.1 Random set based rules with binary weights

Extending the usual nomenclature in fuzzy modelling, three type of fuzzy rules in classification problems can be considered. Let us name "Type 1" fuzzy rules to those that have the form

$$\texttt{if x is } \widetilde{A} \texttt{ then class = } c_k.$$

"Type 2" have one consequent, weighted by a degree of confidence

$$\texttt{if x is } \widetilde{A} \texttt{ then class = } c_k \texttt{ with conf. } p_k$$

and rules not in these two classes (more than one consequent, with or without weights) are "type 3" rules. The interpretability of "type 1" fuzzy rules is higher than those of weighted or "type 3" rules, because they do not carry numerical information in their linguistic expression. Random set based rules are, by definition, equivalent to "Type 3" rules, which are uncommon in fuzzy sets literature. In this section we discuss how to adapt the maximum likelihood estimation discussed before to obtain "Type 1" (we will also call them "not weighted", or "binary") rules.

Two considerations must be taken before:

1. Under the maximum vote scheme (which is the fuzzy inference method equivalent to that of random sets based descriptive classifiers, as we have mentioned) all confidences in the whole rule base can be multiplied by a

constant, and this does not change the description surfaces of the classifier. As a consequence of this, any base comprising "type 3" rules can be written in terms of "type 1" rules. This will be made clear with an example: observe the following three "type 1" fuzzy rules:

if x is $\widetilde{A}$ then class = 1

if x is $\widetilde{A}$ then class = 1

if x is $\widetilde{A}$ then class = 2.

This set of rules is equivalent to this single one:

if x is $\widetilde{A}$ then class = (1 with conf 2s, 2 with conf s)

where the value of the constant "s" depends on the remaining rules in the rule base. It is not difficult to prove that one can always convert a rule base comprising "type 3" rules with rational weights into a "type 1" rule base by splitting each rule and making an adequate number of copies of it. Therefore, the linguistic interpretability of a rule bank in which the antecedents can be repeated is exactly the same as the interpretability of a rule containing "type 3" rules.

2. The addition of a constant to all confidences in one isolated rule does not modify the bank. This means that the rule

   if x is $\widetilde{A}$ then class = (1 with conf s, 2 with conf s)

   can be replaced by

   if x is $\widetilde{A}$ then class = (1 with conf 0, 2 with conf 0)

   and removed from the rule base.

Because of these considerations, we decided that random set based rules equivalent to "type 1" fuzzy rules can belong to two categories:

1. if $\mathbf{x}$ is ϕ_j then class $= c_k$
2. if $\mathbf{x}$ is ϕ_j then class $=$ (1 with conf $1/C$, ..., C with conf $1/C$)

where the second type of rules do not appear in the linguistic expression of the rule base. It is remarked again that we will not admit that two different rules in the base have the same antecedent; it is clear for us that, if this restriction is not enforced, all experiments would conclude that there is not a loss of classification power when weights are binary with respect to that of real valued weights.

4.2 Learning algorithm

Learning the non-weighted rule base is a nonlinear integer programming problem. Standard branch and bound techniques can be applied, as shown in the pseudo code in figure 4, to obtain the best set of "type 1" rules that approximate a base of "type 3" rules.

The crux of the algorithm consists in deciding whether each weighted rule will be replaced by either the rule

$$\text{if } \mathbf{x} \text{ is } \phi_j \text{ then class} = c_k$$

or

$$\text{if } \mathbf{x} \text{ is } \phi_j \text{ then class} = (\ 1 \text{ with conf } 1/C,\ \ldots,\ C \text{ with conf } 1/C),$$

i.e., removed from the base. Let us suppose we replace the first rule of a weighted classifier comprising S' rules by a "type 1" rule; there are $C+1$ different replacements (C binary rules and removing the rule from the base). We can think that each one of these substitutions originates a new weighted subproblem, where the first rule is fixed and the remaining $S'-1$ rules should be modified to find a new maximum likelihood estimation. Let us solve all these $C+1$ weighted problems, write down the final values of the likelihood in each case, and recursively repeat the process for every one of them (the second rule is replaced by a binary one or removed, and so on), finishing when all S' rules have been replaced or removed. If we arrange the result of all experiments in a tree, the leaves are solutions to the integer problem and the internal nodes are solutions to the weighted subproblems.

Obviously, we only need to search a part of this tree, because the likelihoods of intermediate weighted classifiers (the internal nodes of the tree) are lower bounds of the likelihood of the binary weighted classifiers (the leaves of the subtree originated in the internal node). Therefore, as soon as we know the likelihood of any binary solution, we can skip all recursive calls for which the real solution is higher than the likelihood of the binary solution, and prune the search tree as it is shown in the pseudo code in figure 4.

There are three further improvements to the speed of convergence of this algorithm:

1. If we know that the likelihood of the binary solution is in a certain range of the real solution (for example, we usually can expect that the binary classifier log-likelihood is not worse than the real classifier's one +10%) we can skip the recursive calls for which the lower bound is higher than this value, even if a better binary solution has not been reached yet.
2. The order in which the intermediate problems are solved is important: if the problems with a lower bound are solved first, many paths will be removed from the search.
3. If we admit that any binary solution within a certain range of the real solution is precise enough, we can stop the search as soon as this value is reached.

```
best-L ∈ R, best-θ ∈ R^{C×S}, low-bound ∈ R^{C+1}

function learn-bin-weights(θ ∈ R^{C×S}, S' ∈ N) returns (θ' ∈ R^{C×S})
      branch-and-bound(θ, calculate-V(S'))
      θ' = best-θ
end of learn-bin-weights

procedure branch-and-bound(θ ∈ R^{C×S}, V ⊂ 1...S)
   if (V = ∅) then
           if (L(θ) < best-L) then
           best-L=L(θ)
           best-θ = θ
        end if
   else
      r = first element of V
      for k ∈ 0...C
         if (k = 0) θ_ir = 1/C else θ_ir = δ_ir
         low-bound_k = learn-V(θ, V − {r})
         if (low-bound_k < best-L) branch-and-bound(θ, V − {j})
      end for
   end if
end of branch-and-bound
```

Fig. 4. Simplified pseudo code of the numerical algorithm used to approximately solve the set of equations (18), producing at most S' rules with binary weights. Heuristics used to shorten the search (described in section 4.2) are not shown.

5 Numerical examples

5.1 Pure linguistic classification problem

The behavior of the algorithm will be illustrated first with a synthetic example, by means of a data set generated so that the Bayes solution can be described without error by means of a descriptive classifier comprising the following set of type 3 rules:

If x_1 is $\tilde{R}_1$ and x_2 is $\tilde{R}_1$ then $\mathtt{class}_1$=0.90 and $\mathtt{class}_2$=0.10
If x_1 is $\tilde{R}_1$ and x_2 is $\tilde{R}_2$ then $\mathtt{class}_1$=0.85 and $\mathtt{class}_2$=0.15
If x_1 is $\tilde{R}_1$ and x_2 is $\tilde{R}_3$ then $\mathtt{class}_1$=0.60 and $\mathtt{class}_2$=0.40
If x_1 is $\tilde{R}_2$ and x_2 is $\tilde{R}_1$ then $\mathtt{class}_1$=0.40 and $\mathtt{class}_2$=0.60
If x_1 is $\tilde{R}_2$ and x_2 is $\tilde{R}_2$ then $\mathtt{class}_1$=0.80 and $\mathtt{class}_2$=0.20
If x_1 is $\tilde{R}_2$ and x_2 is $\tilde{R}_3$ then $\mathtt{class}_1$=0.40 and $\mathtt{class}_2$=0.60
If x_1 is $\tilde{R}_3$ and x_2 is $\tilde{R}_1$ then $\mathtt{class}_1$=0.20 and $\mathtt{class}_2$=0.80
If x_1 is $\tilde{R}_3$ and x_2 is $\tilde{R}_2$ then $\mathtt{class}_1$=0.10 and $\mathtt{class}_2$=0.90
If x_1 is $\tilde{R}_3$ and x_2 is $\tilde{R}_3$ then $\mathtt{class}_1$=0.00 and $\mathtt{class}_2$=1.00

where the memberships $\widetilde{R}_1$, $\widetilde{R}_2$, $\widetilde{R}_3$ are shown in Figure 5. An algorithm that can generate examples for this problem is shown in Figure 6. Let us generate 1000 examples and apply the algorithms in Figures 3 and 4 to infer the values of the coefficients, with both weighted and not weighted versions.

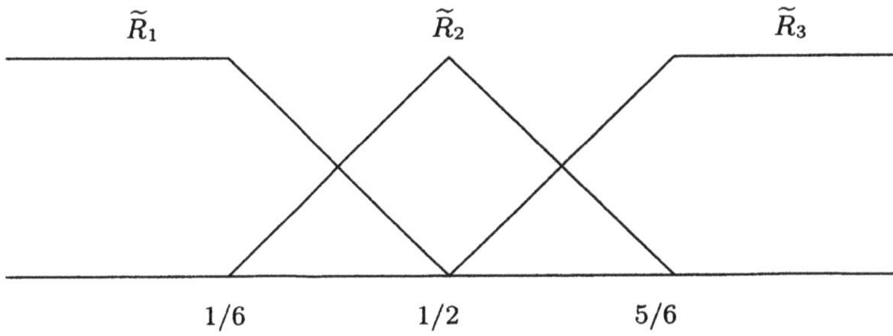

Fig. 5. Membership functions of the example in Section 5.1

x_1=`random(0,1)`; x_2=`random(0,1)`

$$p_1 = 0.90\widetilde{R}_1(x_1)\cdot\widetilde{R}_1(x_2) + 0.85\widetilde{R}_1(x_1)\cdot\widetilde{R}_2(x_2) + 0.60\widetilde{R}_1(x_1)\cdot\widetilde{R}_3(x_2) + 0.40\widetilde{R}_2(x_1)\cdot\widetilde{R}_1(x_2) + 0.80\widetilde{R}_2(x_1)\cdot\widetilde{R}_2(x_2) + 0.40\widetilde{R}_2(x_1)\cdot\widetilde{R}_3(x_2) + 0.20\widetilde{R}_3(x_1)\cdot\widetilde{R}_1(x_2) + 0.15\widetilde{R}_3(x_1)\cdot\widetilde{R}_2(x_2) + 0.00\widetilde{R}_3(x_1)\cdot\widetilde{R}_3(x_2)$$

`if (random(0,1)<` p_1`) output` (x_1, x_2, c_1) `else output` (x_1, x_2, c_2)

Fig. 6. Algorithm used to output a point of the learning sample in the problem discussed in Section 5.1.

0.90	0.40	0.20
0.10	0.60	0.80
0.85	0.80	0.10
0.15	0.20	0.90
0.60	0.40	0
0.40	0.60	1

0.887	0.345	0.198
0.112	0.654	0.801
0.756	0.744	0.100
0.243	0.255	0.899
0.664	0.354	0
0.335	0.645	1

0.5	0.5	0.5
0.5	0.5	0.5
1	0.5	0.5
0	0.5	0.5
0.5	0.5	0
0.5	0.5	1

Fig. 7. True (left), estimated weighted rules (center) and estimated binary rules (right) values for the example explained in Section 5.1.

The inferred rule banks are summarized in Figure 7. The weighted version recovers the original base, while the not weighted one, which is the most precise “type 1” base, (in fact, since it includes rules with consequents ”0.5-0.5” it can be argued that this is not strictly a type 1 classifier; see comments in section 4.1) is clearly suboptimal and comprises only two rules (i.e, only two rules with different values in their consequent part; recall the comments in section 3.1:)

$$\texttt{If } x_1 \texttt{ is } \widetilde{R}_2 \texttt{ and } x_2 \texttt{ is } \widetilde{R}_1 \texttt{ then class}_1$$
$$\texttt{If } x_1 \texttt{ is } \widetilde{R}_3 \texttt{ and } x_2 \texttt{ is } \widetilde{R}_3 \texttt{ then class}_2.$$

This classifier has an estimated error rate of 0.41, while the real solution has an error of 0.26. This example shows us that it is not immediate to pass from the real solution to the best binary solution. In particular, one can not apply an heuristic method to obtain “type 1” rules from “type 3” rules, one by one: such a conversion would depend not only on the single rule being considered, but on rules surrounding it.

5.2 Graphical analysis: Haykin’s two Gaussian problem

With this second example we intend to study the differences in the decision surface between “type 1” and “type 3” rule bases. To be able to do a graphical representation, we are going to analyze the data set proposed in [7]: 4000 points taken from two overlapping Gaussian distributions with different variances. The optimal decision surface is a circle, and the Bayesian test error is 0.185. The error of the linear classifier is 0.24, which is near enough the optimal solution to confuse many rule learning algorithms. The shape of the decision surface in areas with low density of examples (i.e., the left side of the circle) does not contribute too much to the classification error.

In Figure 8, descriptive classifiers are compared when the number of linguistic terms in every partition ranges from 3 to 5. Uniform, unadjusted fuzzy partitions were used. In that Figure we observe that the decision surface of the “type 1” descriptive random set-based classifier is nearer to the “type 3” surface than we could intuitively think. In the worst case (the leftmost one) the difference between “type 1” and “type 3” banks produces less than a 2% increase in the classification error, even while the “type 1” base has two rules less and they all are less complex.

5.3 Significance of the loss of classification power

In third place, to judge whether the loss of power produced when “type 3” bases are downgraded to “type 1” bases, we will study 5 cases: the problem introduced in the preceding section, a multi class synthetic problem similar to “Gauss” but involving five classes (it will be named “Gauss-5”), and three more real-world problems from UCI [14]: Pima, Cancer and Glass.

The experimental framework is as follows: 5x2cv Dietterich’s test [4] will be applied to assess the statistical relevance of the differences between the

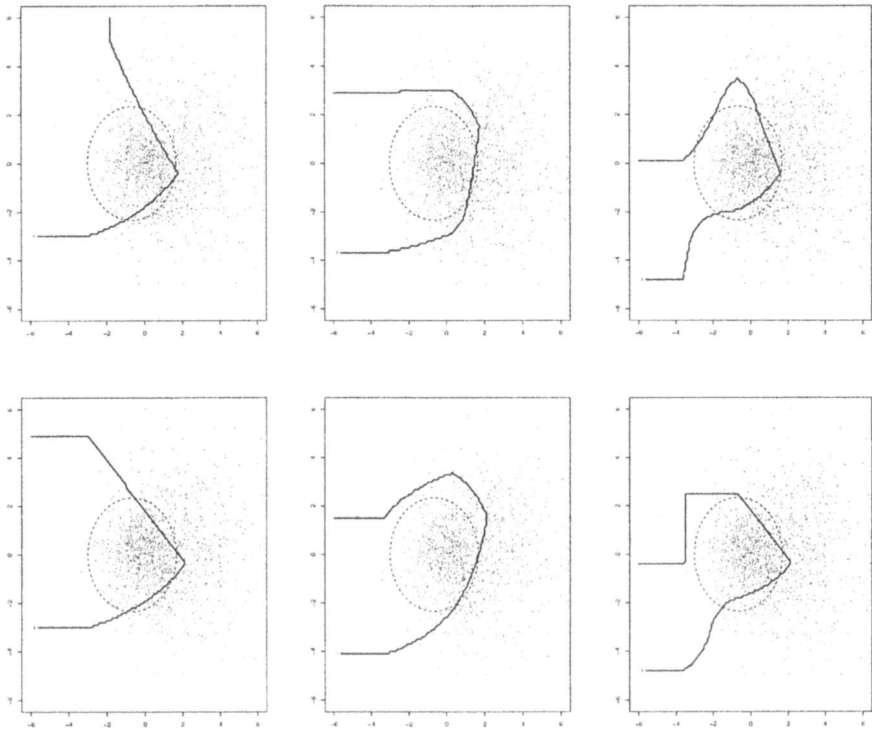

Fig. 8. Effect of the number of terms in the fuzzy partition in the weight removal process. Upper part, from left to right: decision surfaces induced by RSB for classifiers with 3, 4, and 5 terms/feature (9, 16 and 25 "type 3" rules) in the "Gauss" problem. Lower part: the same rule bases, downgraded to "type 1". The dashed line is the optimal decision surface.

two types of rule bases being considered. Data sets are randomly permuted first; the first half of samples is used to train the method, and the second half to test it. Training and test sets are swapped and the learning and test phase repeated. This is repeated for 5 different random permutations. Training errors are discarded, so that box plots only show the dispersion of the error in test sets.

Other statistical (linear, quadratical, nearest neighbor) and artificial intelligence based (neural networks, Wang and Mendel's [3,1], Hong and Lee's [8], Pal and Mandal's [13] and Cordón, Del Jesus' et al. [2] fuzzy classifiers) are included so that the reader can judge the magnitude of the differences between the two methods being considered here. Results of the genetic method in [2] are among the state of the art results in fuzzy classification, and involve tuning of the linguistic partitions in the antecedent; all other fuzzy classification algorithms are based on heuristics and are included as a reference only; besides some of them achieve good results in some data sets, their results are not as consistent as the former genetic method.

	LIN	QUA	NEU	1NN	WM	HL	PM	GIL	KRE	KBI
pima	0.227	0.252	0.255	0.289	0.287	0.301	0.464	0.269	0.238	0.237
cancer	0.044	0.051	0.047	0.048	0.129	0.058	0.087	0.099	0.043	0.043
gauss	0.239	0.190	0.200	0.267	0.477	0.304	0.457	0.205	0.217	0.220
glass	0.403	-	0.439	0.354	0.453	0.503	0.647	0.363	0.392	0.384
gauss5	0.317	0.317	0.321	0.413	0.539	0.344	0.759	0.338	0.328	0.388

Fig. 9. Mean test values of problems in section 5.3

The mean values of test errors are included in Figure 9. Random set based classifiers comprise 200 rules in Pima, Cancer and Glass, and 9 in Gauss and Gauss-5. All features have three linguistic terms but in Cancer data set, where only two values were needed.

The only statistically significant difference between "type 3" and "type 1" data sets is in Gauss-5 (the p-value of the contrast is 0.08, thus we reject the hypotheses of binary and weighted bases producing similar results, with a 92% level). The difference in Gauss has a p-value of 50%. Cancer produced the same results in 9 of the 10 repetitions, thus box plots are roughly the same. Binary results with Pima and Glass seem to be better than the weighted, but the difference is well under the expected deviation of the results.

5.4 Importance of the membership tuning process

Finally, we will study whether an adjust in the memberships recovers the information lost, in the cases in which downgrading to "type 1" rules made a difference; for Gauss, Cancer, Pima and Glass there are not relevant dissimilarities between either the real or the binary solution and the black boxes, thus a membership tuning makes no sense for them.

Let us focus on the problem for which there are significant differences, Gauss-5. This data set comprises 5 classes and there is a maximum of 16 rules in it, thus the weight removal process remove a lot of information and the "type 1" classifier does not perform properly. One may question whether there exists a definition of the membership functions for which the behavior of the "type 1" rule base is comparable to that of black boxes and "type 3" rules. To check this, we have tuned the membership functions, as explained in section 3.2, before launching the weight removal process, and compared the results of weighted rules with unadjusted (uniform partitions) memberships in the antecedents with that of not weighted rules with tuned memberships. Results are plotted in Figure 11. While tuning the membership always improves the

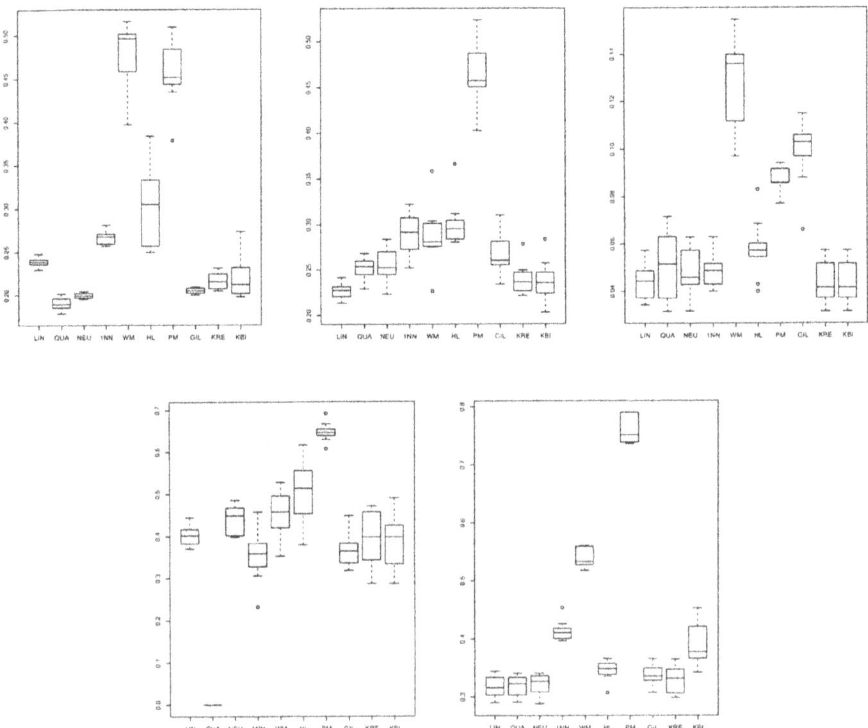

Fig. 10. From left to right and from upper to lower: Box plots of the differences between weighted and not weighted classifiers in gauss, pima, cancer, glass and gauss-5 problems. The columns are: linear, quadratic, neuronal, nearest neighbor, Wang and Mendel's, Hong and Lee's, Pal and Mandal's, Genetic Iterative Learning, Random sets based with "type 3" and "type 1" rules. The bars represent the dispersion of the test results in the 10 repetitions of the experiment.

final results, the gain of classification power does not compensate the loss produced in the weight removal.

6 Concluding remarks and future work

Experimental results have shown that, most of times, there exists very little difference between black boxes and weighted probabilistic rules. These differences decrease with the number of rules, and are statistically significant only when the rule base is rather small. Taking into account that random set based classifiers did not modify the fuzzy memberships in the antecedents, we doubted that the effect of tuning the antecedents was important against the right selection of weights in the rule consequents.

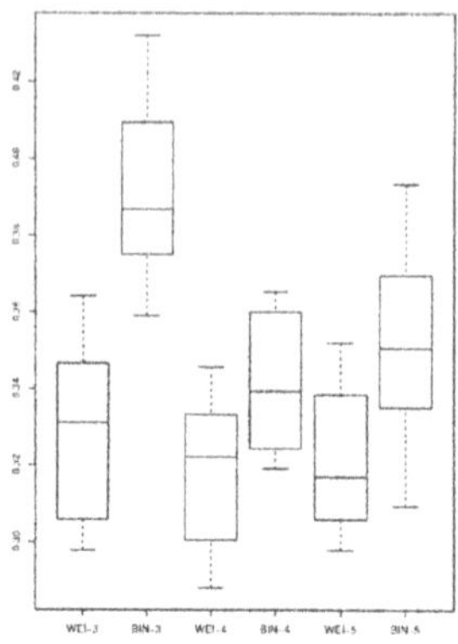

Fig. 11. Comparison between unadjusted memberships + weighted rule bases and adjusted memberships + not weighted rule bases in Gauss-5. From left to right: box plots of weighted and binary solutions of Gauss 5, with 3, 4 and 5 elements/partition. In this case, tuning the memberships does not recover the classification power lost in the weight removal process. The bars represent the dispersion of the test results in the 10 repetitions of the experiment.

Against our initial thought, the results of this study show that the use of weighted or "type 3" does not uniformly produce results significantly better than those obtained by simpler (and easier to interpret) rules without weights. It is remarkable that neither weights in the rules nor multi consequent rules achieved significant improvements in the representation power in real world data sets (where the rule base comprised more than one hundred rules.) In the set of experiments that we performed, the tradeoff between precision and interpretability was best achieved when memberships were uniform and confidences were not used in rule bases with size from moderate to large; by the contrary, both weighted rules and adjusted memberships improved the results of type 1 rules in small bases. In this last case, using weights on uniform partitions produces a gain of precision similar to that achieved when both the definition of the linguistic terms are adjusted, and type 1 rules used.

Acknowledgments

The authors wish to thank the anonymous reviewers for their effort revising this chapter and their valuable suggestions for future works.

References

1. Cordón, O., Del Jesus, M. J., Herrera, F. "A proposal on reasoning methods in fuzzy rule-based classification systems". International Journal of Approximate Reasoning **20**(1), pp. 21-45, 1999.

2. Cordón O., del Jesus M. J., Herrera F. y Lozano M. (1999) Mogul: A methodology to obtain genetic fuzzy rule-based systems under the iterative rule learning approach. *International Journal of Intelligent Systems* 14(9).
3. Chi, Z., Yan, H., Pham, T. *Fuzzy Algorithms: With Applications to Image Processing and Pattern Recognition.* World Scientific. 1996.
4. Dietterich, G. "Approximate Statistical Tests for Comparing Supervised Classification Learning Algorithms". Neural Computation, 10(**7**), pp 1895-1924. 1998
5. Geman, S., Bienenstock, E., Doursat, R. "Neural networks and the bias/variance dilemma". Neural Computation, 4, pp. 1-58. 1992.
6. Hand, D. J. *Discrimination and Classification.* Wiley. 1981
7. Haykin, S. *Neural Networks.* Prentice Hall, 1999.
8. Hong, T. P., Lee, C. Y. Induction of fuzzy rules and membership functions from training examples. *Fuzzy Sets and Systems* 84. pp 33-47. 1996.
9. Ishibuchi, H. and Nakashima, T. "Effect of rule weights in fuzzy rule-based classification systems," *Proc. of 9th International Conference on Fuzzy Systems*, pp 59-64 (San Antonio, May 7-10, 2000).
10. Ishibuchi, H., Nakashima, T.: Effect of Rule Weights in Fuzzy Rule-Based Classification Systems, *IEEE Trans. on Fuzzy Systems*, vol. 9, no. 4, pp. 506-515, August 2001.
11. Nelder, J.A. and Mead, R., A simplex method for function minimization, Computer J., 7 (1965), 308-313.
12. Nauck, D. and Kruse, R. "How the learning of rule weights affects the interpretability of fuzzy systems". *Proc. of the 7th IEEE International Conference on Fuzzy Systems*, pp. 1235-1240 (Anchorage, May 4-9, 1998).
13. Pal, S. K., Mandal, D. P. "Linguistic recognition system based in approximate reasoning". *Information Sciences* **61**, pp. 135-161. 1992.
14. Prechelt, L. "PROBEN1 – A set of benchmarks and benchmarking rules for neural network training algorithms". Tech. Rep. 21/94, Fakultät für Informatik, Universität Karlsruhe, 1994.
15. Sánchez, L., Casillas, J., Cordón, O., Del Jesus, M. J. "Some relationships between fuzzy and random set-based classifiers and models". Accepted for publication in IJAR, 2001.
16. Trillas, E., Alsina, C., Terricabras, J. *Introducción a la lógica borrosa.* Ariel Matemática. 1995.
17. Wang, L. X., Mendel, J. "Generating fuzzy rules by learning from examples". IEEE Trans. on Systems, Man and Cybernetics, **25**(2), pp. 353-361, 1992.
18. Zadeh, L.A. "The concept of a linguistic variable and its application to approximate reasoning". Information Science, Part I: vol. 8, pp. 199-249, 1975; Part II: vol. 8, pp. 301-357, 1975; Part III: vol. 9, pp. 43-80, 1975.
19. Zadeh, L.A. "Fuzzy Languages and Their Relation to Human and Machine Intelligence", in *Fuzzy Sets, Fuzzy Logic and Fuzzy Systems*, Klir, Yuan, eds. pp 148-179. World Scientific, 1996.

www.ingramcontent.com/pod-product-compliance
Ingram Content Group UK Ltd.
Pitfield, Milton Keynes, MK11 3LW, UK
UKHW021858190726
13853UKWH00003B/1321

* 9 7 8 3 6 4 2 5 3 5 3 6 9 *